# Current Topics in Microbiology and Immunology

## Volume 331

Ethel-Michele de Villiers • Harald zur Hausen
Editors

# TT Viruses

## The Still Elusive Human Pathogens

Springer

*Editor*
Prof. Dr. Ethel-Michele de Villiers
Deutsches Krebsforschungszentrum Abt.
Tumorvirus-Charakterisierung
Im Neuenheimer Feld 242
69120 Heidelberg
Germany

Prof. Dr. mult. Harald zur Hausen
Deutsches Krebsforschungszentrum (DKFZ)
Im Neuenheimer Feld 280
69120 Heidelberg
Germany

ISBN 978-3-540-70971-8 e-ISBN 978-3-540-70972-5

DOI 10.1007/978-3-540-70972-5

Current Topics in Microbiology and Immunology ISSN 0070-217x

Library of Congress Catalog Number: 2008935114



*Cover design*: WMX Design GmbH, Heidelberg, Germany

Printed on acid-free paper

9 8 7 6 5 4 3 2 1

springer.com

# Preface

Eleven years ago the circular DNA of a novel single-stranded virus was cloned and partially characterized by Nishizawa, Okamoto and their colleagues. According to the initials of the patient from whom the isolate originated, the virus was named TT virus. This name has been subsequently changed by the International Committee on Taxonomy of Viruses (ICTV) into Torque teno virus, permitting the further use of the abbreviation TTV. Although initially suspected to play a role in non-A–E hepatitis, subsequent studies have failed to support this notion.

Within a remarkably short period of time it became clear that TT viruses are widely spread globally, infect a large proportion of all human populations studied thus far, and represent an extremely heterogeneous group of viruses, now labelled as Anelloviruses. TT virus-like infections have also been noted in various animal species. The classification of this virus group turns out to be difficult. Their DNA contains between 2,200 and 3,800 nucleotides, while related so-called TT miniviruses and a substantial proportion of intragenomic recombinants further complicate attempts to combine these viruses into a unifying phylogenetic concept.

Although studied in many laboratories, the most interesting medical question concerning their possible pathogenic role in humans still remains unanswered today. We know of a substantial number of other infections that persist for life within infected individuals: members of the herpesvirus group, including the Epstein-Barr virus, and also polyomaviruses, such as BK and JC viruses, may serve as examples. In virtually all of these instances these viruses are able to induce human diseases, at least in some patients: e.g. Epstein-Barr virus may cause infectious mononucleosis and in immunosuppressed patients B cell lymphomas. BK and JC virus infections do not seem to cause acute conditions; immunosuppression, however, may result in BK virus-induced haemorrhagic cystitis or, in JC, virus-caused progressive multifocal leukoencephalopathy. Thus, it is not unreasonable to suspect that persistence of at least some specific TT virus genotypes may result in some infected individuals in a definable pathogenicity. We believe that this volume provides the first hints in support of this view.

When the editors were approached by Peter Vogt to compile a specific volume of *Current Topics in Microbiology and Immunology* on TT viruses, they were pleased and quickly agreed. For comparative purposes we also invited contributions on structurally related single-stranded DNA viruses, like chicken anaemia and plant

Gemini viruses. The response of invited contributors was exceedingly good and all of them delivered their contributions in time. We gratefully appreciate their help in compiling this volume, which seems to represent the first comprehensive documentation of this interesting virus group. We are also grateful to Springer, and specifically to Anne Clauss, for gently accompanying the editing and production process of this volume.

Heidelberg, 8 June 2008
Ethel-Michele de Villiers
Harald zur Hausen

# Contents

# List of Contributors

Mauro Bendinelli
Retrovirus Center and Department of Experimental Pathology, University of Pisa 37, Via San Zeno I-56127 Pisa, Italy, bendinelli@biomed.unipi.it

P. Biagini
UMR CNRS 6578 Equipe Emergence et co-évolution virale, Etablissement Français du Sang Alpes-Méditerranée et Université de la Méditerranée, 27, Bd. Jean Moulin, 13005 Marseille, France, pbiagini-ets-ap@gulliver.fr, philippe.biagini@univmed.fr

Maarten H de Smit
Department of Molecular Genetics, Leiden Institute of Chemistry, Leiden University, Einsteinweg 55, 2333 CC Leiden, The Netherlands, m.smit@chem.leidenuniv.nl

Ethel-Michele de Villiers
Division for the Characterisation of Tumour Viruses, Deutsches Krebsforschungszentrum, Im Neuenheimer Feld 242, 69120 Heidelberg, Germany, e.devilliers@dkfz-heidelberg.de

Harald zur Hausen
Deutsches Krebsforschungszentrum, Im Neuenheimer Feld 280, 69120 Heidelberg, Germany, h.zurhausen@dkfz-heidelberg.de

Klaus Hedman
Department of Virology, Haartman Institute, University of Helsinki and Helsinki University Central Hospital Laboratory, Finland, klaus.hedman@helsinki.fi

Shigeo Hino
Medical Scanning, 4-3-1F, Kanda-Surugadai, Chiyoda, Tokyo 101-0062 Japan, shg.hino@gmail.com

Holger Jeske
Institute of Biology, Department of Molecular Biology and Plant Virology, University of Stuttgart, Pfaffenwaldring 57, 70550 Stuttgart, Germany, holger.jeske@bio.uni-stuttgart.de

Laura Kakkola
Department of Virology, Haartman Institute, University of Helsinki and Helsinki University Central Hospital Laboratory, Finland, laura.kakkola@helsinki.fi

Fabrizio Maggi
Virology Unit, Pisa University Hospital 35-37, Via San Zeno I-56127 Pisa, Italy, maggif@biomed.unipi.it

Mathieu HM Noteborn
Department of Molecular Genetics, Leiden Institute of Chemistry, Leiden University, Einsteinweg 55, 2333 CC Leiden, The Netherlands, m.noteborn@chem.leidenuniv.nl

H. Okamoto
Division of Virology, Department of Infection and Immunity, Jichi Medical University School of Medicine, 3311-1 Yakushiji, Shimotsuke-Shi, Tochigi-Ken 329-0498, Japan, hokamoto@jichi.ac.jp

David Pintel
Department of Molecular Microbiology and Immunology, University of Missouri-Columbia, USA, PintelD@health.missouri.edu

Afiono Agung Prasetyo
Division of Virology, Faculty of Medicine, Tottori University, Yonago 683-8503, Japan

Jianming Qiu
Department of Microbiology, Molecular Genetics and Immunology, University of Kansas Medical Center, USA, jqiu@kumc.edu

Karel A. Schat
Department of Microbiology and Immunology, College of Veterinary Medicine, Cornell University, Ithaca, NY 14853, USA, kas24@cornell.edu

Maria Söderlund-Venermo
Department of Virology, Haartman Institute, University of Helsinki and Helsinki University Central Hospital Laboratory, Finland, maria.soderlund-venermo@helsinki.fi

# History of Discoveries and Pathogenicity of TT Viruses

**H. Okamoto**

**Contents**



**Abstract** Since 1997, groups of novel nonenveloped DNA viruses with a circular, single-stranded (negative sense) DNA genome of 3.6–3.9 kb, 3.2 kb, or 2.8–2.9 kb in size have been discovered and designated Torque teno virus (TTV), Torque teno midi virus (TTMDV), and Torque teno mini virus (TTMV), respectively, in the floating genus *Anellovirus*. These three anelloviruses frequently and ubiquitously infect humans, and the infections are characterized by lifelong viremia and great genetic variability. Although TTV infection has been epidemiologically suggested to be associated with many diseases including liver diseases, respiratory disorders, hematological disorders, and cancer, there is no direct causal evidence for links between TTV infection and specific clinical diseases. The pathogenetic role of

H. Okamoto
Division of Virology, Department of Infection and Immunity, Jichi Medical University School of Medicine, 3311-1 Yakushiji, Shimotsuke-Shi, Tochigi-Ken, 329-0498, Japan
hokamoto@jichi.ac.jp

E.-M. de Villiers, H. zur Hausen (eds.), *TT Viruses: The Still Elusive Human Pathogens,* 

TTMV and TTMDV infections remains unknown. The changing ratio of the three anelloviruses to each other over time, relative viral load, or combination of different genotype(s) of each anellovirus may be associated with the pathogenicity or the disease-inducing potential of these three human anelloviruses. To clarify their disease association, polymerase chain reaction (PCR) systems for accurately detecting, differentiating, and quantitating all of the genotypes and/or genogroups of TTV, TTMDV, and TTMV should be established and standardized, as should methods to detect past infections and immunological responses to anellovirus infections.

## History of Discoveries

### *Discovery of Original TT Virus*

In 1997, while searching for an as-yet-unidentified hepatitis viruses, Nishizawa et al. found a novel DNA virus in a Japanese patient (initials T.T.) with posttransfusion hepatitis of unknown etiology (Nishizawa et al. 1997). The patient was 58 years old, and had received 35 units of blood during heart surgery. He had elevated alanine aminotransferase (ALT) levels 9–11 weeks after the surgery (peak, 180 IU/l at 10 weeks after transfusion). Representational difference analysis (RDA) (Lisitsyn et al. 1993) was performed for the specific amplification of nucleic acid sequences present in the serum of the patient during the period of his acute hepatitis, but which were absent before transfusion. After three courses of subtraction, a broad but clear band 0.5 kb in size was visualized on electrophoresis, and subjected to molecular cloning. Among the 36 clones obtained, varying in size from 281 to 564 bp, 9 clones of 500 bp were similar to each other, whose sequence was detectable only during the period of acute hepatitis in the index patient. A representative clone (N22) with the consensus sequence showed poor homology to any of the 1,731,752 sequences deposited in DNA databases as of 2 October 1997 (Nishizawa et al. 1997).

The N22 clone was found to originate from the genome of a nonenveloped, single-stranded DNA virus based on data using a PCR method with N22-derived primers RD037 and RD038 in the first round and RD051 and RD052 in the second round. The virus was provisionally named “TT” virus (TTV) after the initials of the index patient (Nishizawa et al. 1997). In brief, since the N22 sequence was not amplified from any of four human genomic DNA samples, a nonhost origin of N22 was attested. Furthermore, since the N22 sequence fractionated in sucrose gradient at 1.26 g/cm$^3$ and was resistant to DNase I, it was concluded to be encapsidated and thereby of viral origin. Furthermore, serum-derived TTV DNA was sensitive to mung bean nuclease but resistant to RNase A and restriction enzymes. Hence, TTV was believed to be a DNA virus that had a single-stranded genome (Nishizawa et al. 1997; Okamoto et al. 1998b). Since the density of Tween 80-treated TTV remained unchanged in sucrose gradient, TTV was understood to be a nonenveloped virus (Okamoto et al. 1998b).

The N22 sequence was extended to 3,739 nt in the prototype TTV isolate (TA278) obtained from a 34-year-old male blood donor with an elevated ALT level of 106 IU/l and containing TTV DNA in high titer ($10^5$ copies/ml) detectable by PCR with nested N22 primers, but its extreme 5′- and 3′-end sequences remained undetermined at that time (Okamoto et al. 1998b). In 1999, the presence of a GC-rich sequence of approximately 120 nt was reported (Miyata et al. 1999; Mushahwar et al. 1999), leading to the recognition of the circular nature of the TTV genome with negative polarity. The entire genomic length of the TA278 isolate was finally determined to be 3,853 nt, with the unique stem-and-loop structures in the GC-rich region, which would play a pivotal role in viral replication (Okamoto et al. 1999b, c).

## Virological Characterization of TTV

Subsequent studies revealed the following characteristics of TTV. The buoyant density in cesium chloride (CsCl) was found to be 1.31–1.33 g/cm$^3$ for TTV in serum and 1.33–1.35 g/cm$^3$ for TTV in feces (Okamoto et al. 1998a). TTV particles in the circulation were bound to immunoglobulin G (IgG), forming immune complexes (Itoh et al. 2000). Therefore, TTV-associated particles with a diameter of 30–32 nm recovered from the sera of infected humans were observed as aggregates of various sizes on electron microscopy. In contrast, TTV particles in feces exist as free virions. TTV particles of genotype 1a with a diameter of 30–32 nm and banding at 1.33–1.35 g/cm$^3$ have been visualized in fecal supernatant by immune electron microscopy using $\gamma$-globulins from human plasma containing TTV genotype 1a-specific antibodies (Itoh et al. 2000; Tsuda et al. 1999).

Kamahora et al. (2000) analyzed the messenger RNAs (mRNAs) transcribed from a plasmid containing the whole genome construct of TTV in COS1 cells. They recovered three spliced mRNAs of 3.0 kb, 1.2 kb, and 1.0 kb with common 5′- and 3′-termini, and showed that the splicing sites link distant open reading frames (ORFs) to create two new ORFs capable of encoding 286 amino acids (aa) and 289aa. Such spliced mRNAs of TTV have also been observed in actively replicating cells including bone marrow cells in infected humans (Okamoto et al. 2000d). The proposed genomic organization of the prototype TTV isolate (TA278; Accession No. AB017610) is illustrated in Fig. 1.

## Discovery of Many TTV-Like Variants in Humans

After the discovery of the original TTV isolate, with the use of primers based on a conserved untranslated region, many TTV variants with marked genetic variability were identified (Hallett et al. 2000; Hijikata et al. 1999b; Khudyakov et al. 2000; Muljono et al. 2001; Okamoto et al. 1999a, c, d, 2000d, 2001; Peng et al. 2002;

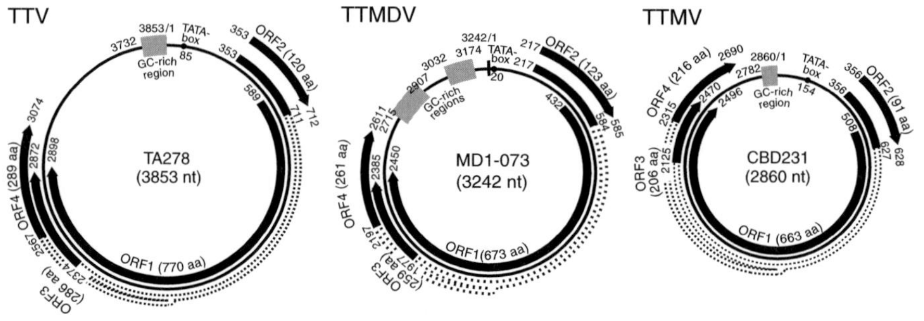


**Fig. 1** Genomic organization of the prototype TTV, TTMDV, and TTMV isolates. The *circumference* of each circle represents the relative size of the genome. The *closed arrows* represent ORFs (ORF1 to ORF4). The *open boxes* located between an *upstream closed box* and *downstream closed arrow* in ORF3 and ORF4, which encode joint proteins, represent areas corresponding to introns in the mRNA (Kamahora et al. 2000; Okamoto et al. 2000c). The *shaded box* indicates the GC-rich stretch and the *small closed circle* represents the position of the TATA box. (Reproduced from Ninomiya et al. 2007a, with permission)

Takahashi et al. 2000a; Ukita et al. 2000) and segregated into at least 39 genotypes with a difference of greater than 30% or five major genetic groups with a difference of greater than 50% difference (Okamoto et al. 2004; Peng et al. 2002; Fig. 2). Independently, the SEN virus (SENV) was discovered by RDA (patent application WO 00/28039, 2000). However, it soon became apparent that it represented different genotypes of TTV (Tanaka et al. 2001).

In addition to ubiquitous distribution of many TTV-like variants in various tissues and body fluids of humans, multiple genotypes of TTV may be found within an infected individual, often with different genotype combinations predominating in different tissues (Okamoto et al. 2001). This suggests that certain virus genotypes might be better adapted to particular cell or tissue types.

## *Discovery of Torque Teno Mini Virus*

In 2000, a small virus that was distantly related to TTV was accidentally discovered by PCR of human plasma samples using TTV-specific primers that partially matched homologous sequences but generated a noticeably shorter amplicon than expected for TTV, and was provisionally named as TTV-like mini virus (TLMV) (Takahashi et al. 2000b). The genome of TLMV consists of a circular, single-stranded DNA of approximately 2,800–2,900 nt with negative polarity (Fig. 1). The size of a TLMV virion has been estimated to be less than 30 nm in diameter (Takahashi et al. 2000b). TLMV resembles TTV in genomic structure, and also contains an arginine-rich N-terminus as well as Rep-motifs in the ORF1 region and a chicken anemia virus (CAV)-like motif in the ORF2 region (Biagini et al. 2001b, 2007; Okamoto

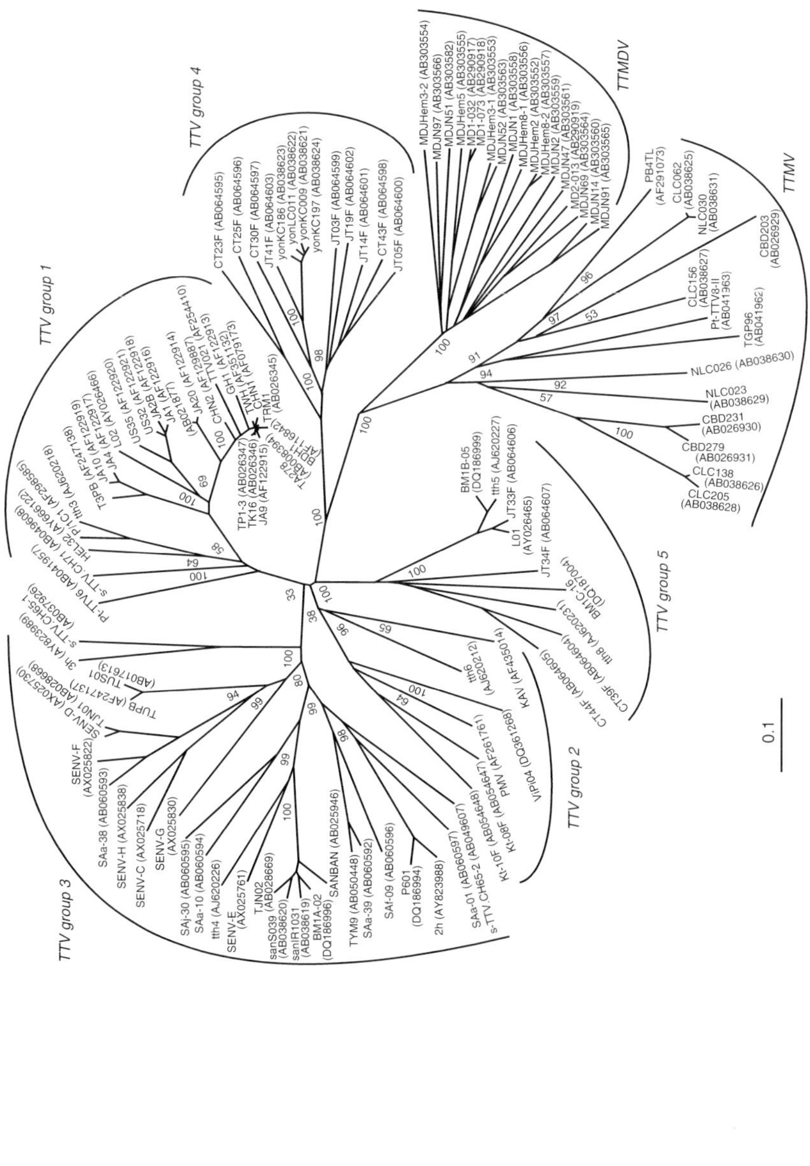


**Fig. 2** Phylogenetic tree based on the entire nucleotide sequence of ORF1 of the 81 TTV, 18 TTMDV, and 13 TTMV isolates by the neighbor-joining method (Saitou and Nei 1987). The percentage of bootstrap values generated from 1,000 samplings of the data is shown near the nodes. The *scale bar* represents the number of nucleotide substitutions per position. (Reproduced from Ninomiya et al. 2008, with permission)

et al. 2000a; Takahashi et al. 2000b). TLMV is also highly divergent: the first three TLMV sequences reported by Takahashi et al. (2000b) differed from each other by 42% at the nucleotide level and by 67% at the amino acid level.

TLMV was found to be distributed worldwide among healthy individuals (Biagini et al. 2001b; Niel et al. 2001). The prevalence of TLMV DNA among blood donors is reported to be 48%–72% (Biagini et al. 2006b; Moen et al. 2002; Niel et al. 2001). TLMV has been isolated from various body fluids and tissues, such as plasma/serum, peripheral blood mononuclear cells (PBMC), feces, saliva, bone marrow, spleen, and cervical swabs (Biagini et al. 2001b; Fornai et al. 2001; Thom et al. 2003; Vasconcelos et al. 2002).

Recently, the International Committee on Taxonomy of Viruses (ICTV) officially designated TTV and TLMV as Torque teno virus (TTV) and Torque teno mini virus (TTMV), respectively, deriving from the Latin terms *torque* meaning "necklace" and *tenuis* meaning "thin", and classified them into a novel floating genus, *Anellovirus*. These terms were chosen to reflect the organizational arrangement of the TTV genome, without changing the abbreviation TTV (Biagini et al. 2005).

## *Discovery of Torque Teno Midi Virus*

By means of the DNase-sequence independent single primer amplification (SISPA) method, two new TTV-like viruses named small anellovirus 1 (SAV1) and small anellovirus 2 (SAV2) were isolated from the sera of patients with acute viral infection syndrome (Jones et al. 2005). SAV1 possessed a genomic DNA of 2,249 nt with three putative ORFs, while SAV2 had a genomic DNA of 2,635 nt with five ORFs. These two viruses (collectively, SAVs) were provisionally classified as anelloviruses on the basis of the circular nature of the genomic DNA and the presence of regions homologous to TTV and TTMV in the largest ORF (ORF1) and noncoding region. The SAV ORF2 region was shown to possess a similar CAV-like motif as TTVs and TTMVs (Andreoli et al. 2006). Similar to TTVs and TTMVs, SAV isolates showed wide genomic variation of up to 41%. SAV has also been isolated from various body fluids and tissues, including saliva and PBMC (Biagini et al. 2006a) as well as nasopharyngeal aspirates (Chung et al. 2007). Similar to TTVs and TTMVs, SAVs were found to be common among healthy individuals and were present in 20% of French blood donors (Biagini et al. 2006a) and in 34.5% of Korean children (Chung et al. 2007). In addition, using a combined rolling-circle amplification (RCA) and SISPA approach, isolates related to SAV, but with even shorter genomes (2,002 nt and 2,454 nt) have been identified (Biagini et al. 2007); they differed from SAVs by approximately 40%.

Recently, in the process of amplifying the SAV sequence in human sera, amplicons longer than expected were obtained, and the full-length clones were 3,242–3,253 nt, with all of the characteristics of TTV-like viruses (Fig. 1). Most importantly, the previously described SAVs were found to be deletion mutants or artifacts generated during amplification of these longer isolates. These newly identified isolates were named Torque teno midi virus (Ninomiya et al. 2007a). Upon analyzing 15 additional TTMDV sequences over the entire genome (Fig. 2), it was found that they form a large clade of isolates differing in length (3,175–3,230 nt) and in sequence (up to 33% divergence at the nucleotide level and 61% divergence at the amino acid level of ORF1; Ninomiya et al. 2007b). In addition to other TTV-like characteristics, three Rep-motifs were identified in the ORF1 region, as well as putative stem-loop structures in the GC-rich region. Therefore, TTMDV is provisionally classified as the third group in the genus *Anellovirus* (Ninomiya et al. 2007a, b).

## Multiple Infections of Three Human Anelloviruses (TTV, TTMDV, and TTMV) in Humans

The development of methods for specific detection of human anelloviruses has been made more difficult by the discovery of the third human anellovirus, TTMDV. In fact, TTMDV DNA can be erroneously amplified by previously reported TTMV PCR assays (Biagini et al. 2006b; Vasconcelos et al. 2002). Biagini et al. (2006a) reported that TTMDV/SAV showed a 20% prevalence among French blood donors, which is comparable to the 9% frequency of TTMDV/SAV DNA among Italian blood donors (Andreoli et al. 2006). The selection of PCR primers and the length of the genomic region for PCR amplification crucially influence the detection of TTV, which has an extremely divergent genome (Biagini et al. 2001a; Itoh et al. 1999; Okamoto et al. 1999d). Recent surveys using primers specific for individual genotypes or genogroups of TTV, or those that differentiate TTV from TTMV sequences indicated that approximately 90% of study populations (generally healthy adults) were viremic for TTV or TTMV, with co-infection of TTV and TTMV in 44% (Biagini et al. 2006b). Therefore, it is likely that the actual prevalence of TTV, TTMV, or TTMDV DNA is higher than was previously reported.

### *PCR Assays for Differential Detection of Three Human Anelloviruses*

Despite marked divergence with a difference of greater than 50% among TTV genomes (Khudyakov et al. 2000; Okamoto et al. 2004; Peng et al. 2002; Thom et al. 2003), a difference of up to 33% among TTMDV genomes (Ninomiya et al. 2007b), and a difference of up to 40% among TTMV genomes (Biagini et al. 2001, 2007; Okamoto et al. 2000a; Takahashi et al. 2000b), there exists a highly conserved area of 130 nt located just downstream of the TATA-box in the TTV, TTMDV, and TTMV genomes. Taking advantage of this particular genomic area that is conserved among known anelloviruses, virus species-specific PCR assays were developed in which the genomic DNA of all three anelloviruses is amplified by first-round PCR with universal primers; TTV DNA, TTMDV DNA, or TTMV DNA is separately amplified by each of the three second-round PCR assays with species-specific primers (Ninomiya et al. 2008). All 257 molecular clones of the PCR products amplified by universal primers obtained from three subjects (subjects D1 to D3) who were co-infected with TTV, TTMDV, and TTMV were classifiable into TTV, TTMDV, or TTMV by the three differential PCR assays. The reliability of these assays for classification was confirmed by phylogenetic analysis (Fig. 3). When the newly developed PCR assays were applied to serum samples from adults in the general population in Japan, high prevalence rates of TTV DNA (100%), TTMV DNA (82%), and TTMDV DNA (75% or 75/100) were found.

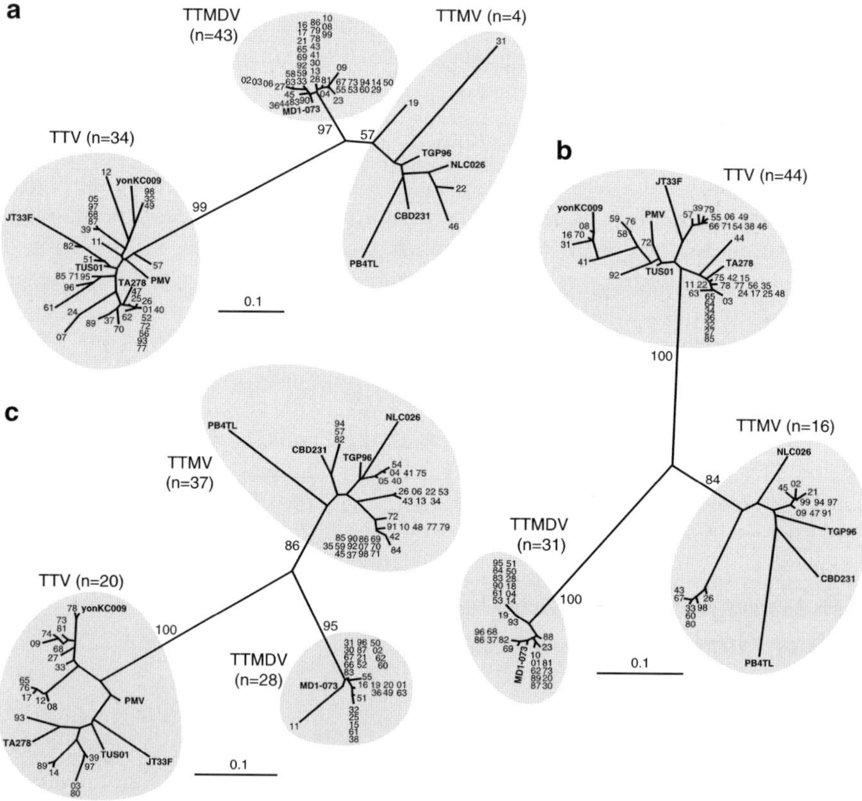


**Fig. 3 a–c** Phylogenetic trees constructed by the neighbor-joining method based on the partial nucleotide sequences (85–97 nt) of TTV, TTMDV, and TTMV isolates obtained from subject D1 (**a**), subject D2 (**b**), and subject D3 (**c**). The representative TTV, TTMDV, and TTMV isolates whose full-length sequence is known are indicated in *boldface*. Bootstrap values are indicated for the major nodes as a percentage obtained from 1,000 resamplings of the data. *Bar*, 0.1 nucleotide substitutions per site. (Reproduced from Ninomiya et al. 2008, with permission)

## *Early Acquisition of Dual or Triple Infection of three Human Anelloviruses During Infancy*

Ten cord blood samples, 44 serum samples from infants (<1 year of age), and 261 serum samples from individuals age 1 year or older were tested for TTV, TTMDV, and TTMV DNA. Although none of the 10 umbilical cord blood samples had the genomic DNA of any of TTV, TTMDV, and TTMV, rapidly rising prevalence rates of TTV, TTMDV, and TTMV infections were noted over the subsequent months (Table 1), reaching a rate of 100% within the first 2 years of life (Ninomiya et al. 2008). This finding corroborated earlier studies reporting that TTV infection prevails in childhood (Kazi et al. 2000; Lin et al. 2002; Toyoda et al. 1999; Vasconcelos et al. 2002). Therefore, maternal transmission may play only a minor role in early acquisition of TTV and TTMV as well as TTMDV in infants, and early acquisition via horizontal infection during infancy may be common to all three anelloviruses (Ninomiya et al. 2008). TTV and TTMV genomes have been detected in feces (Biagini et al. 2001b; Okamoto et al. 1998a), saliva (Biagini et al. 2001b; Deng et al. 2000; Gallian et al. 2000; Ishikawa et al. 1999; Vasconcelos et al. 2002), and breast milk (Gerner et al. 2000; Toyoda et al. 1999), suggesting transmission of anelloviruses including TTMDV via horizontal routes.

Due to the frequent dual or triple infection of TTV, TTMDV, and TTMV noted in infants and adults, it remains unknown whether each of the three anelloviruses can replicate and maintain the persistent carrier state independently or in concert with one or two of the other anelloviruses. However, the presence of single infection of TTV or TTMV in some infected hosts suggests that TTV, TTMDV, and TTMV can independently infect susceptible hosts and replicate in their tissues or organs (Ninomiya et al. 2008).

**Table 1** Age-specific prevalence of TTV, TTMDV, and TTMV DNAs. (Reproduced from Ninomiya et al. 2008, with permission)

| Age (years) and sample type | No. of samples tested | No. of samples (%) with: TTV DNA | TTMDV DNA | TTMV DNA |
|---|---|---|---|---|
| <1 | | | | |
| Cord blood | 10 | 0 | 0 | 0 |
| Serum[a] | 44 | 35 (79.5) | 22 (50.0) | 33 (75.0) |
| Subtotal | 54 | 35 (64.8) | 22 (40.7) | 33 (61.1) |
| ≥1 (serum) | | | | |
| 1 | 28 | 28 (100) | 28 (100) | 28 (100) |
| 2–4 | 29 | 29 (100) | 28 (96.6) | 28 (96.6) |
| 5–9 | 42 | 41 (97.6) | 36 (85.7) | 41 (97.6) |
| 10–19 | 62 | 61 (98.4) | 48 (77.4) | 55 (88.7) |
| 20–49 | 32 | 32 (100) | 22 (68.8) | 24 (75.0) |
| 50–81 | 68 | 68 (100) | 53 (77.9) | 58 (85.3) |
| Subtotal | 261 | 259 (99.2) | 215 (82.4) | 234 (89.7) |
| Total | 315 | 294 (93.3) | 237 (75.2) | 267 (84.8) |

[a]The serum samples were from infants <1 year of age (9–364 days old)

## Pathogenesis and Clinical Manifestations

### *Pathogenesis*

Recently, two cell culture systems supporting virus replication and virion formation of transfected cloned TTV have been reported, but in neither case was efficient virus propagation achieved (Kakkola et al. 2007; Leppik et al. 2007). Infection with TTV is characterized by persistent lifelong viremia in humans, with circulating levels of up to $10^6$ copies/ml in the general population (Hu et al. 2005; Pistello et al. 2001). As an indication of its great replicative capacity in vivo, a study on the kinetics of clearance of TTV viremia suggested that a daily production rate of $>10^{10}$ virions is required to maintain the observed levels of viremia (Maggi et al. 2001b). TTV replicates in the liver, as documented by in situ hybridization and/or quantitative PCR (Ohbayashi et al. 2001; Rodríguez-Iñigo et al. 2000), as well as the detection of double-stranded replicative intermediates in the liver (Okamoto et al. 2000e) and a high level of TTV excreted in bile (Luo et al. 2000; Nakagawa et al. 2000; Ukita et al. 1999). Excretion of TTV in bile may be the main source of TTV in the gastrointestinal tract and its fecal shedding. Replication of TTV is not restricted to the liver. High viral loads, double-stranded replicative forms of TTV DNA, and mRNA transcripts have also been detected in lung tissues (Okamoto et al. 2001), pancreas (Okamoto et al. 2001), bone marrow (Fanci et al. 2004; Kikuchi et al. 2000; Okamoto et al. 2000b), spleen (Okamoto et al. 2001; Jelcic et al. 2004), and other lymphoid tissues (Kakkola et al. 2004). TTV DNA is frequently detectable in PBMCs (Barril et al. 2000; Lopez-Alcorocho et al. 2000; Okamoto et al. 1999a, 2000c; Okamura et al. 1999). In these lymphoid cells, TTV shows a very broad tropism with viral DNA detected not only in T and B lymphocytes, monocytes, and natural killer (NK) cells (Maggi et al. 2001; Takahashi et al. 2002; Zhong et al. 2002) but also in granulocytes and other polymorphonuclear cells (Maggi et al. 2001a; Takahashi et al. 2002).

### *Host's Immune Response*

As described above, TTV infection is likely acquired early in infancy, which may lead to substantial immune tolerance, as is well known for other viruses such as hepatitis B virus. The persistent nature of infection and co-infection of multiple TTV variants in the circulation, representing repeated rounds of concurrent infection, may suggest the presence of mechanisms of immune evasion that have evolved in TTV to establish persistent infection in immunocompetent individuals. However, antibodies against native TTV virions (Tsuda et al. 1999) or recombinant ORF1 proteins (Handa et al. 2000; Ott et al. 2000) are detected in viremic and nonviremic individuals, and TTV particles in the circulation are often bound to IgG, forming immune complexes (Itoh et al. 2000). Nevertheless, at present there is no evidence to indicate an association with diseases evoked by the deposition of immune complex, such as glomerulonephritis.

TTV viral loads have been shown to increase in human immunodeficiency virus (HIV)-infected patients who are progressing toward AIDS, and a high TTV viral load was associated with a low CD4 cell count, indicating a potential role of the immune system in controlling TTV replication (Christensen et al. 2000; Shibayama et al. 2001; Thom and Petrik 2007; Touinssi et al. 2001; Zhong et al. 2002). Although it remains unclear what role the immune system plays in the natural course of TTV infection, TTV may act as an opportunistic pathogen in immunocompromised hosts, analogous to human cytomegalovirus in AIDS patients.

## *Disease Associations*

Although TTV, TTMDV, and TTMV are potentially related to many diseases, conflicting opinions exist on their disease-causing potential due to their nearly universal presence in human populations (i.e., lack of controls). It is possible that certain genotypes/genogroups of TTV, TTMDV, and TTMV may be specifically pathogenic. Interestingly, the expression of genotype 1–ORF1 in transgenic mice, leading to production of a spliced protein, caused pathological changes in the kidneys. Expression of the TTV protein seemed to interfere with differentiation of renal epithelial cells (Yokoyama et al. 2002). It has recently been suggested that subgenomic fragments of TTV identified in human sera could have some role in diseases as is the case with plant geminiviruses (Leppik et al. 2007). Unfortunately, however, there are only a few reports to support the disease-inducing potential of TTV.

### Possible Association of TTV with Liver Diseases

When the original TTV isolate was discovered, TTV (genotype 1) detectable by N22 PCR was found in three of five patients with posttransfusion acute hepatitis of unknown etiology, and the presence of TTV genotype 1 was closely associated with the serum ALT level (Nishizawa et al. 1997). When a serum sample obtained from an 11-month-old infant with acute hepatitis of unknown etiology who had been transiently infected with genotype 1 TTV ($10^5$ copies/ml) was inoculated intravenously into a naive chimpanzee, TTV DNA was transiently detected in the chimpanzee at 5–15 weeks postinoculation (PI), with the titer peaking at 12–13 weeks PI (Tawara et al. 2000). This viremia was accompanied by an abrupt elevation of the serum $\alpha$-glutathione-*S*-transferase level and mild elevation of ALT level. Histological changes in biopsied liver samples (ballooning degeneration of hepatocytes) were observed in association with the reduction in TTV DNA titer and the appearance of IgM-class and IgG-class anti-TTV (genotype 1) antibodies, suggesting that TTV genotype 1 has hepatitis-inducing capacity. Shibata et al. (2000) reported that TTV genotype 1 may play a role in the pathogenesis of non-A, -B, or -C fulminant hepatic failure (FHF), since the TTV-positive rate was significantly higher among the group with non-A, -B, or -C FHF (6/7 or 86%) than among the group with non-A, -B, or -C acute hepatitis (4/17 or 24%; $p=0.005$). Tajiri et al. (2001) presented three infants diagnosed with idiopathic neonatal hepatitis and intrahepatic fatty degeneration and whose livers were infected with TTV detectable by N22 PCR. Several other studies also suggested that TTV of genotype 1 may be more pathogenic than other genotypes of TTV in children with liver disease of unknown etiology (Okamura et al. 2000; Sugiyama et al. 2000). On the other hand, infection with TTV genotype 12 or 16, which was described as SENV-D and SENV-H, respectively, was found to be much more prevalent among patients with transfusion-associated non-A to -E hepatitis than among transfused patients without hepatitis in the United States (92% vs 24%, $p<0.001$) (Umemura et al. 2001). Foschini et al. (2001) reported an Italian case of TTV (genotype 13)-related acute recurrent hepatitis, with clinicopathological findings reinforcing the suggestion that TTV can be responsible for a mild form of liver disease. Other investigators (Charlton et al. 1998; Ikeda et al. 1999; Kanda et al. 1999; Okamura et al. 2000; Tanaka et al. 1998, 2000; Tuveri et al. 2000) also showed an association between the prevalence of TTV and/or TTV load and various hepatic disorders. It was also suggested that persistent TTV infections could contribute to cryptogenic hepatic failure in hemophiliacs (Takayama et al. 1999).

However, contradictory results showing that TTV is not associated with ALT levels or with any form of hepatitis (posttransfusion, chronic idiopathic, acute or fulminant) have also been presented (Hijikata et al. 1999a; Hsieh et al. 1999; Naoumov et al. 1998; Niel et al. 1999; Prati et al. 1999; Viazov et al. 1998). Additionally for SENV, it was reported that infection is not related to hepatitis or other liver disease (Akiba et al. 2005; Kao et al. 2002; Schroter et al. 2003; Umemura et al. 2001).

Tokita et al. (2002b) reported that a high TTV viral load was independently associated with the complication of hepatocellular carcinoma (HCC) and that it may have prognostic significance in patients with hepatitis C virus (HCV)-related chronic liver disease. There are two possible explanations for the findings in this report. One explanation is that high TTV viremia has an adverse effect on the progression of chronic liver disease in concert with concurrent HCV infection and may be associated with the development of HCC. Zein et al. (1999) reported that TTV infection was more prevalent among patients with advanced HCV-associated liver disease (decompensated cirrhosis and HCC) than among those with stable disease (chronic hepatitis and compensated cirrhosis). Moriyama et al. (2001) reported that the score of irregular regeneration of hepatocytes among TTV-infected cirrhotic patients with chronic hepatitis C was higher than that among patients who were not infected with TTV. These findings suggest that TTV plays a role in the development of cirrhosis and subsequent complications. However, another explanation is possible. A correlation between high TTV titer and a low CD4 T cell count among patients infected with HIV type 1, and the possible prognostic significance of TTV viral load in immunocompromised patients, has been reported (Christensen et al. 2000; Shibayama et al. 2001). Therefore, it is likely that an impaired immune system or suppression of the immune system is involved in elevated TTV viremia in HCC patients.

At present, despite evidence for hepatic replication of TTV, TTV does not fulfill the criteria for being a hepatitis virus. For TTV to be characterized as a hepatitis virus, direct causal evidence of cytopathology or specific inflammatory changes associated with replication as well as statistical difference in comparison with controls in terms of TTV prevalence, loads, sequence variation, genotype distribution, or co-infection among liver disease patients have to be demonstrated in future studies.

### Possible Association of TTV with Respiratory Diseases

It has been suggested that TTV infection has a potential role in children with respiratory diseases. Importantly, TTV replication has been shown to occur in lung tissues (Okamoto et al. 2001; Bando et al. 2001). Infection with TTV coincided with mild rhinitis in a neonate (Biagini et al. 2003), and children hospitalized with acute respiratory disease or with bronchiectasis showed higher TTV viral loads than controls (Maggi et al. 2003b, c; Pifferi et al. 2006). In addition, children with high TTV loads in nasal specimens were shown to have worse spirometric values, and TTV was suggested to contribute to the pathogenesis of asthma (Pifferi et al. 2005). Although the precipitating factors of idiopathic pulmonary fibrosis have not been elucidated, Bando et al. (2001) first reported the influence of TTV infection on the disease activity and prognosis of idiopathic pulmonary fibrosis. Furthermore, the association between TTV infection and the complication of lung cancer in patients with idiopathic pulmonary fibrosis has been reported (Bando et al. 2008).

These observations raise interesting questions about the pathophysiological significance of TTV in the respiratory tract of infected humans. However, it remains undetermined if TTV is the cause or the result of the disease. Interestingly, it was suggested that TTV replication could twist the immunobalance toward the T helper 2 cell (Th2) response that is known to have a role in the pathogenesis of asthma (Pifferi et al. 2005).

### Possible Association of TTV with Hematological Disorders

Hepatitis-associated aplastic anemia mainly occurs after acute non-A, non-B, non-C hepatitis (Brown et al. 1997). A high level of TTV replication in bone marrow has been suggested as being responsible for hepatitis-associated aplastic anemia of unexplained etiology (Kikuchi et al. 2000). A possible association between TTV infection and aplastic anemia has also been suggested by others (Miyamoto et al. 2000). However, contradictory results were also presented indicating that TTV is not associated with post-hepatitis aplastic anemia (Poovorawan et al. 2001; Safadi et al. 2001).

As for the association of TTV with hematopoietic malignancies, TTV DNA was detected in lymphocytes circulating in the lymph nodes of patients with B-cell lymphomas and those with Hodgkin's disease (Garbuglia et al. 2003). It was postulated that TTV could somehow modulate the infected T cells and thus play some role in the pathogenesis of lymphomas.

### Possible Association of TTV with Cancer

In addition to hepatocellular carcinoma, lung cancer, and hematopoietic malignancies, the possible involvement of TTV infection in other malignancies or malignant changes has been suggested. TTV DNA has been detected in a wide variety of neoplastic tissues (de Villiers et al. 2002). Co-infection of TTV genotype 1 and human papillomavirus was related to poor outcome of laryngeal carcinoma (Szladek et al. 2005). However, similar to other small DNA viruses such as parvoviruses and circoviruses, there is no plausible causal association of TTV infection with tumorigenesis or malignant transformation of cells.

## Conclusions

Although TTV was discovered relatively recently in 1997 (Nishizawa et al. 1997), it seems to be a well-adapted virus of humans that has been a persistent source of infection since the distant past. TTV infections are extremely prevalent even in healthy individuals. The high prevalence of TTV is not usual per se among viruses; for example, papillomaviruses and herpesviruses are frequently found in healthy individuals. However, TTV differs from all other known viruses in its ability to sustain lifelong viremia, i.e., to actively replicate and continuously produce virus in the blood for decades, even in healthy individuals. Due to its global distribution and persistent viremia in human populations, there is no definitive causal association of TTV infection with the diseases that have been investigated. It may be possible that TTVs per se do not cause any disease and do not have any adverse effect whatsoever on human health (Griffiths 1999; Simmonds et al. 1999). On the other hand, as with opportunistic pathogens, disease may appear only under exceptional circumstances. In some virus infections, the viral load is a critical determinant of development of disease. It has been suggested that TTV could be a commensal in normal conditions, incapable of exceeding the threshold of a disease-causing load (Griffiths 1999), and if TTV is a genuine symbiont, the virus should benefit the host, which is an intriguing aspect hitherto unexplored with TTV.

At present, however, we cannot rule out the possibility that some isolates/genotypes could be more pathogenic than others (Maggi et al. 2003a, 2007; Okamura et al. 2000; Sugiyama et al. 2000; Tokita et al. 2002a), as is well known for human papillomaviruses, either alone or co-infecting with other TTV strains or other pathogens, that they have an effect on the outcome or progression of some disease(s), and that the level of TTV in tissue and/or in the bloodstream could affect any of the disease conditions.

Our newly developed PCR method with high sensitivity and reliability has revealed frequent dual or triple infection of these three anelloviruses, even in infants (Ninomiya et al. 2008). The pathogenetic role of TTMV and TTMDV infections remains unknown. The changing ratio of the three anelloviruses to each other over time, their relative viral load, or the combination of different genotypes of each anellovirus may be associated with the pathogenicity or the disease-inducing potential of these three human anelloviruses. In this context, further efforts are warranted to develop methods to separately or simultaneously quantify the genomic DNA of the three anelloviruses and to clarify their disease association. PCR systems for detecting, differentiating, and quantitating all of the genotypes and/or genogroups of TTV, TTMDV, and TTMV should be established and standardized, as should methods to detect past infections and immunological responses to anellovirus infections.

## References

Akiba J, Umemura T, Alter HJ, et al (2005) SEN virus: epidemiology and characteristics of a transfusion-transmitted virus. Transfusion 45:1084–1088

Andreoli E, Maggi F, Pistello M, et al (2006) Small anellovirus in hepatitis C patients and healthy controls. Emerg Infect Dis 12:1175–1176

Bando M, Ohno S, Oshikawa K, et al (2001) Infection of TT virus in patients with idiopathic pulmonary fibrosis. Respir Med 95:935–942

Bando M, Takahashi M, Ohno S, et al (2008) Torque teno virus DNA titre elevated in idiopathic pulmonary fibrosis with primary lung cancer. Respirology 13:263–269

Barril G, Lopez-Alcorocho JM, Bajo A, et al (2000) Prevalence of TT virus in serum and peripheral mononuclear cells from a CAPD population. Perit Dial Int 20:65–68

Biagini P, Gallian P, Attoui H, et al (2001a) Comparison of systems performance for TT virus detection using PCR primer sets located in non-coding and coding regions of the viral genome. J Clin Virol 22:91–99

Biagini P, Gallian P, Attoui H, et al (2001b) Genetic analysis of full-length genomes and subgenomic sequences of TT virus-like mini virus human isolates. J Gen Virol 82:379–383

Biagini P, Charrel RN, de Micco P, et al (2003) Association of TT virus primary infection with rhinitis in a newborn. Clin Infect Dis 36:128–129

Biagini P, Todd D, Bendinelli M, et al (2005) Anellovirus. In: Fauquet CM, Mayo MA, Maniloff J, Desselberger U, Ball LA (eds) Virus taxonomy: classification and nomenclature of viruses, eight report of the international committee on taxonomy of viruses. Elsevier/Academic Press, London, pp 335–341

Biagini P, de Micco P, de Lamballerie X (2006a) Identification of a third member of the *Anellovirus* genus ("small anellovirus") in French blood donors. Arch Virol 151:405–408

Biagini P, Gallian P, Cantaloube JF, et al (2006b) Distribution and genetic analysis of TTV and TTMV major phylogenetic groups in French blood donors. J Med Virol 78:298–304

Biagini P, Uch R, Belhouchet M, et al (2007) Circular genomes related to anelloviruses identified in human and animal samples by using a combined rolling-circle amplification/sequence-independent single primer amplification approach. J Gen Virol 88:2696–2701

Brown KE, Tisdale J, Barrett AJ, et al (1997) Hepatitis-associated aplastic anemia. N Engl J Med 336:1059–1064

Charlton M, Adjei P, Poterucha J, et al (1998) TT-virus infection in North American blood donors, patients with fulminant hepatic failure, and cryptogenic cirrhosis. Hepatology 28:839–842

Christensen JK, Eugen-Olsen J, SŁrensen M, et al (2000) Prevalence and prognostic significance of infection with TT virus in patients infected with human immunodeficiency virus. J Infect Dis 181:1796–1799
Chung JY, Han TH, Koo JW, et al (2007) Small anellovirus infections in Korean children. Emerg Infect Dis 13:791–793
de Villiers EM, Schmidt R, Delius H, et al (2002) Heterogeneity of TT virus related sequences isolated from human tumour biopsy specimens. J Mol Med 80:44–50
Deng X, Terunuma H, Handema R, et al (2000) Higher prevalence and viral load of TT virus in saliva than in the corresponding serum: another possible transmission route and replication site of TT virus. J Med Virol 62:531–537
Fanci R, De Santis R, Zakrzewska K, et al (2004) Presence of TT virus DNA in bone marrow cells from hematologic patients. New Microbiol 27:113–117
Fornai C, Maggi F, Vatteroni ML, et al (2001) High prevalence of TT virus (TTV) and TTV-like minivirus in cervical swabs. J Clin Microbiol 39:2022–2024
Foschini MP, Morandi L, Macchia S, et al (2001) TT virus-related acute recurrent hepatitis. Histological features of a case and review of the literature. Virchows Arch 439:752–755
Gallian P, Biagini P, Zhong S, et al (2000) TT virus: a study of molecular epidemiology and transmission of genotypes 1, 2 and 3. J Clin Virol 17:43–49
Garbuglia AR, Iezzi T, Capobianchi MR, et al (2003) Detection of TT virus in lymph node biopsies of B-cell lymphoma and Hodgkin's disease, and its association with EBV infection. Int J Immunopathol Pharmacol 16:109–118
Gerner P, Oettinger R, Gerner W, et al (2000) Mother-to-infant transmission of TT virus: prevalence, extent and mechanism of vertical transmission. Pediatr Infect Dis J 19:1074–1077
Griffiths P (1999) Time to consider the concept of a commensal virus? Rev Med Virol 9:73–74
Hallett RL, Clewley JP, Bobet F, et al (2000) Characterization of a highly divergent TT virus genome. J Gen Virol 81:2273–2279
Handa A, Dickstein B, Young NS, et al (2000) Prevalence of the newly described human circovirus, TTV, in United States blood donors. Transfusion 40:245–251
Hijikata M, Iwata K, Ohta Y, et al (1999a) Genotypes of TT virus (TTV) compared between liver disease patients and healthy individuals using a new PCR system capable of differentiating 1a and 1b types from others. Arch Virol 144:2345–2354
Hijikata M, Takahashi K, Mishiro S (1999b) Complete circular DNA genome of a TT virus variant (isolate name SANBAN) and 44 partial ORF2 sequences implicating a great degree of diversity beyond genotypes. Virology 260:17–22
Hsieh SY, Wu YH, Ho YP, et al (1999) High prevalence of TT virus infection in healthy children and adults and in patients with liver disease in Taiwan. J Clin Microbiol 37:1829–1831
Hu YW, Al-Moslih MI, Al Ali MT, et al (2005) Molecular detection method for all known genotypes of TT virus (TTV) and TTV-like viruses in thalassemia patients and healthy individuals. J Clin Microbiol 43:3747–3754
Ikeda H, Takasu M, Inoue K, et al (1999) Infection with an unenveloped DNA virus (TTV) in patients with acute or chronic liver disease of unknown etiology and in those positive for hepatitis C virus RNA. J Hepatol 30:205–212
Ishikawa T, Hamano Y, Okamoto H (1999) Frequent detection of TT virus in throat swabs of pediatric patients. Infection 27:298
Itoh K, Takahashi M, Ukita M, et al (1999) Influence of primers on the detection of TT virus DNA by polymerase chain reaction. J Infect Dis 180:1750–1751
Itoh Y, Takahashi M, Fukuda M, et al (2000) Visualization of TT virus particles recovered from the sera and feces of infected humans. Biochem Biophys Res Commun 279:718–724
Jelcic I, Hotz-Wagenblatt A, Hunziker A, et al (2004) Isolation of multiple TT virus genotypes from spleen biopsy tissue from a Hodgkin's disease patient: genome reorganization and diversity in the hypervariable region. J Virol 78:7498–7507
Jones MS, Kapoor A, Lukashov VV, et al (2005) New DNA viruses identified in patients with acute viral infection syndrome. J Virol 79:8230–8236

Kakkola L, Kaipio N, Hokynar K, et al (2004) Genoprevalence in human tissues of TT-virus genotype 6. Arch Virol 149:1095–1106
Kakkola L, Tommiska J, Boele LC, et al (2007) Construction and biological activity of a full-length molecular clone of human Torque teno virus (TTV) genotype 6. FEBS J 274:4719–4730
Kamahora T, Hino S, Miyata H (2000) Three spliced mRNAs of TT virus transcribed from a plasmid containing the entire genome in COS1 cells. J Virol 74:9980–9986
Kanda T, Yokosuka O, Ikeuchi T, et al (1999) The role of TT virus infection in acute viral hepatitis. Hepatology 29:1905–1908
Kao JH, Chen W, Chen PJ, et al (2002) Prevalence and implication of a newly identified infectious agent (SEN virus) in Taiwan. J Infect Dis 185:389–392
Kazi A, Miyata H, Kurokawa K, et al (2000) High frequency of postnatal transmission of TT virus in infancy. Arch Virol 145:535–540
Khudyakov YE, Cong ME, Nichols B, et al (2000) Sequence heterogeneity of TT virus and closely related viruses. J Virol 74:2990–3000
Kikuchi K, Miyakawa H, Abe K, et al (2000) Indirect evidence of TTV replication in bone marrow cells, but not in hepatocytes, of a subacute hepatitis/aplastic anemia patient. J Med Virol 61:165–170
Leppik L, Gunst K, Lehtinen M, et al (2007) In vivo and in vitro intragenomic rearrangement of TT viruses. J Virol 81:9346–9356
Lin HH, Kao JH, Lee PI, Chen DS (2002) Early acquisition of TT virus in infants: possible minor role of maternal transmission. J Med Virol 66:285–290
Lisitsyn N, Lisitsyn N, Wigler M (1993) Cloning the differences between two complex genomes. Science 259:946–951
Lopez-Alcorocho JM, Mariscal LF, de Lucas S, et al (2000) Presence of TTV DNA in serum, liver and peripheral blood mononuclear cells from patients with chronic hepatitis. J Viral Hepat 7:440–447
Luo K, Liang W, He H, et al (2000) Experimental infection of nonenveloped DNA virus (TTV) in rhesus monkey. J Med Virol 61:159–164
Maggi F, Fornai C, Zaccaro L, et al (2001a) TT virus (TTV) loads associated with different peripheral blood cell types and evidence for TTV replication in activated mononuclear cells. J Med Virol 64:190–194
Maggi F, Pistello M, Vatteroni M, et al (2001b) Dynamics of persistent TT virus infection, as determined in patients treated with alpha interferon for concomitant hepatitis C virus infection. J Virol 75:11999–12004
Maggi F, Marchi S, Fornai C, et al (2003a) Relationship of TT virus and Helicobacter pylori infections in gastric tissues of patients with gastritis. J Med Virol 71:160–165
Maggi F, Pifferi M, Fornai C, et al (2003b) TT virus in the nasal secretions of children with acute respiratory diseases: relations to viremia and disease severity. J Virol 77:2418–2425
Maggi F, Pifferi M, Tempestini E, et al (2003c) TT virus loads and lymphocyte subpopulations in children with acute respiratory diseases. J Virol 77:9081–9083
Maggi F, Andreoli E, Riente L, et al (2007) Torquetenovirus in patients with arthritis. Rheumatology (Oxford) 46:885–886
Miyamoto M, Takahashi H, Sakata I, et al (2000) Hepatitis-associated aplastic anemia and transfusion-transmitted virus infection. Intern Med 39:1068–1070
Miyata H, Tsunoda H, Kazi A, et al (1999) Identification of a novel GC-rich 113-nucleotide region to complete the circular, single-stranded DNA genome of TT virus, the first human circovirus. J Virol 73:3582–3586
Moen EM, Huang L, Grinde B (2002) Molecular epidemiology of TTV-like mini virus in Norway. Arch Virol 147:181–185
Moriyama M, Matsumura H, Shimizu T, et al (2001) Histopathologic impact of TT virus infection on the liver of type C chronic hepatitis and liver cirrhosis in Japan. J Med Virol 64:74–81
Muljono DH, Nishizawa T, Tsuda F, et al (2001) Molecular epidemiology of TT virus (TTV) and characterization of two novel TTV genotypes in Indonesia. Arch Virol 146:1249–1266

Mushahwar IK, Erker JC, Muerhoff AS, et al (1999) Molecular and biophysical characterization of TT virus: evidence for a new virus family infecting humans. Proc Natl Acad Sci U S A 96:3177–3182

Nakagawa N, Ikoma J, Ishihara T, et al (2000) Biliary excretion of TT virus (TTV). J Med Virol 61:462–467

Naoumov NV, Petrova EP, Thomas MG, et al (1998) Presence of a newly described human DNA virus (TTV) in patients with liver disease. Lancet 352:195–197

Niel C, Lampe E (2001) High detection rates of TTV-like mini virus sequences in sera from Brazilian blood donors. J Med Virol 65:199–205

Niel C, de Oliveira JM, Ross RS, et al (1999) High prevalence of TT virus infection in Brazilian blood donors. J Med Virol 57:259–263

Ninomiya M, Nishizawa T, Takahashi M, et al (2007a) Identification and genomic characterization of a novel human torque teno virus of 3.2 kb. J Gen Virol 88:1939–1944

Ninomiya M, Takahashi M, Shimosegawa T, et al (2007b) Analysis of the entire genomes of fifteen torque teno midi virus variants classifiable into a third group of genus Anellovirus. Arch Virol 152:1961–1975

Ninomiya M, Takahashi M, Nishizawa T, et al (2008) Development of PCR assays with nested primers specific for differential detection of three human anelloviruses and early acquisition of dual or triple infection during infancy. J Clin Microbiol 46:507–514

Nishizawa T, Okamoto H, Konishi K, et al (1997) A novel DNA virus (TTV) associated with elevated transaminase levels in posttransfusion hepatitis of unknown etiology. Biochem Biophys Res Commun 241:92–97

Ohbayashi H, Tanaka Y, Ohoka S, et al (2001) TT virus is shown in the liver by in situ hybridization with a PCR-generated probe from the serum TTV-DNA. J Gastroenterol Hepatol 16:424–428

Okamoto H, Akahane Y, Ukita M, et al (1998a) Fecal excretion of a nonenveloped DNA virus (TTV) associated with posttransfusion non-A-G hepatitis. J Med Virol 56:128–132

Okamoto H, Nishizawa T, Kato N, et al (1998b) Molecular cloning and characterization of a novel DNA virus (TTV) associated with posttransfusion hepatitis of unknown etiology. Hepatol Res 10:1–16

Okamoto H, Kato N, Iizuka H, et al (1999a) Distinct genotypes of a nonenveloped DNA virus associated with posttransfusion non-A to G hepatitis (TT virus) in plasma and peripheral blood mononuclear cells. J Med Virol 57:252–258

Okamoto H, Nishizawa T, Ukita M (1999b) A novel unenveloped DNA virus (TT virus) associated with acute and chronic non-A to G hepatitis. Intervirology 42:196–204

Okamoto H, Nishizawa T, Ukita M, et al (1999c) The entire nucleotide sequence of a TT virus isolate from the United States (TUS01): comparison with reported isolates and phylogenetic analysis. Virology 259:437–448

Okamoto H, Takahashi M, Nishizawa T, et al (1999d) Marked genomic heterogeneity and frequent mixed infection of TT virus demonstrated by PCR with primers from coding and noncoding regions. Virology 259:428–436

Okamoto H, Nishizawa T, Tawara A, et al (2000a) Species-specific TT viruses in humans and nonhuman primates and their phylogenetic relatedness. Virology 277:368–378

Okamoto H, Nishizawa T, Tawara A, et al (2000b) TT virus mRNAs detected in the bone marrow cells from an infected individual. Biochem Biophys Res Commun 279:700–707

Okamoto H, Takahashi M, Kato N, et al (2000c) Sequestration of TT virus of restricted genotypes in peripheral blood mononuclear cells. J Virol 74:10236–10239

Okamoto H, Takahashi M, Nishizawa T, et al (2000d) Replicative forms of TT virus DNA in bone marrow cells. Biochem Biophys Res Commun 270:657–662

Okamoto H, Ukita M, Nishizawa T, et al (2000e) Circular double-stranded forms of TT virus DNA in the liver. J Virol 74:5161–5167

Okamoto H, Nishizawa T, Takahashi M, et al (2001) Heterogeneous distribution of TT virus of distinct genotypes in multiple tissues from infected humans. Virology 288:358–368

Okamoto H, Nishizawa T, Takahashi M (2004) Torque teno virus (TTV): molecular virology and clinical implications. In: Mushahwar IK (ed) Viral hepatitis: molecular biology, diagnosis, epidemiology and control. Elsevier, Amsterdam
Okamura A, Yoshioka M, Kubota M, et al (1999) Detection of a novel DNA virus (TTV) sequence in peripheral blood mononuclear cells. J Med Virol 58:174–177
Okamura A, Yoshioka M, Kikuta H, et al (2000) Detection of TT virus sequences in children with liver disease of unknown etiology. J Med Virol 62:104–108
Ott C, Duret L, Chemin I, et al (2000) Use of a TT virus ORF1 recombinant protein to detect anti-TT virus antibodies in human sera. J Gen Virol 81:2949–2958
Peng YH, Nishizawa T, Takahashi M, et al (2002) Analysis of the entire genomes of thirteen TT virus variants classifiable into the fourth and fifth genetic groups, isolated from viremic infants. Arch Virol 147:21–41
Pifferi M, Maggi F, Andreoli E, et al (2005) Associations between nasal torquetenovirus load and spirometric indices in children with asthma. J Infect Dis 192:1141–1148
Pifferi M, Maggi F, Caramella D, et al (2006) High torquetenovirus loads are correlated with bronchiectasis and peripheral airflow limitation in children. Pediatr Infect Dis J 25:804–808
Pistello M, Morrica A, Maggi F, et al (2001) TT virus levels in the plasma of infected individuals with different hepatic and extrahepatic pathology. J Med Virol 63:189–195
Poovorawan Y, Tangkijvanich P, Theamboonlers A, et al (2001) Transfusion transmissible virus TTV and its putative role in the etiology of liver disease. Hepatogastroenterology 48:256–260
Prati D, Lin YH, De Mattei C, et al (1999) A prospective study on TT virus infection in transfusion-dependent patients with beta-thalassemia. Blood 93:1502–1505
Rodríguez-Iñigo E, Casqueiro M, Bartolomé J, et al (2000) Detection of TT virus DNA in liver biopsies by in situ hybridization. Am J Pathol 156:1227–1234
Safadi R, Or R, Ilan Y, et al (2001) Lack of known hepatitis virus in hepatitis-associated aplastic anemia and outcome after bone marrow transplantation. Bone Marrow Transplant 27:183–190
Saitou N, Nei M (1987) The neighbor-joining method: a new method for reconstructing phylogenetic trees. Mol Biol Evol 4:406–425
Schroter M, Laufs R, Zollner B, et al (2003) A novel DNA virus (SEN) among patients on maintenance hemodialysis: prevalence and clinical importance. J Clin Virol 27:69–73
Shibata M, Morizane T, Baba T, et al (2000) TT virus infection in patients with fulminant hepatic failure. Am J Gastroenterol 95:3602–3606
Shibayama T, Masuda G, Ajisawa A, et al (2001) Inverse relationship between the titre of TT virus DNA and the CD4 cell count in patients infected with HIV. Aids 15:563–570
Simmonds P, Prescott LE, Logue C, et al (1999) TT virus—part of the normal human flora? J Infect Dis 180:1748–1750
Sugiyama K, Goto K, Ando T, et al (2000) TT virus infection in Japanese children: isolates from genotype 1 are overrepresented in patients with hepatic dysfunction of unknown etiology. Tohoku J Exp Med 191:233–239
Szladek G, Juhasz A, Kardos G, et al (2005) High co-prevalence of genogroup 1 TT virus and human papillomavirus is associated with poor clinical outcome of laryngeal carcinoma. J Clin Pathol 58:402–405
Tajiri H, Tanaka T, Sawada A, et al (2001) Three cases with TT virus infection and idiopathic neonatal hepatitis. Intervirology 44:364–369
Takahashi K, Hijikata M, Samokhvalov EI, et al (2000a) Full or near full length nucleotide sequences of TT virus variants (Types SANBAN and YONBAN) and the TT virus-like mini virus. Intervirology 43:119–123
Takahashi K, Iwasa Y, Hijikata M, et al (2000b) Identification of a new human DNA virus (TTV-like mini virus, TLMV) intermediately related to TT virus and chicken anemia virus. Arch Virol 145:979–993
Takahashi M, Asabe S, Gotanda Y, et al (2002) TT virus is distributed in various leukocyte subpopulations at distinct levels, with the highest viral load in granulocytes. Biochem Biophys Res Commun 290:242–248

Takayama S, Miura T, Matsuo S, et al (1999) Prevalence and persistence of a novel DNA TT virus (TTV) infection in Japanese haemophiliacs. Br J Haematol 104:626–629

Tanaka H, Okamoto H, Luengrojanakul P, et al (1998) Infection with an unenveloped DNA virus (TTV) associated with posttransfusion non-A to G hepatitis in hepatitis patients and healthy blood donors in Thailand. J Med Virol 56:234–238

Tanaka Y, Hayashi J, Ariyama I, et al (2000) Seroepidemiology of TT virus infection and relationship between genotype and liver damage. Dig Dis Sci 45:2214–2220

Tanaka Y, Primi D, Wang RY, et al (2001) Genomic and molecular evolutionary analysis of a newly identified infectious agent (SEN virus) and its relationship to the TT virus family. J Infect Dis 183:359–367

Tawara A, Akahane Y, Takahashi M, et al (2000) Transmission of human TT virus of genotype 1a to chimpanzees with fecal supernatant or serum from patients with acute TTV infection. Biochem Biophys Res Commun 278:470–476

Thom K, Petrik J (2007) Progression towards AIDS leads to increased Torque teno virus and Torque teno minivirus titers in tissues of HIV infected individuals. J Med Virol 79:1–7

Thom K, Morrison C, Lewis JC, et al (2003) Distribution of TT virus (TTV), TTV-like minivirus, and related viruses in humans and nonhuman primates. Virology 306:324–333

Tokita H, Murai S, Kamitsukasa H, et al (2002a) TT virus of certain genotypes may reduce the platelet count in patients who achieve a sustained virologic response to interferon treatment for chronic hepatitis C. Hepatol Res 23:105–114

Tokita H, Murai S, Kamitsukasa H, et al (2002b) High TT virus load as an independent factor associated with the occurrence of hepatocellular carcinoma among patients with hepatitis C virus-related chronic liver disease. J Med Virol 67:501–509

Touinssi M, Gallian P, Biagini P, et al (2001) TT virus infection: prevalence of elevated viraemia and arguments for the immune control of viral load. J Clin Virol 21:135–141

Toyoda H, Naruse M, Yokozaki S, et al (1999) Prevalence of infection with TT virus (TTV), a novel DNA virus, in healthy Japanese subjects, newborn infants, cord blood and breast milk. J Infect 38:198–199

Tsuda F, Okamoto H, Ukita M, et al (1999) Determination of antibodies to TT virus (TTV) and application to blood donors and patients with post-transfusion non-A to G hepatitis in Japan. J Virol Methods 77:199–206

Tuveri R, Jaffredo F, Lunel F, et al (2000) Impact of TT virus infection in acute and chronic, viral- and non viral-related liver diseases. J Hepatol 33:121–127

Ukita M, Okamoto H, Kato N, et al (1999) Excretion into bile of a novel unenveloped DNA virus (TT virus) associated with acute and chronic non-A-G hepatitis. J Infect Dis 179:1245–1248

Ukita M, Okamoto H, Nishizawa T, et al (2000) The entire nucleotide sequences of two distinct TT virus (TTV) isolates (TJN01 and TJN02) remotely related to the original TTV isolates. Arch Virol 145:1543–1559

Umemura T, Yeo AE, Sottini A, et al (2001) SEN virus infection and its relationship to transfusion-associated hepatitis. Hepatology 33:1303–1311

Vasconcelos HC, Cataldo M, Niel C (2002) Mixed infections of adults and children with multiple TTV-like mini virus isolates. J Med Virol 68:291–298

Viazov S, Ross RS, Varenholz C, et al (1998) Lack of evidence for an association between TTV infection and severe liver disease. J Clin Virol 11:183–187

Yokoyama H, Yasuda J, Okamoto H, et al (2002) Pathological changes of renal epithelial cells in mice transgenic for the TT virus ORF1 gene. J Gen Virol 83:141–150

Zein NN, Arslan M, Li H, et al (1999) Clinical significance of TT virus infection in patients with chronic hepatitis C. Am J Gastroenterol 94:3020–3027

Zhong S, Yeo W, Tang M, et al (2002) Frequent detection of the replicative form of TT virus DNA in peripheral blood mononuclear cells and bone marrow cells in cancer patients. J Med Virol 66:428–434

# Classification of TTV and Related Viruses (Anelloviruses)

**P. Biagini**

**Contents**



**Abstract** Ten years after the identification of the first partial sequences of Torque teno virus (TTV), more than 200 full-length related genomes have been characterized in humans and in several animal species. As suspected in the earlier stages of their description, a considerable genetic variability characterizes TTV and related viruses, the current members of the floating genus *Anellovirous*. Since information related to anelloviruses diversity is in constant evolution, the challenge in their taxonomic classification is to take into account all pertinent parameters, along with the taxonomic situation of other viruses having circular single-stranded DNA genomes. Past, present and future phylogenetic and taxonomic considerations are exposed.

## Introduction

Since the characterization of the first Torque teno virus (TTV) clone at the end of 1997, the number of partial and complete sequences related to this new class of viruses (characterizing the genus *Anellovirus*) has dramatically increased during the past 10 years. Thus, a query of the GeneBank database reveals more than 5,200

P. Biagini
UMR CNRS 6578 Equipe <<Emergence et co-évolution virale>>,
Etablissement Français du Sang Alpes-Méditerranée et Université de la Méditerranée,
27, Bd. Jean Moulin, 13005 Marseille, France
pbiagini-ets-ap@gulliver.fr / philippe.biagini@univmed.fr

matching nucleotide results, as of April 2008, including more than 200 complete sequences identified in human and in several animal species. As suspected in the earlier stages of their description, a considerable genetic diversity characterizes *Anellovirus* members. The challenge in taxonomic classification of TTV and related viruses is to take into account all pertinent parameters, along with the taxonomic situation of other viruses having circular single-stranded DNA genomes. This article presents an overview of such approaches and perspectives.

## Identification of TTV and Related Viruses

TTV was initially discovered in 1997 by means of a subtractive technique (representational difference analysis, RDA) performed on paired sera from a Japanese patient (initials T.T.) hospitalized for a heart operation (Nishizawa et al. 1997). Driver and tester samples used for the RDA procedure were collected before and during an alanine aminotransferase (ALT) peak noted in the serum of the patient. The 500-nt clone (N22) obtained was used to design a polymerase chain reaction (PCR) system, giving a 396-bp product which was also detectable in the sera of two additional patients. As demonstrated in the original publication, the three partial sequences showed up to 13.2% and 8% of divergence at the nucleotide and amino acids level, respectively. This was the first indication that a noticeable genetic diversity characterized this new class of viruses.

Additional data rapidly followed with the analysis of novel partial sequences (~222 bp) in the N22 zone: two distinct genotypes of TTV were proposed in 1998 (Okamoto et al. 1998), followed by the identification of a third genotype early in 1999 (Biagini et al. 1999; Mushahwar et al. 1999). At that time, it was proposed that isolates could be assigned to distinct genotypes if their sequence divergence was above 27%–30%. Based on extensive PCR studies located on the N22 region, new genogroups were progressively identified the same year, adding up to 16 TTV genotypes described (Okamoto et al. 1999b); this value was further extended to 23 genotypes 1 year later (Okamoto and Mayumi 2001a).

The next step in the understanding of TTV genetic diversity was brought by the characterization of highly divergent full-length sequences. Thus, isolates TUS01 and SANBAN, along with related variants identified in 1999–2000 (including SEN virus), depicted new complete sequences harbouring a considerable divergence (>40%, nt level) when compared to the reference sequence TA278/TTV-1a (Fig. 1; Hijikata et al. 1999; Miyata et al. 1999; Okamoto et al. 1999a; Biagini et al. 2000). Such identifications reinforced the view that TTV members exhibited a considerable genetic diversity, which was still barely revealed, and thus allowed investigators to raise the description of such viral diversity to major TTV phylogenetic groups. A third major group was rapidly identified following the characterization of the PMV isolate (Hallett et al. 2000). This new virus was detected in the serum of a patient (PM) with acute non-A–E hepatitis. The same year, Takahashi and colleagues identified a new variant of TTV designated as YONBAN (meaning the fourth, in Japanese),

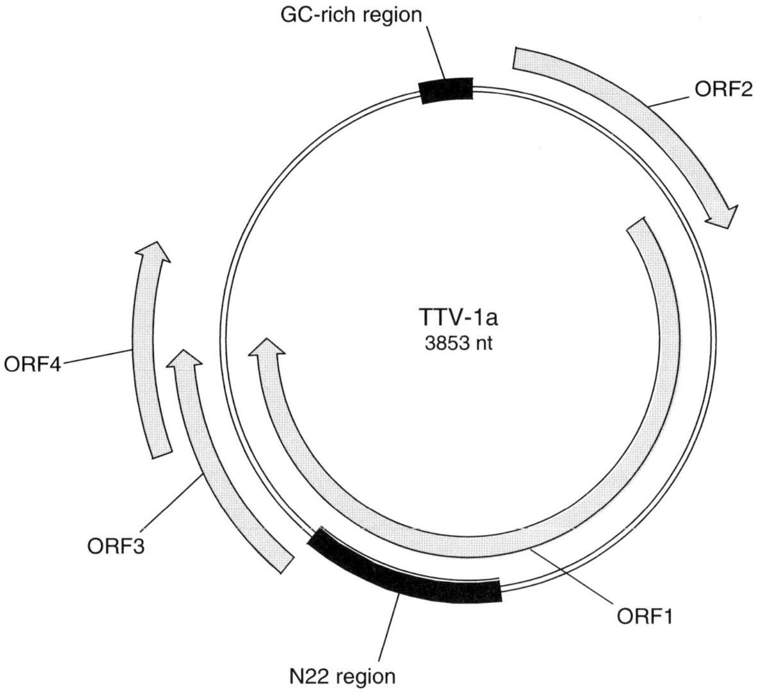


**Fig. 1** Genome organization of TTV prototype (isolate TTV-1a). The *arrows* represent major ORFs (longer than 50aa). *GC-rich* and *N22* regions are indicated

the prototype of the fourth major group (Takahashi et al. 2000b). At the end of 2001, a fifth major phylogenetic group was identified following the characterization of several full-length TTV isolates in serum samples from infants (Fig. 2; Peng et al. 2002). The corresponding prototype isolate was designated JT33F.

Despite their high genetic divergence, all of these isolates show a conserved genetic organization with a coding region containing the major open reading frame, ORF1, an overlapping ORF2 and several ORFs, and an untranslated region containing a GC-rich zone (Fig. 1); they also exhibit genomes of comparable sizes (~3,800 nt).

The characterization of shorter viral genomes in 2000 revealed that the heterogeneity of these newly discovered viruses was not solely restricted to sequence diversity but may also be extended to genome length (Takahashi et al. 2000a, 2000b). Indeed, TTV-like mini viruses (TTMV, initially TLMV) harbour genomes of about 2,800 nt (prototype CBD231), with low sequence homology to other TTV members except in a short portion of the genome (~130 nt) located upstream from the ORF1. However, they present a genetic organization similar to that observed for TTVs. Despite a limited number of complete sequences described at present, TTMV members present a high genetic heterogeneity, at least of the same magnitude of the one identified for TTV (Takahashi et al. 2000b; Biagini et al. 2001). A clustering in at least four major phylogenetic groups was initially proposed (Biagini et al.

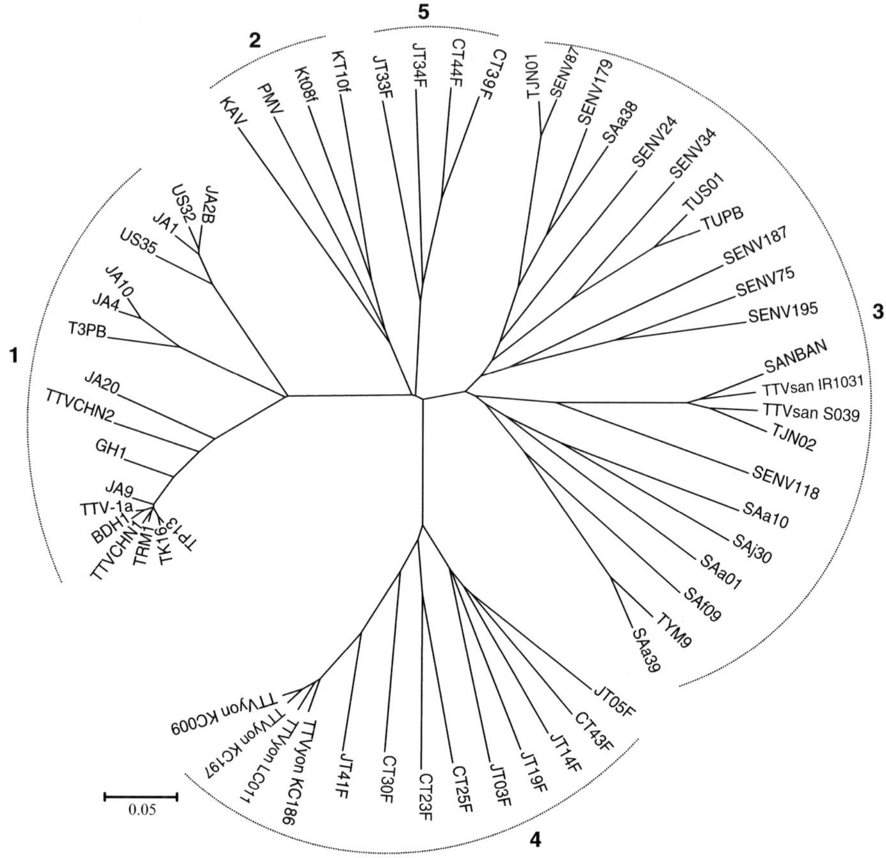


**Fig. 2** Neighbour-joining phylogenetic tree built with nearly full-length sequences of TTV (situation at the end of 2001). Classification of TTV variants into five distinct clusters is shown

2006), but the identification of highly divergent sequences using a sequence-independent approach (RCA-SISPA) recently increased this pattern of divergence (Biagini et al. 2007).

A new representative member of the genus *Anellovirus* was identified in 2005 following the characterization by Jones and colleagues of two highly divergent circular sequences (~2,200 nt and ~2,600 nt) in the plasma of patients with acute viral infection syndrome (Jones et al. 2005). As for TTMV, these isolates, provisionally named "small anelloviruses" (SAV, 1 and 2), harboured a conserved genetic organization like other anelloviruses, but showed an extremely short non-coding region. New full-length sequences related to SAV were further characterized in 2007: one study described the smallest SAV genome identified to date in humans (2,002 nt) (Biagini et al. 2007), while two other studies described numerous isolates with identical genomic length (~3,200 nt, acronym TTMDV), intermediate between those of TTV and TTMV (Ninomiya et al. 2007a, 2007b).

Interestingly, it was demonstrated that SAV1/SAV2 and TTMDV isolates are closely related, differing mainly by an extension of the non-coding region in the latter, and further suggested that SAV isolates were deletion mutants of TTMDV forms. Whether or not these conclusions are applicable to the various isolates of SAV identified to date is a question which still needs to be answered. In the same way, the recent demonstration of in vivo intragenomic rearrangement of some TTV isolates had also generated new questions related to the diversity of anelloviruses (Leppik et al. 2007).

Anellovirus isolates have been also identified in various animal species. In 1999, Verschoor and colleagues characterized for the first time partial TTV sequences in sera from both common chimpanzees and pygmy chimpanzees (Verschoor et al. 1999). The following years, full-length sequences were progressively detected in other non-human primates (macaques, tamarins and douroucoulis), tupaias, pets (cats, dogs) and farm animals such as pigs (Leary et al. 1999; Abe et al. 2000; Cong et al. 2000; Okamoto et al. 2000a, 2001b, 2002; McKeown et al. 2004; Bigarre et al. 2005; Niel et al. 2005; Biagini et al. 2007). Viral genomic sequences identified in these hosts were generally highly divergent when compared to thosc obtained from human biological samples, but exhibited a genetic organization typical to anelloviruses. Moreover, collected data confirmed that the genetic variability among anelloviruses in infected animal species is high, potentially of the same order of magnitude as the one already identified in humans. For example, two isolates exhibiting 46% of nucleotide sequence divergence (ORF1) were characterized recently in one saliva sample from a domestic cat (Biagini et al. 2007).

## Other Circular Single-Stranded DNA Viruses

Viruses with a circular single-stranded DNA genome have been characterized in various non-human hosts, i.e. animals, plants and bacteria, well before the discovery of anelloviruses. On the basis of molecular and structural considerations, they are grouped into several taxonomic families, as defined by the current official classification of the International Committee on Taxonomy of Viruses (ICTV) (Fig. 3):

1. Viruses infecting bacteria are grouped into two viral families, *Inoviridae* (genera *Inovirus* and *Plectrovirus*) and *Microviridae* (genera *Bdellomicrovirus*, *Chlamydiamicrovirus*, *Microvirus*, and *Spiromicrovirus*) (Day and Hendrix 2005; Fane 2005). Inoviruses and microviruses harbour a monopartite genome.
2. Families *Geminiviridae* and *Nanoviridae* contain viruses infecting plants (Stanley et al. 2005; Vetten et al. 2005). The former is composed of four genera, *Begomovirus*, *Curtovirus*, *Mastrevirus*, and *Topocuvirus*, whereas two genera, *Babuvirus* and *Nanovirus* describe the latter. Geminiviruses harbour a mono- or bipartite genome with an ambisense organization. Nanoviruses have a genome consisting of 6–10 components and present a unisense organization.
3. Members of the family *Circoviridae* infect animal hosts (avian and porcine species) and are taxonomically grouped into two genera named *Circovirus*

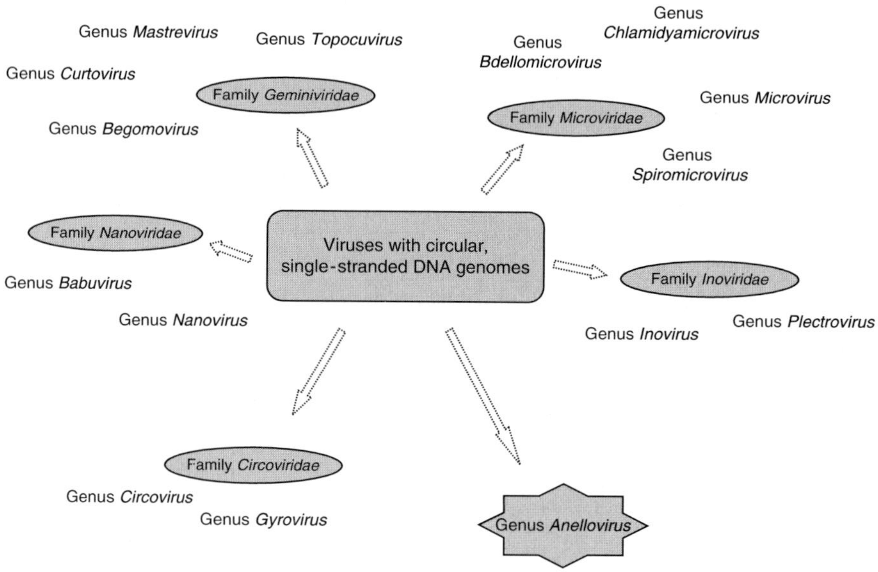


**Fig. 3** Current official taxonomic description of viruses with circular, single-stranded DNA genomes (ICTV Report 2005)

(type-species *porcine circovirus-1*, PCV-1) and *Gyrovirus* (type-species *chicken anaemia virus*, CAV) (Todd et al. 2005). Circoviruses have a monopartite genome, with an ambisense (genus *Circovirus*) or unisense organization (genus *Gyrovirus*).

Despite their classification into separate families, *Circoviridae*, *Geminiviridae* and *Nanoviridae* members present some conserved regions in their genomes which coincide globally with highly conserved functions such as replication (Rep protein). An important feature of viruses infecting animals is that CAV exhibits no close phylogenetic relationship to the other species in the genus *Circovirus*, as revealed by comparison of Rep proteins among all members of the *Circoviridae* family. Results of electron microscopic studies demonstrating that viruses in the genera *Gyrovirus* and *Circovirus* are not structurally related strengthen such divergence (Crowther et al. 2003).

## Phylogenetic and Taxonomic Considerations

### *Past*

Early in the description of TTV and related species, the distinct biophysical and molecular characteristics exhibited by this new class of viruses suggested their taxonomic grouping into a new genus or family. Indeed, sequence comparisons with other viruses having circular single-stranded DNA genome globally failed to show consistent similarities apart from the presence of some conserved motifs related to the Rep protein (ORF1). In 1999–2000 the terms *Circinoviridae* and *Paracircoviridae* were cited in the literature (Mushahwar et al. 1999; Takahashi et al. 2000b), but the official taxonomic nomenclature was determined by the Study Group (Current) ICTV *Circoviridae-Anellovirus* Study Group: P. Biagini (Chair), M. Bendinelli, S. Hino, L. Kakkola, A. Mankertz, C. Niel, H. Okamoto, S. Raidal, C.G. Teo, D. Todd, and D. McGeoch (Chair of the Vertebrate Virus Subcommittee) in charge of this classification, based on the virus classification system approved by ICTV:

1. The creation of the genus *Anellovirus*, unattached to any existing viral family (i.e. *floating*), was established at the end of 2001. The term *Anello* was chosen since it is derived from Latin *Anello*, "the ring", and relates to the circular nature of their genome.
2. The viruses identified at that time, i.e. TTV and TLMV, were officially designated:

   - Torque teno virus (TTV) replacing the name "TT virus" or "transfusion transmitted virus"
   - Torque teno mini virus (TTMV) replacing the name "TT virus-like mini virus"

   The term Torque teno is derived from Latin Torques", *the necklace*, and Latin *Tenuis*, "thin", and relates to the circular, single-stranded nature of its genome.

3. The designation of anellovirus isolates was established as: TTV-isolate name (or TTMV-isolate name).

The description of the genus *Anellovirus* in the VIII$^{th}$ report of the ICTV was effective in 2004 and corresponded to the first step in the official classification of anelloviruses (Fig. 3; Biagini et al. 2005).

At that time, the creation of a new genus instead of a new viral family was favoured due to the relatively limited information collected on the other TTV-related viruses. However, the creation of this floating genus was considered as a prerequisite for the future creation of a specific family hosting several genera.

Comparisons of sequences with a functional/structural entity are logically used in the official taxonomic classification of viruses. Early studies dealing with the phylogeny of anelloviruses proposed to perform phylogenetic analyses on short nucleotide sequences (<200 nt) located in the historical N22 region or in the untranslated portion of the genome. However, these approaches appear to be unreliable for precise taxonomic studies, the latter approach being also biased by the occurrence of recombination events which are statistically more frequent in this location than in the rest of the genome (Worobey et al. 2000; Lemey et al. 2002; Manni et al. 2002).

So, nearly full-length sequences or sequences deduced from the largest ORF (ORF1) were used for the description of anelloviruses diversity. The ORF1 encompasses at least 60% of the entire genome of the various anelloviruses described, and is presumed to encode the putative capsid protein and replication-associated protein (Rep). Corresponding transcripts have been identified in both in vivo and in vitro studies confirming its functionality (Kamahora et al. 2000; Okamoto et al. 2000b).

Based on these considerations, the first official report on anelloviruses taxonomy listed the various anellovirus isolates identified at that time. The diversity and phylogenetic relationships harboured by TTV sequences (the most-described isolates) were also depicted by the description of five major phylogenetic groups separated by a sequence divergence of about 45% (nt level) (Fig. 2). Finally, the extreme genetic divergence also identified between TTV, TTMV and animal isolates, along with potential similarities with other taxa, were considered as supports for future taxonomic evolutions.

## ***Present and Future***

In contrast to the lack of precise morphologic and structural data, information related to anellovirus diversity is in constant evolution. Since the first official taxonomic report, a growing number of sequences have been submitted in databases, and new TTV-related forms have been identified, i.e. SAV/TTMDV (Fig. 4). According to new data collected in the literature, and on the basis of the work of the Study Group, evolution of anelloviruses classification is currently under evaluation by the ICTV.

The first step in such an evolution would concern the creation of a distinct and structured taxonomic family comprising a large number of viruses that are extremely variable in sequences, but with similar type of genome organization. Such a taxonomic family has to be defined by several genera and species identified using precise demarcation criteria (Ball 2005). It should be based on the analysis of the entire ORF1 (Fig. 5a), which could be considered as representative of the whole genome identity as demonstrated by distance matrix comparisons (Fig. 5b).

As general guidelines on this taxonomic approach:

- The term *Anelloviridae* would be the logical evolution of the precedent designation.
- Nine to ten genera provisionally named
    - *Alphatorquevirus*
    - *Betatorquevirus*
    - *Gammatorquevirus*, ...

  should make up this new family.
- More than 40 type-species would be identifiable.
- A precise and unambiguous naming of the various species listed should be performed.

A subsequent step would be to take into account similarities identified with CAV, the type-species of genus *Gyrovirus*.

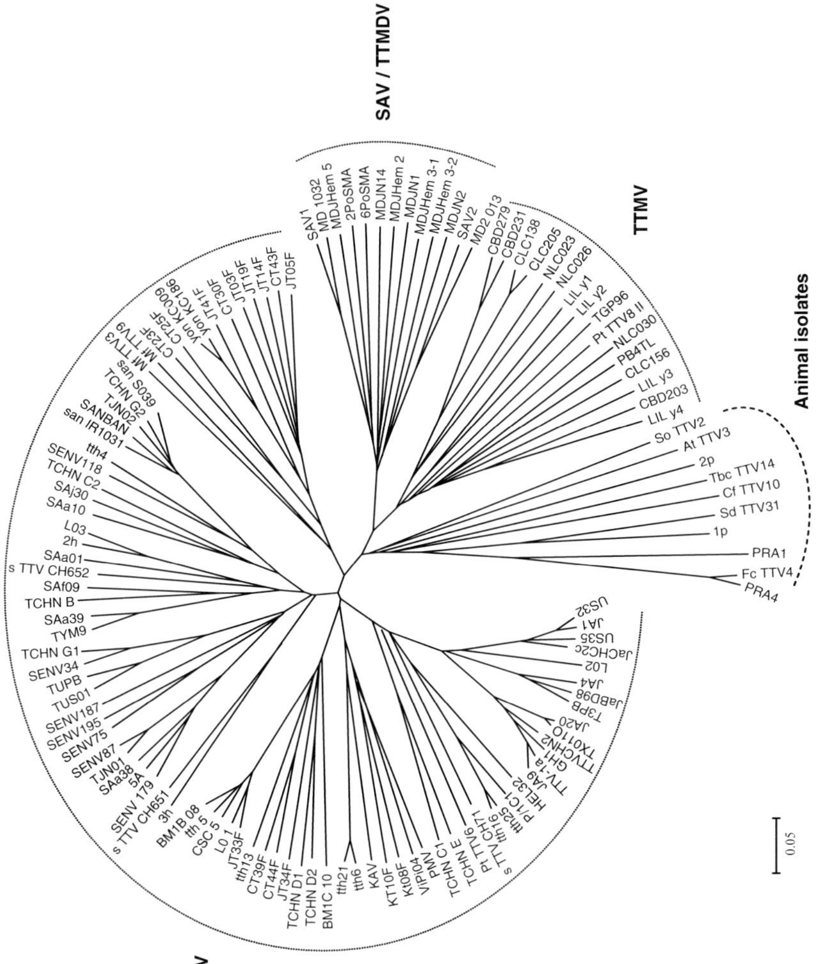


**Fig. 4** Neighbour-joining phylogenetic tree built with representative human and animal anellovirus sequences (ORF1, nt). (Situation at the end of 2007)

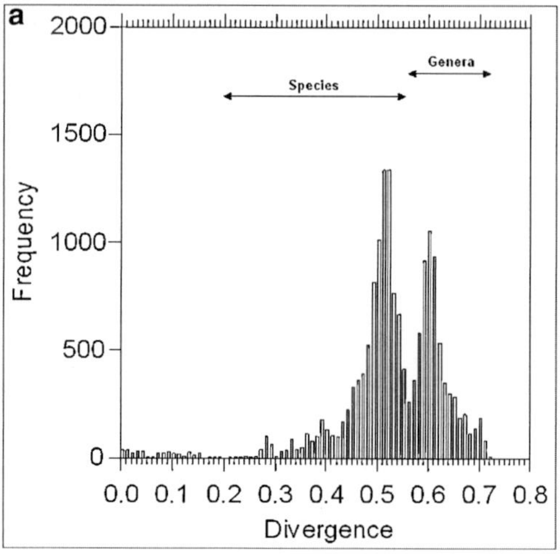


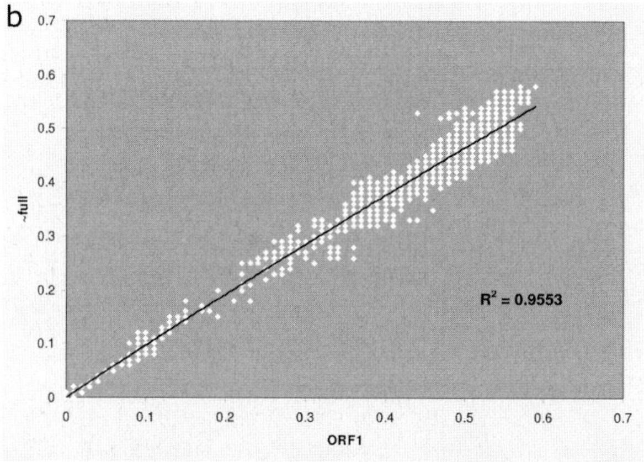


**Fig. 5** **a** Frequency distribution of pairwise-distance comparisons (divergence) among human and animal anellovirus ORF1 sequences. Peaks possibly corresponding to various genera and species are identified. **b** Example of pairwise-distance matrices comparison: ~full-length sequences vs ORF1 (Torque teno virus isolates). Correlation coefficient (*R*2) is shown

Features in common are:

- Comparable genomic organizations. (Both harbour negative sense genomes, with overlapping ORFs; Simmonds 2002).
- Similar transcription profiles with spliced transcripts (Kamada et al. 2006).
- A non-coding region containing G+C rich sequences and some conserved patterns (Takahashi et al. 2000b).
- Some conserved motifs related to the Rep protein on the $ORF1_{TTVs}/VP1_{CAV}$, and one highly conserved motif ($WX_7HX_3CXCX_5H$) located on the $ORF2_{TTVs}/VP2_{CAV}$ (Takahashi et al. 2000b; Biagini et al. 2001; Andreoli et al. 2006).

- Supposed identical structural and functional roles (capsid and replication functions) for $ORF1_{TTVs}/VP1_{CAV}$ (Crowther et al. 2003).
- Protein phosphatase activity demonstrated for $ORF2_{TTVs}/VP2_{CAV}$ (Peters et al. 2002).
- The ability of $\text{“ORF3”}_{TTVs}/VP3_{CAV}$ to induce apoptosis in some human transformed cells (Kooistra et al. 2004).

Based on these considerations, a review of CAV's taxonomic position could be considered, leading to the re-classification of gyroviruses to the future *Anelloviridae* family (for example by the creation of distinct subfamilies).

## References

Abe K, Inami T, Ishikawa K, et al (2000) TT virus infection in nonhuman primates and characterization of the viral genome: identification of simian TT virus isolates. J Virol 74:1549–1553

Andreoli E, Maggi F, Pistello M, et al (2006) Small anellovirus in hepatitis C patients and healthy controls. Emerg Infect Dis 12:1175–1176

Ball LA (2005) The universal taxonomy of viruses in theory and practise. In: Virus taxonomy, VIIIth report of the International Committee for the Taxonomy of Viruses. Elsevier/Academic Press, London, pp 3–8

Biagini P, Gallian P, Attoui H, et al (1999) Determination and phylogenetic analysis of partial sequences from TT virus isolates. J Gen Virol 80:419–424

Biagini P, Attoui H, Gallian P, et al (2000) Complete sequences of two highly divergent European isolates of TT virus. Biochem Biophys Res Commun 271:837–841

Biagini P, Gallian P, Attoui H, et al (2001) Genetic analysis of full-length genomes and subgenomic sequences of TT virus-like mini virus human isolates. J Gen Virol 82:379–383

Biagini P, Todd D, Bendinelli M, et al (2005) Anellovirus. In: Virus taxonomy, VIIIth report of the International Committee for the Taxonomy of Viruses. Elsevier/Academic Press, London, pp 335–341

Biagini P, Gallian P, Cantaloube JF, et al (2006) Distribution and genetic analysis of TTV and TTMV major phylogenetic groups in French blood donors. J Med Virol 78:298–304

Biagini P, Uch R, Belhouchet M, et al (2007) Circular genomes related to anelloviruses identified in human and animal samples by using a combined rolling-circle amplification/sequence-independent single primer amplification approach. J Gen Virol 88:2696–2701

Bigarre L, Beven V, de Boisseson C, et al (2005) Pig anelloviruses are highly prevalent in swine herds in France. J Gen Virol 86:631–635

Cong ME, Nichols B, Dou XG, et al (2000) Related TT viruses in chimpanzees. Virology 274:343–355

Crowther RA, Berriman JA, Curran WL, et al (2003) Comparison of the structures of three circoviruses: chicken anemia virus, porcine circovirus type 2, and beak and feather disease virus. J Virol 77:13036–13041

Day LA, Hendrix RW (2005) Inoviridae. In: Virus taxonomy, VIIIth report of the International Committee for the Taxonomy of Viruses. Elsevier/Academic Press, London, pp 279–287

Fane B (2005) Microviridae. In: Virus taxonomy, VIIIth report of the International Committee for the Taxonomy of Viruses. Elsevier/Academic Press, London, pp 289–299

Hallett RL, Clewley JP, Bobet F, et al (2000) Characterization of a highly divergent TT virus genome. J Gen Virol 81:2273–2279

Hijikata M, Takahashi K, Mishiro S (1999) Complete circular DNA genome of a TT virus variant (isolate name SANBAN) and 44 partial ORF2 sequences implicating a great degree of diversity beyond genotypes. Virology 260:17–22

Jones MS, Kapoor A, Lukashov VV, et al (2005) New DNA viruses identified in patients with acute viral infection syndrome. J Virol 79:8230–8236

Kamada K, Kuroishi A, Kamahora T, et al (2006) Spliced mRNAs detected during the life cycle of chicken anemia virus. J Gen Virol 87:2227–2233

Kamahora T, Hino S, Miyata H (2000) Three spliced mRNAs of TT virus transcribed from a plasmid containing the entire genome in COS1 cells. J Virol 74:9980–9986

Kooistra K, Zhang YH, Henriquez NV, et al (2004) TT virus-derived apoptosis-inducing protein induces apoptosis preferentially in hepatocellular carcinoma-derived cells. J Gen Virol 85:1445–1450

Leary TP, Erker JC, Chalmers ML, et al (1999) Improved detection systems for TT virus reveal high prevalence in humans, non-human primates and farm animals. J Gen Virol 80:2115–2120

Lemey P, Salemi M, Bassit L, et al (2002) Phylogenetic classification of TT virus groups based on the N22 region is unreliable. Virus Res 85:47–59

Leppik L, Gunst K, Lehtinen M, et al (2007) In vivo and in vitro intragenomic rearrangement of TT viruses. J Virol 81:9346–9356

Manni F, Rotola A, Caselli E, et al (2002) Detecting recombination in TT virus: a phylogenetic approach. J Mol Evol 55:563–572

McKeown NE, Fenaux M, Halbur PG, et al (2004) Molecular characterization of porcine TT virus, an orphan virus, in pigs from six different countries. Vet Microbiol 104:113–117

Miyata H, Tsunoda H, Kazi A, et al (1999) Identification of a novel GC-rich 113-nucleotide region to complete the circular, single-stranded DNA genome of TT virus, the first human circovirus. J Virol 73:3582–3586

Mushahwar IK, Erker JC, Muerhoff AS, et al (1999) Molecular and biophysical characterization of TT virus: evidence for a new virus family infecting humans. Proc Natl Acad Sci U S A 96:3177–3182

Niel C, Diniz-Mendes L, Devalle S (2005) Rolling-circle amplification of Torque teno virus (TTV) complete genomes from human and swine sera and identification of a novel swine TTV genogroup. J Gen Virol 86:1343–1347

Ninomiya M, Nishizawa T, Takahashi M, et al (2007a) Identification and genomic characterization of a novel human torque teno virus of 3.2 kb. J Gen Virol 88:1939–1944

Ninomiya M, Takahashi M, Shimosegawa T, et al (2007b) Analysis of the entire genomes of fifteen torque teno midi virus variants classifiable into a third group of genus Anellovirus. Arch Virol 152:1961–1975

Nishizawa T, Okamoto H, Konishi K, et al (1997) A novel DNA virus (TTV) associated with elevated transaminase levels in posttransfusion hepatitis of unknown etiology. Biochem Biophys Res Commun 241:92–97

Okamoto H, Mayumi M (2001a) TT virus: virological and genomic characteristics and disease associations. J Gastroenterol 36:519–529

Okamoto H, Nishizawa T, Kato N, et al (1998) Molecular cloning and characterization of a novel DNA virus (TTV) associated with posttransfusion hepatitis of unknown etiology. Hepatol Res 10:1–16

Okamoto H, Nishizawa T, Ukita M, et al (1999a) The entire nucleotide sequence of a TT virus isolate from the United States (TUS01): comparison with reported isolates and phylogenetic analysis. Virology 259:437–448

Okamoto H, Takahashi M, Nishizawa T, et al (1999b) Marked genomic heterogeneity and frequent mixed infection of TT virus demonstrated by PCR with primers from coding and noncoding regions. Virology 259:428–436

Okamoto H, Nishizawa T, Tawara A, et al (2000a) Species-specific TT viruses in humans and nonhuman primates and their phylogenetic relatedness. Virology 277:368–378

Okamoto H, Nishizawa T, Tawara A, et al (2000b) TT virus mRNAs detected in the bone marrow cells from an infected individual. Biochem Biophys Res Commun 279:700–707

Okamoto H, Nishizawa T, Takahashi M, et al (2001b) Genomic and evolutionary characterization of TT virus (TTV) in tupaias and comparison with species-specific TTVs in humans and non-human primates. J Gen Virol 82:2041–2050

Okamoto H, Takahashi M, Nishizawa T, et al (2002) Genomic characterization of TT viruses (TTVs) in pigs, cats and dogs and their relatedness with species-specific TTVs in primates and tupaias. J Gen Virol 83:1291–1297

Peng YH, Nishizawa T, Takahashi M, et al (2002) Analysis of the entire genomes of thirteen TT virus variants classifiable into the fourth and fifth genetic groups, isolated from viremic infants. Arch Virol 147:21–41

Peters MA, Jackson DC, Crabb BS, et al (2002) Chicken anemia virus VP2 is a novel dual specificity protein phosphatase. J Biol Chem 277:39566–39573

Simmonds P (2002) TT virus infection: a novel virus-host relationship. J Med Microbiol 51:455–458

Stanley J, Bisaro DM, Briddon RW, et al (2005) Geminiviridae. In: Virus taxonomy, VIIIth report of the International Committee for the Taxonomy of Viruses. Elsevier/Academic Press, London, pp 301–326

Takahashi K, Hijikata M, Samokhvalov EI, et al (2000a) Full or near full length nucleotide sequences of TT virus variants (types SANBAN and YONBAN) and the TT virus-like mini virus. Intervirology 43:119–123

Takahashi K, Iwasa Y, Hijikata M, et al (2000b) Identification of a new human DNA virus (TTV-like mini virus, TLMV) intermediately related to TT virus and chicken anemia virus. Arch Virol 145:979–993

Todd D, Bendinelli M, Biagini P, et al (2005) Circoviridae. In: Virus taxonomy, VIIIth report of the International Committee for the Taxonomy of Viruses. Elsevier/Academic Press, London, pp 327–334

Verschoor EJ, Langenhuijzen S, Heeney JL (1999) TT viruses (TTV) of non-human primates and their relationship to the human TTV genotypes. J Gen Virol 80:2491–2499

Vetten HJ, Chu PWG, Dale JL, et al (2005) Nanoviridae. In: Virus taxonomy, VIIIth report of the International Committee for the Taxonomy of Viruses. Elsevier/Academic Press, London, pp 343–352

Worobey M (2000) Extensive homologous recombination among widely divergent TT viruses. J Virol 74:7666–7670

# TT Viruses in Animals

**H. Okamoto**

**Contents**



**Abstract** Infection with TT virus (Torque teno virus, TTV), a small, nonenveloped virus with a circular, single-stranded DNA genome classified in the floating genus *Anellovirus*, is not restricted to humans. Using highly conserved primers derived from the untranslated region of the human TTV genome, a variety of TTV-like viruses have been found circulating in nonhuman primates such as chimpanzees, macaques, and tamarins. TTV variants in nonhuman primates are species-specific, although some genetic groups of human and chimpanzee TTVs cluster to make human/chimpanzee clades. TTVs from macaques and tamarins are increasingly divergent from TTV variants infecting humans and chimpanzees. TTV-like mini virus (TTMV) infections have also been detected in chimpanzees, with genotypes distinct but interspersed with human TTMV genotypes. Pets are also naturally infected with species-specific TTVs, and several isolates have been found in cats and dogs. In addition, other mammals such as tupaias and pigs have species-specific

H. Okamoto
Division of Virology, Department of Infections and Immunity, Jichi Medical University School of Medicine, 3311-1 Yakushiji, Shimotsuke-Shi, Tochigi-Ken, 329-0498, Japan
hokamoto@jichi.ac.jp

E.-M. de Villiers, H. zur Hausen (eds.), *TT Viruses: The Still Elusive Human Pathogens,* 

TTVs: swine TTVs are found among pigs worldwide. The genomic organization and proposed transcriptional profiles of TTVs infecting nonhuman primate and other mammalian species are similar to those of human TTVs, and co-evolution of TTVs with their hosts has been suggested. To date, TTVs infecting nonhuman primates and other mammalian species have been under-examined. It is likely that essentially all animals are naturally infected with species-specific TTVs.

## Introduction

In 1997, by means of representational difference analysis (RDA) (Lisitsyn et al. 1993), the first "TT" virus (TTV) isolate was identified from a Japanese patient (with initials T.T.) with posttransfusion hepatitis of unknown etiology (Nishizawa et al. 1997). Subsequent studies revealed that TTV and TTV-like viruses with marked genetic variability are nonenveloped, single-stranded (negative sense), circular DNA viruses with a genomic length of 3.6–3.9 kb (Nishizawa et al. 1997; Miyata et al. 1999; Mushahwar et al. 1999; Okamoto et al. 1998b, 1999b, c, 2004; Peng et al. 2002). In 2000, a small virus was discovered by conventional polymerase chain reaction (PCR) of human plasma samples using TTV-specific primers that partially matched homologous sequences but generated an unexpectedly shorter amplicon, and was provisionally named as TTV-like mini virus (TLMV) (Takahashi et al. 2000b). The genome of TLMV consists of a circular, single-stranded DNA of approximately 2.8–2.9 kb with negative polarity. Recently, the International Committee on Taxonomy of Viruses proposed that the abbreviation "TT" of "TTV" stands for "Torque teno," deriving from the Latin terms "torque" meaning "necklace" and "tenuis" meaning "thin." These terms reflect the arrangement of the genomic organization of TTV and TLMV. TTV and TLMV have been renamed as Torque teno virus (TTV), without changing its abbreviation, and Torque teno mini virus (TTMV), respectively (Biagini et al. 2005). TTV and TTMV have a common presumed genomic organization with four open reading frames (ORFs, ORF1–ORF4). Three distinct messenger RNAs (mRNAs)—of 2.9–3.0 kb, 1.2 kb, and 1.0 kb with common 5′ and 3′ termini that are transcribed from the genomic, minus-strand of TTV DNA—have been observed in culture cells transfected with recombinant TTV DNA as well as in bone marrow cells obtained from an infected human (Kamahora et al. 2000; Okamoto et al. 2000c). These three mRNAs have in common a short splicing of approximately 100 nt. The 1.2-kb and 1.0-kb mRNAs possess an additional splicing of approximately 1,700 nt and 1,900 nt, respectively, leading to the creation of two novel ORFs (ORF3 and ORF4).

TTV has an extremely wide range of sequence divergence, and the TTV genomes are tentatively classified into at least 39 genotypes with sequence divergence of more than 30% from one another, or into five major phylogenetic groups (groups 1–5) with sequence divergence of more than 50% from one another (Hallett et al. 2000; Hijikata et al. 1999; Khudyakov et al. 2000; Muljono et al. 2001; Okamoto et al. 1999b, c, 2000c, 2001b, 2004; Peng et al. 2002; Takahashi et al. 2000a; Ukita et al. 2000). Complete sequences have thus far been determined for 17 TTMV genomes, and high genetic divergence has also been noted among the TTMV genomes (Biagini et al. 2001b, 2007; Okamoto et al. 2000b; Takahashi et al. 2000b). Co-infection of individuals with multiple TTV and TTMV variants is frequent (Hino and Miyata 2007; Niel et al. 2000; Ninomiya et al. 2008; Okamoto et al. 1999a, d, 2001b; Takayama et al. 1999).

Infection of TTVs is not restricted to human hosts. Although many aspects of TTV infection including the precise host range and cross-species infection remain to be elucidated, increasing lines of evidence indicate that nonhuman primates, tupaias (tree shrews: *Tupaia belangeri chinensis*), and livestock and some companion animals are infected with TTVs (Abe et al. 2000; Brassard et al. 2008; Cong et al. 2000; Leary et al. 1999; Okamoto et al 2000a, b, 2001a, 2002; Romeo et al. 2000; Thom et al. 2003; Verschoor et al. 1999). The entire nucleotide sequences of species-specific TTVs that infect nonhuman primates such as the chimpanzee (*Pan troglodytes*), Japanese macaque (*Macaca fuscata*), cotton-top tamarin (*Saguinus oedipus*) and douroucouli (*Aotus trivirgatus*) and tupaias as well as domestic animals including pig (*Sus domesticus*), cat (*Felis catus*) and dog (*Canis familiaris*), have been determined (Biagini et al. 2007; Inami et al. 2000; Niel et al. 2005; Okamoto et al. 2000b, 2001a, 2002; Table 1). Furthermore, TTV DNA has been detected in serum samples obtained from domesticated farm animals such as chickens, cows, and sheep (Brassard et al. 2008; Leary et al. 1999). However, the TTVs in these farm animals have not been fully characterized as yet.

**Table 1** TT virus isolates obtained from nonhuman primates and other mammalian species whose entire nucleotide sequence is known

| Animal species | Isolate | Genomic length (nt) | Accession No. | Reference |
|---|---|---|---|---|
| Chimpanzee | CH65–1 | 3,899 | AB037826 | Inami et al. 2000 |
| | Pt-TTV6 | 3,690 | AB041957 | Okamoto et al. 2000b |
| | Pt-TTV8-II | 2,785 | AB041963 | Okamoto et al. 2000b |
| Japanese macaque | Mf-TTV3 | 3,798 | AB041958 | Okamoto et al. 2000b |
| | Mf-TTV9 | 3,763 | AB041959 | Okamoto et al. 2000b |
| Tamarin | So-TTV2 | 3,371 | AB041960 | Okamoto et al. 2000b |
| Douroucouli | At-TTV3 | 3,718 | AB041961 | Okamoto et al. 2000b |
| Tupaia | Tbc-TTV14 | 2,199 | AB057358 | Okamoto et al. 2001a |
| Pig | Sd-TTV31 | 2,878 | AB076001 | Okamoto et al. 2002 |
| | Sd-TTV1p | 2,872 | AY823990 | Niel et al. 2005 |
| | Sd-TTV2p | 2,735[a] | AY823991 | Niel et al. 2005 |
| Dog | Cf-TTV10 | 2,797 | AB076002 | Okamoto et al. 2002 |
| Cat | Fc-TTV4 | 2,064 | AB076003 | Okamoto et al. 2002 |
| | PRA1 | 2,019 | EF538877 | Biagini et al. 2007 |
| | PRA2 | 2,065 | EF538878 | Biagini et al. 2007 |

[a]Contains a total of 27 undetermined nucleotides

## Methods of Isolating TTV in Animals

When PCR was performed using the three PCR methods for amplifying TTV in human samples with primers based on the untranslated region (UTR) [NG133 and NG147 for the first round, and NG134 and NG132 for the second round (nUTR-PCR v1)] (Okamoto et al. 1999d, 2000a), N22 primers (NG059 and NG063 for the first round, and NG061 and NG063 for the second round) (Okamoto et al. 1998b) or set B primers (forward 1 and reverse 1 for the first round, and forward 2 and reverse 2 for the second round) (Leary et al. 1999), no amplification signals were observed in the serum samples from any of the nonprimate animals tested (Okamoto et al. 2000a). This suggested the absence of cross-species infection of human TTVs in animals including pigs, cats, and dogs. However, these results do not exclude the possibility of the presence of species-specific TTVs in each animal species.

Determination and comparative analysis of the full-length genomic sequence of TTV and TTMV isolates indicated the extremely wide range of sequence divergence among isolates, as described above. However, when the TTV and TTMV sequences from humans and nonhuman primates were aligned over the entire genome so as to obtain maximal homology, a specific region of approximately 130 nt located just downstream of a common internal promoter, i.e., the TATA-box (ATATAA), in each anellovirus that contains highly conserved areas at both ends, was recognized (Okamoto et al. 2000b). Taking advantage of this highly conserved UTR sequence, PCR with primers NG343 and NG344 that had been derived from the well-conserved areas (Okamoto et al. 2001a, 2002; Fig. 1), was performed on sera from nonprimate animals. As a result, tupaias and domestic animals including pigs, cats, and dogs were found to carry species-specific TTVs at extremely high frequencies (Okamoto et al. 2001a, 2002). Using primers derived from the amplified short UTR sequence that were made to be specific to each isolate, the entire genomic sequence of animal TTV was amplified by long-distance inverted PCR and determined. By utilizing this method, the full-length sequence was determined for TTVs isolated from a tupaia (Tbc-TTV14), pig (Sd-TTV31), cat (Fc-TTV4), and dog (Cf-TTV10) (Okamoto et al. 2001a, 2002) (Table 1). Therefore, the primers (NG343 and NG344) used for the initial detection of tupaia, swine, canine, and feline TTVs would be instrumental in extended research on TTVs in unexamined animal species in an attempt to further define their virological characteristics and evolutionary relationships.

Recently, two new methods using the rolling-circle amplification (RCA)-based sequence-independent technique were applied to isolate TTVs from infected animals. In brief, by using the technique of multiply primed RCA (Esteban et al. 1993; Dean et al. 2001), circular genomes of papillomavirus were successfully amplified (Rector et al. 2004). In this method, a random hexamer primer anneals to multiple sites on a DNA template; these sites are isothermally extended by the phi29 DNA polymerase (Garmendia et al. 1992; Esteban et al. 1993), therefore producing multiple copies of the complete genome, without the need for prior knowledge of their sequences. This approach permitted the amplification and cloning of a circovirus (Johne et al. 2006) and geminivirus (Haible et al. 2006) as well as TTV in eight

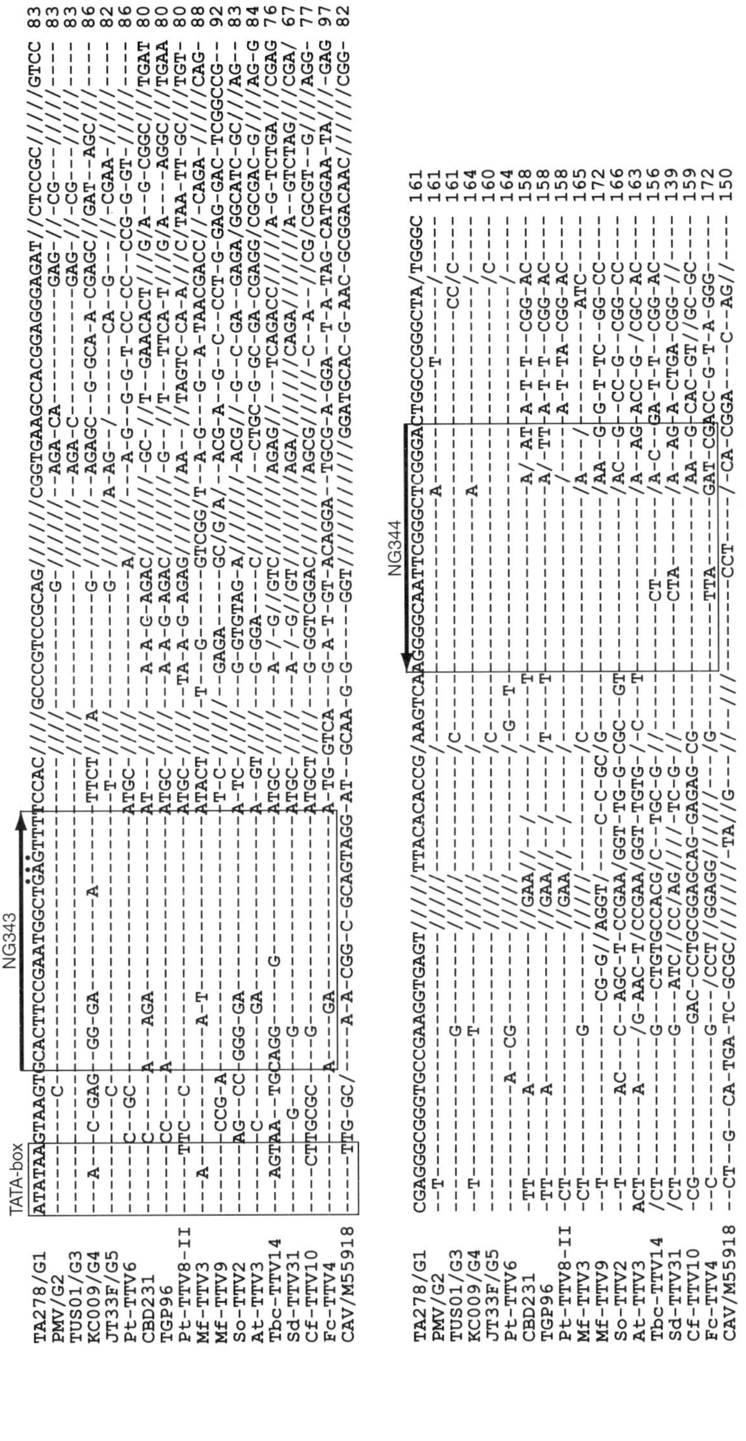


**Fig. 1** Alignment of the nucleotide sequence of TTVs and TTMVs obtained from humans and nonhuman primates, TTVs obtained from a tupaia and domestic animals as well as chicken anemia virus (CAV) (M55918). The sequences corresponding to primers NG343 (sense) and NG344 (antisense) are *boxed*. (Reproduced from Okamoto et al. 2002, with permission)

human and four pig serum samples (Niel et al. 2005). All TTV samples gave amplification products of high molecular weight (>30 kb). By restriction endonuclease digestion, these products generated DNA fragments whose sizes were consistent with those of human TTV (3.8 kb) and swine TTV (2.9 kb) genomes. Two TTV isolates [2 h (accession No. AY823988) and 3 h (AY823989)] derived from a single acquired immunodeficiency syndrome (AIDS) patient, as well as two swine TTV isolates (Sd-TTV1p and Sd-TTV2p) derived from a single pig, were characterized by nucleotide sequencing over the entire genome (Table 1).

Furthermore, a combined RCA and sequence-independent single primer amplification (SISPA) approach was applied to four samples of human plasma and one sample of saliva from a cat in an attempt to obtain circular genomes (Biagini et al. 2007). This approach is based on the use of endonuclease restriction of target sequences previously converted to double-stranded DNA, followed by nonspecific linker ligation and PCR amplification (Reyes and Kim 1991; Allander et al. 2001), and permitted the characterization of nine anelloviruses including two cat TTV genomes obtained from the cat's saliva (PRA1 and PRA4; Table 1). These results highlight the potential of the RCA-SISPA method for detecting species-specific anelloviruses in as yet unexamined animals.

## TTV in Nonhuman Primates

In the search for viruses resembling human TTV, when serum samples from nonhuman primates were subjected to PCR with UTR primers (nUTR-PCR v1) deduced from well-conserved areas in the untranslated region, TTV DNA was detected frequently in serum samples from chimpanzees, Japanese macaques, red-bellied tamarins, cotton-top tamarins, and douroucoulis. Analysis of the amplification products of 90–106 nt revealed TTV DNA sequences specific for each species, with decreasing similarity to human TTV in the order of chimpanzee, Japanese macaque, and tamarin/douroucouli TTVs (Okamoto et al. 2000b). Full-length viral sequences amplified by PCR with inverted nested primers deduced from the UTR of TTV DNA from each species showed that all TTVs from nonhuman primates were circular with a genomic length of 3.5–3.8 kb, which was comparable to or slightly shorter than human TTV (Okamoto et al. 2000b; Fig. 2). Additionally, it was found that chimpanzees had TTMVs of 2.8 kb, similar to humans (Okamoto et al. 2000b).

### *TTV in Chimpanzees*

The entire nucleotide sequence has been determined for two chimpanzee TTV isolates (Pt-TTV6 and CH65–1) consisting of 3,690 nt and 3,899 nt, respectively (Table 1), which differ from each other by 43% and from reported human TTVs by more than 50% (Okamoto et al. 2000b; Inami et al. 2000). However, when serum samples were subjected to PCR with primers deduced from the human TTV sequence of a coding region (ORF1) spanning 222–231 nt (the N22 region), which is capable of detecting human TTV strains of group 1, they were detected in 49 (47%) of the 104 chimpanzees tested (Okamoto et al. 2000a). This result corroborates those of Verschoor et al. (1999), who detected TTV DNAs in 60 (49%) of 123 chimpanzees of the *Pan troglodytes verus* species and in 4 (67%) of 6 chimpanzees of the *P. paniscus* species by PCR with the N22 primers. In addition, sera from chimpanzees

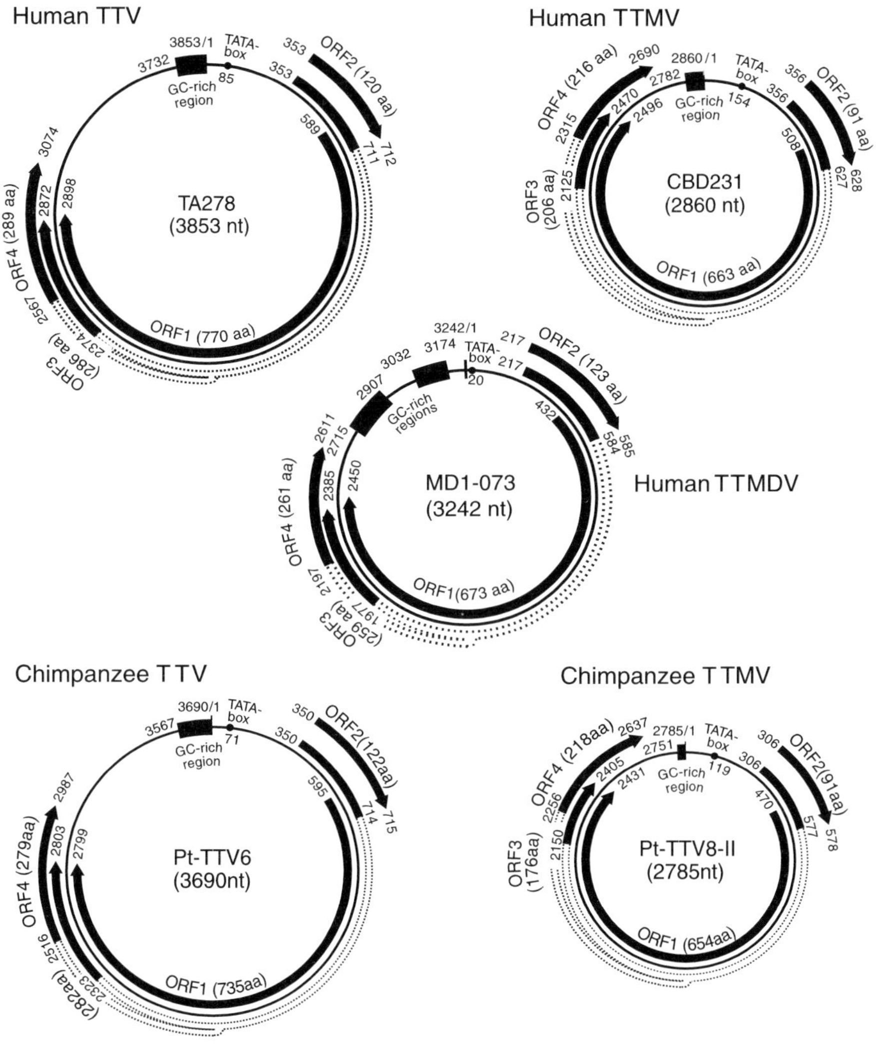


**Fig. 2** Genomic organization of the 12 anelloviruses from humans, nonhuman primates, a tupaia, and domestic animals. The *circumference of each circle* represents the relative size of the genome. *Arrows* represent ORFs, and *closed boxes* indicate the GC-rich regions

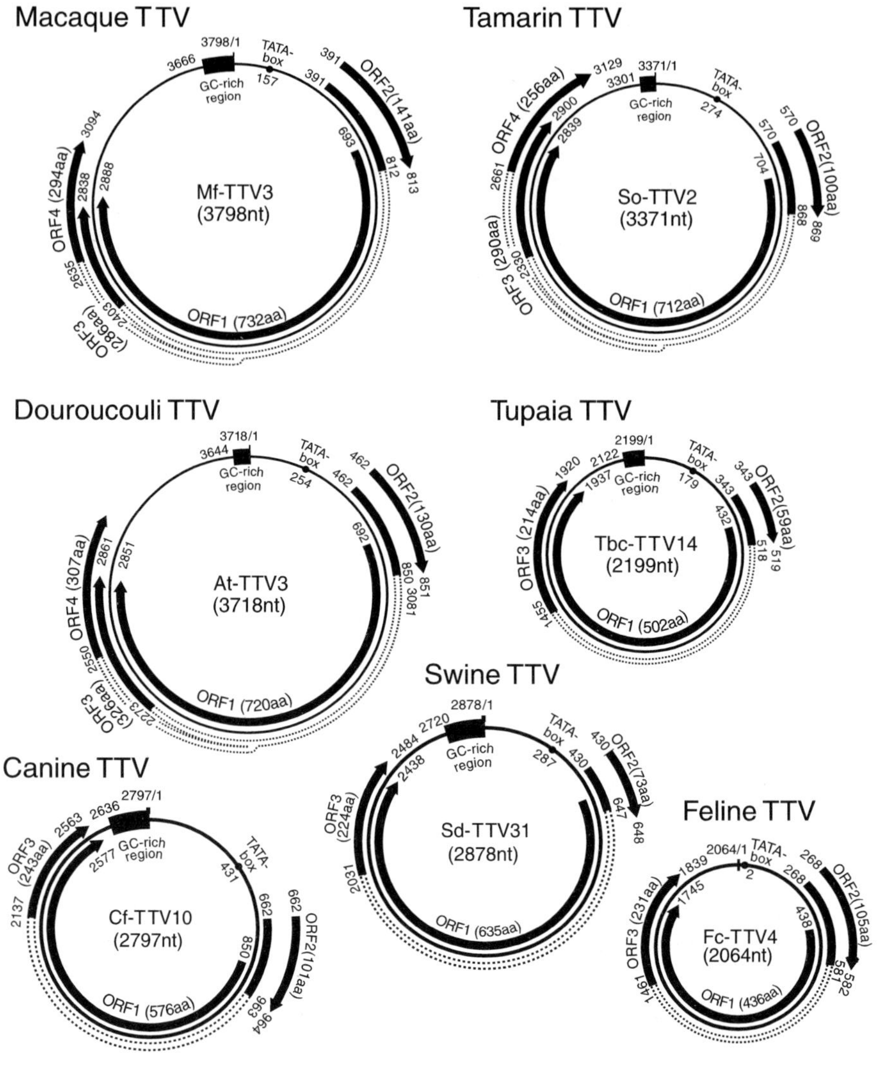


**Fig. 2** (continued)

were tested for TTV sequences resembling the human TTV genome by using PCR primers in the UTR (nUTR-PCR v1) (Okamoto et al. 1999d), which are capable of detecting human TTVs of groups 1–4 (but not most group 5 TTVs). Sera from 102 (98%) of 104 chimpanzees tested positive for TTV DNA by UTR PCR.

Sequence analysis of the N22 region (222–231 nt) of chimpanzee TTV (Pt-TTV) DNAs disclosed four genetic groups differing by 36%–50% from one another; they were 35%–53% divergent from human TTVs (Okamoto et al. 2000a). The four genetic groups of Pt-TTV, however, did not cluster in a clade separate from human TTVs in a phylogenetic tree. Human TTV genotypes intermingled with genetic groups of Pt-TTV. Furthermore, some Pt-TTVs had UTR sequences which were very similar to that of human TTV. These results are in line with the report of Verschoor et al. (1999); they observed that *P. troglodytes verus* TTV strains cluster with human TTV of a certain genotype while *P. paniscus* TTV strains cluster with that of another genotype. Interestingly, it has been reported that simian TTV infection also occurs in humans, whereby TTV may be of zoonotic origin (Iwaki et al. 2003). Hence, it is possible that some Pt-TTVs belong to unknown genotypes of human TTV.

It is likely that cross-species infection occurred from chimpanzee to humans via animal handling, particularly for viruses transmitted through contamination of blood, represented by retroviruses. Interspecies infection would be relatively easy for TTV, which is excreted in the feces for a possible fecal-oral transmission route (Okamoto et al. 1998a). It would be worthwhile to test the sera of individuals in Africa, where the chimpanzees were caught, for the presence of Pt-TTV and to analyze it phylogenetically. One of the 104 chimpanzees tested was infected with human TTV of genotype 1a (Okamoto et al. 2000a), confirming cross-species infection as reported by Mushahwar et al. (1999) and Tawara et al. (2000). This chimpanzee was among the 53 that participated in transmission experiments with human hepatitis viruses in the past. When the other 53 chimpanzees were tested for the presence of antibody against TTV of genotype 1a by immunoprecipitation (Tsuda et al. 1999), it was detected in 11 chimpanzees. Anti-TTV of genotype 1a was detected significantly more frequently among the chimpanzees used in transmission experiments than among those not used in these experiments [8/28 (29%) and 3/35 (9%), respectively; $p$=0.038]. Hence, at least 11 more chimpanzees had been infected with TTV of genotype 1a, most probably through human materials contaminated with TTV of this genotype, and resolved the infection by raising humoral antibodies. Notably, being different from other nonhuman primates, chimpanzees are also frequently infected with chimpanzee TTMV of 2.8 kb, represented by Pt-TTV8-II whose entire nucleotide sequence is known (Okamoto et al. 2000b). Recently, a novel human virus with a circular DNA genome of 3.2 kb, tentatively designated as Torque teno midi virus (TTMDV), that is classifiable into a third group in the genus *Anellovirus*, was identified from human sera (Ninomiya et al. 2007a, b), and frequent dual or triple infection of TTV, TTMDV, and TTMV has been recognized in humans (Ninomiya et al. 2008). Therefore, it is very likely that chimpanzees are also infected with TTMDV. Clarifying the presence of TTMDV in chimpanzees may be valuable in determining the origin, nature, and transmission of human anelloviruses.

## *TTV in Lower-Order Primates*

Abe et al. (2000) tested 400 serum samples from 24 different species of nonhuman primates for the presence of TTV DNA by their own UTR PCR assay. TTV DNA was detected in 87 (89%) of 98 chimpanzees and 3 (14%) of 21 crab-eating macaques (*Macaca fascicularis*), but not in any of 22 other species including Old World monkeys, New World monkeys, and Prosimii. However, by means of PCR (nUTR-PCR v1) (Okamoto et al. 1999d), TTV DNA was detected in sera from 9 (90%) of 10 Japanese macaques, 4 (100%) of 4 red-bellied tamarins, 5 (83%) of 6 cotton-top tamarins, and all 5 (100%) of 5 douroucoulis (Okamoto et al. 2000a). Thus, TTV DNA is highly prevalent in nonhuman primates other than chimpanzees as well. Unlike the detection of TTV DNA in human sera, TTV DNAs in sera from these monkeys were amplified by the first round (NG133 and NG147) but not the second round (NG134 and NG132) of PCR, suggesting the marked genomic heterogeneity of TTVs specific to monkeys. Amplification products of the first round of UTR PCR measuring 90–106 bp (primer sequences at both ends excluded) were sequenced for all 23 animals testing positive, including 9 Japanese macaques, 9 red-bellied or cotton-top tamarins and 5 douroucoulis, and were analyzed phylogenetically. The tree revealed phylogenetic differences of TTV depending on the species (Fig. 3). Due to these phylogenetic differences, TTV isolated from Japanese macaques (*Macaca fuscata*) is referred to as Mf-TTV, that from cotton-top tamarins (*Saguinus oedipus*) is referred to as So-TTV, and that from douroucoulis (*Aotus trivirgatus*) is referred to as At-TTV. The entire nucleotide sequence has been determined for the prototype Mf-TTV, So-TTV, and At-TTV (Okamoto et al. 2000b; Table 1, Fig. 2). Supporting the notion that monkeys are infected with TTVs with lower similarity to those in humans, TTV DNA was not detectable by N22 PCR in any of the animals of Japanese macaques ($n$=10) and tamarins ($n$=10) tested (Okamoto et al. 2000a). Verschoor et al. (1999) also could not detect TTV DNA by

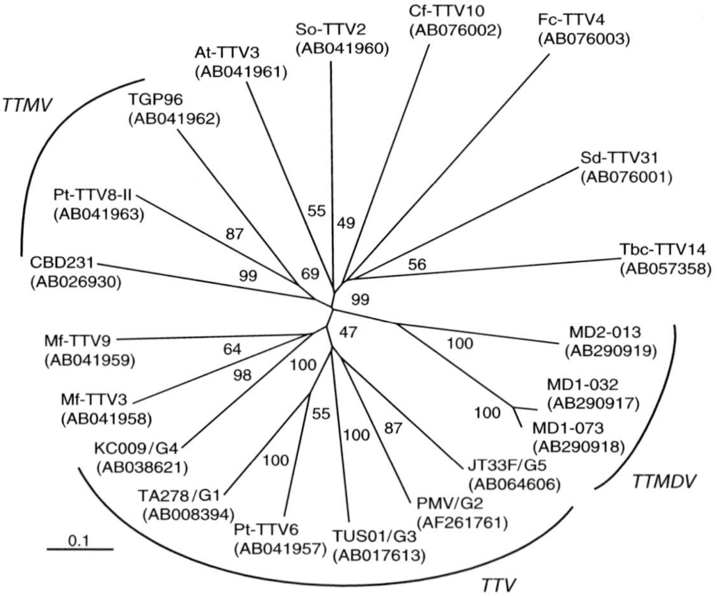


**Fig. 3** Phylogenetic trees constructed based on the amino acid sequences of ORF1 by the neighbor-joining method (Saitou and Nei 1987). The percentage of bootstrap values generated from 1,000 samplings of the data is shown near the nodes. The *scale bar* represents the number of amino acid substitutions per position. (Reproduced from Ninomiya et al. 2007a, with permission)

N22 PCR in nonchimpanzee primates including orangutans (*Pongo pygmaeus*) (*n*=22), rhesus macaques (*M. mulatta*) (*n*=25), long-tailed macaques (*M. fascicularis*) (*n*=10), cotton-top tamarins (*n*=10), common marmosets (*Callithrix jacchus*) (*n*=10), and common squirrel monkeys (*Saimiri sciureus*) (*n*=10), indicating that human TTV can cross-infect chimpanzees but not other animals. Of note, however, TTV sequences identified in gibbons bred in captivity were found to be closely related to human sequences, and it was hypothesized that TTVs have been transmitted to gibbons from animal handlers (Noppornpanth et al. 2001).

## TTV in Nonprimate Animals

TTV has also been detected in nonprimate animal species, including companion animals (e.g., cats and dogs), livestock (e.g., cattle, chickens, pigs, and sheep), wild animals (wild boars), and tupaias (Brassard et al. 2008; Leary et al. 1999; Martinez et al. 2006; Okamoto et al. 2001a, 2002).

Leary et al. (1999) first reported frequent detection of TTV DNA in farm animals, including 19% of chickens, 20% of pigs, 25% of cows, and 30% of sheep, by their improved PCR methods with set B primers. On the basis of their findings, they speculated that domesticated farm animals would serve as a source of TTV infection in humans. It is possible that animals could, in part, serve as reservoirs of human TTVs. This scenario is supported by the results showing human TTVs in camels in the United Arab Emirates (Al-Moslih et al. 2007). However, other investigators have not verified the results on farm animals (Brassard et al. 2008; Okamoto et al. 2002; Thom et al. 2003). Specifically, TTV DNA was not detectable by either PCR with the set B primers or UTR PCR (nUTR-PCR v1) in domestic animals including pigs, cows, and horses (Okamoto et al. 2002; H. Okamoto, unpublished observations). Whether human TTVs can infect animals other than chimpanzees is a matter of great concern, because it would influence the epidemiology of TTV infection in humans.

Using PCR with primers NG343 and NG344 derived from a highly conserved area in the noncoding region of the TTV genome (Fig. 1), the TTV sequences in serum samples obtained from tupaias (tree shrews), pigs, cats, and dogs were identified, and the entire genomic sequence was determined for a representative isolate from each animal species. Currently available data indicate that even mammals other than primates are naturally infected with species-specific TTVs with small genomic size, and suggest wide distribution of TTVs with extremely divergent genomic sequence and length in animals.

### *TTV in Tupaias*

Tupaias share characteristics with both primates and insectivores, and have been classified in a single order called Scandentia, rather than in the order Primates or the order Insectivora (Martin, 1990). TTV DNA was not detectable by either N22

PCR or UTR PCR (nUTR-PCR v1) in any of the sera from the 19 tupaias tested. As described above, however, when PCR was performed using the primers NG343 and NG344, TTV DNA was detected in the sera of 5 (26%) of the 19 tupaias (Okamoto et al. 2001a). The PCR-amplified products from the 5 tupaias were subjected to sequence analysis. Based on a common 84-bp sequence in the 5 sequences, nested primers (NG478–NG479 for the first round and NG480–NG481 for the second round) were newly designed to sensitively detect TTV DNA in the sera from the tupaias. Upon performing PCR with the nested primers, 15 (79%) of the tupaias, including the above-mentioned 5, were found to be positive for TTV DNA, indicating that TTV infection is prevalent in tupaias (Okamoto et al. 2001a).

One tupaia TTV isolate (Tbc-TTV14) consisting of only 2,199 nt had three ORFs, spanning 1,506 nt (ORF1), 177 nt (ORF2), and 642 nt (ORF3) (Fig. 2), which were in the same orientation as the ORFs of the TTVs and TTMVs in humans. ORF3 was presumed to arise from a splicing of TTV mRNA, similar to reported human TTVs whose spliced mRNAs have been identified, and encode a joint protein of 214 amino acids (aa) with a Ser-, Lys-, and Arg-rich sequence at the C terminus. Comparison of the complete DNA sequence of tupaia TTV (Tbc-TTV14) with species-specific TTVs and TTMVs from humans and nonhuman primates indicated that the TTV that infects tupaias has the simplest viral genome (2,199 nt) among the TTVs and TTMVs thus far identified from humans and nonhuman primates, and that although the tupaia TTV genome differs from all other TTV and TTMV genomes by more than 50% at both the nucleotide and amino acid levels, its putative genomic organization and transcription profile are similar to those of TTVs and TTMVs (Okamoto et al. 2001a).

## *TTV in Pigs*

Upon performing PCR with primers NG343 and NG344 (Fig. 1), nine (82%) of 11 pigs (*Sus domesticus*) in Japan were found to be positive for TTV DNA (Okamoto et al. 2002). Based on the sequence of the PCR-amplified products, which measured 69–99 bp (primer sequences at both ends excluded), two pairs of inverse primers were designed to amplify the entire swine TTV genomes. The swine TTV genome was found to be approximately 2.9 kb in size in all 9 samples. Swine TTV strains could be divided into two genetic groups (Sd-TTV31 in one group and Sd-TTV83 in the second group). Then the PCR-amplified product from each of the swine serum samples with the strongest PCR signal was molecularly cloned, and the TTV clone named Sd-TTV31 was sequenced over the entire genome. The Sd-TTV31 isolate had a circular genomic structure of 2,878 nt; it was deduced to be single-stranded from the results of the PCR upon digestion with S1 nuclease or Mung bean nuclease, similar to the genomic DNAs of known TTVs and TTMVs.

The Sd-TTV31 isolate has three ORFs encoding 635aa (ORF1), 73aa (ORF2) and 224aa (ORF3), but lacks ORF4 similar to tupaia TTV (Fig. 2). ORF3 is presumed to arise from a splicing of TTV mRNA, similar to human prototype TTV. Although the nucleotide sequence of Sd-TTV31 differs by more than 50% from TTVs of 3.6–3.9 kb, TTMDVs of 3.2 kb, and TTMVs of 2.8–2.9 kb isolated from humans and nonhuman primates as well as tupaia TTVs of 2.2 kb, it resembled known anelloviruses with regard to genomic organization and presumed transcriptional profile, rather than animal circoviruses (chicken anemia virus, porcine circovirus, psittacine beak and feather disease virus) of 1.7–2.3 kb (Todd et al. 2000).

To date, the full-length genomic sequence of three swine TTV isolates have been published (Niel et al. 2005; Okamoto et al. 2002; Table 1). Two of the strains, Sd-TTV31 and Sd-TTV1p, share relatively high nucleotide sequence identity (70%), whereas strain Sd-TTV2p shows approximately 44% sequence identity to the other two strains (Niel et al. 2005). Due to the wide genetic diversity of these isolates, it has recently been suggested that Sd-TTV2p would be the prototype of a novel genogroup 2, whereas Sd-TTV31 would be the prototype of genogroup 1 (Niel et al. 2005; Okamoto et al., 2002). The Sd-TTV83 isolate, belonging to the second group of swine TTV in our previous study (Okamoto et al. 2002), was found to comprise 2,796 nt and was 94% identical to Sd-TTV2p, indicating that Sd-TTV83 is classifiable into genogroup 2 and that swine genogroup 2 TTV also prevails in Japan (H. Okamoto, unpublished observations).

TTV infection in pigs is distributed worldwide (Brassard et al. 2008; Kekarainen et al. 2008). Swine TTV detectable by genogroup 1-specific PCR is prevalent at a frequency ranging from 24% to 100% in serum samples collected from different geographical regions, including Canada, China, France, Italy, Korea, Spain, Thailand, and the United States (Bigarré et al. 2005; Kekarainen et al. 2006; Martelli et al. 2006; McKeown et al. 2004). The prevalence of swine TTV genogroup 2 has been reported to be as high as 77% in Spanish pigs (Kekarainen et al. 2006). Also in Spain, 58% and 66% of wild boars were found to be infected with swine TTV of genogroup 1 and 2, respectively (Martinez et al. 2006). Since genogroup 2 TTV has also been identified in farm pigs in Brazil and Japan as described above, both TTV genogroups 1 and 2 may be present on most swine farms worldwide.

Swine TTV has not yet been shown to be pathogenic; however, its role during co-infection with other pathogens remains unknown. Post-weaning multisystemic wasting syndrome (PMWS), an economically important swine disease worldwide, has been linked etiologically with porcine circovirus type 2 (PCV2) (Chae 2005), a member of the family Circoviridae, genus *Circovirus*. PCV2, similar to swine TTV, is a ubiquitous virus of pigs, indicating that PMWS is a multifactorial disease in which PCV2 infection is necessary, but not sufficient, to trigger the clinical outcome (Allan et al. 2007). Therefore, besides PCV2, other factors are needed for the full expression of PMWS in most cases. In pigs, a higher prevalence of TTV infection was observed in the sera from pigs with PMWS than in the sera from pigs without PMWS (97% vs 78%). Of interest, pigs with PMWS were more likely to be infected with swine TTV genogroup 2 than the nonaffected pigs (Kekarainen et al. 2006). However, the biological importance of this finding remains to be elucidated. The high prevalence of swine TTV raises a possible risk in xenotransplantation for immunosuppressed organ recipients (McKeown et al. 2004).

## *TTV in Cats and Dogs*

Pets have also been shown to have TTVs. Several isolates have been found in cats and dogs (Biagini et al. 2007; Okamoto et al. 2002). Using PCR with primers NG343 and NG344 (Fig. 1), the TTV sequence was identified in serum samples obtained from 3 (36%) of 8 dogs and 3 (43%) of 7 cats in Japan, and the entire genomic sequence was determined for each representative isolate (Okamoto et al. 2002). Two TTV isolates (Cf-TTV10 from a dog and Fc-TTV4 from a cat) consisted of 2,797 and 2,064 nt, respectively, and both had three ORFs encoding 436–576aa (ORF1), 101–105aa (ORF2) and 231–243aa (ORF3), but lacked ORF4 similar to tupaia TTV and swine TTV (Fig. 2). ORF3 was presumed to arise from a splicing of TTV mRNA. Although the nucleotide sequence of Cf-TTV10 and Fc-TTV4 differed by more than 50% from each other and from previously reported anelloviruses that had been isolated from humans and nonhuman primates as well as tupaias and pigs, they resembled known anelloviruses with regard to genomic organization and presumed transcriptional profile.

Upon comparison of the nucleotide sequence of the NG343/NG344 amplicons, two genetic groups of TTV with intergroup differences of 23%–25% and 56%–57% were identified among each of the dogs and cats, respectively, similar to swine TTV. Two feline TTV isolates (PRA1 and PRA4) obtained from the saliva sample of a cat in France, that are separated by a genetic distance of 46%, clustered within the group formed by the unique TTV species identified so far for cats. Identification of these two feline isolates indicates the presence of a highly divergent feline TTV (PRA1, 54% identities with Fc-TTV4) classifiable into a new phylogenetic branch, and that of an isolate very close to the Japanese isolate (PRA4, 95% identities with Fc-TTV4), confirming the absence of geographical cluster, as previously demonstrated for human and swine anelloviruses as well as the existence of co-infections in pets. Therefore, the genetic variability among TTVs in infected cats is probably very high, as already suggested for pigs (Bigarré et al. 2005; Niel et al. 2005), and potentially of the same order of magnitude as that already identified in humans.

## Conclusions

Exposure to TTVs is not restricted to humans, and various animal species carry their own TTVs. Based on the current information on TTVs in animals as described herein, it would be likely that essentially all animals have species-specific TTVs, and that each mammalian species may harbor a complex "viral flora" of TTV as diverse as those found in humans. Thus far, however, TTVs infecting nonhuman primates and other mammalian species have not been fully examined. Further studies on the presence and characteristics of TTVs in unexamined animal species as well as viral cell and tissue tropism, host's immune responses to these viruses, viral persistence and potential outcome of infection are warranted to better understand the evolution and pathogenesis of TTVs in humans and other animal species.

## References


Abe K, Inami T, Ishikawa K, et al (2000) TT virus infection in nonhuman primates and characterization of the viral genome: identification of simian TT virus isolates. J Virol 74:1549–1553

Al-Moslih MI, Perkins H, Hu YW (2007) Genetic relationship of Torque teno virus (TTV) between humans and camels in United Arab Emirates (UAE). J Med Virol 79:188–191

Allan GM, Caprioli A, McNair I, et al (2007) Porcine circovirus 2 replication in colostrum-deprived piglets following experimental infection and immune stimulation using a modified live vaccine against porcine respiratory and reproductive syndrome virus. Zoonoses Public Health 54:214–222

Allander T, Emerson SU, Engle RE, et al (2001) A virus discovery method incorporating DNase treatment and its application to the identification of two bovine parvovirus species. Proc Natl Acad Sci U S A 98:11609–11614

Biagini P, Gallian P, Attoui H, et al (2001) Genetic analysis of full-length genomes and subgenomic sequences of TT virus-like mini virus human isolates. J Gen Virol 82:379–383

Biagini P, Todd D, Bendinelli M, et al (2005) Genus *Anellovirus*. In: Fauquet CM, Mayo MA, Maniloff J, et al (eds) Virus taxonomy: classification and nomenclature of viruses, eight report of the international committee on taxonomy of viruses. Elsevier/Academic Press, London, pp 335–341

Biagini P, Uch R, Belhouchet M, et al (2007) Circular genomes related to anelloviruses identified in human and animal samples by using a combined rolling-circle amplification/sequence-independent single primer amplification approach. J Gen Virol 88:2696–2701

Bigarré L, Beven V, de Boisseson C, et al (2005) Pig anelloviruses are highly prevalent in swine herds in France. J Gen Virol 86:631–635

Brassard J, Gagne MJ, Lamoureux L, et al (2008) Molecular detection of bovine and porcine Torque teno virus in plasma and feces. Vet Microbiol 126:271–276

Chae C (2005) A review of porcine circovirus 2-associated syndromes and diseases. Vet J 169:326–336

Cong ME, Nichols B, Dou XG, et al (2000) Related TT viruses in chimpanzees. Virology 274:343–355

Dean FB, Nelson JR, Giesler TL, et al (2001) Rapid amplification of plasmid and phage DNA using Phi 29 DNA polymerase and multiply-primed rolling circle amplification. Genome Res 11:1095–1099

Esteban JA, Salas M, Blanco L (1993) Fidelity of phi 29 DNA polymerase. Comparison between protein-primed initiation and DNA polymerization. J Biol Chem 268:2719–2726

Garmendia C, Bernad A, Esteban JA, Blanco L, Salas M (1992) The bacteriophage phi 29 DNA polymerase, a proofreading enzyme. J Biol Chem 267:2594–2599

Haible D, Kober S, Jeske H (2006) Rolling circle amplification revolutionizes diagnosis and genomics of geminiviruses. J Virol Methods 135:9–16

Hallett RL, Clewley JP, Bobet F, et al (2000) Characterization of a highly divergent TT virus genome. J Gen Virol 81:2273–2279

Hijikata M, Takahashi K, Mishiro S (1999) Complete circular DNA genome of a TT virus variant (isolate name SANBAN) and 44 partial ORF2 sequences implicating a great degree of diversity beyond genotypes. Virology 260:17–22

Hino S, Miyata H (2007) Torque teno virus (TTV): current status. Rev Med Virol 17:45–57

Inami T, Obara T, Moriyama M, et al (2000) Full-length nucleotide sequence of a simian TT virus isolate obtained from a chimpanzee: evidence for a new TT virus-like species. Virology 277:330–335

Iwaki Y, Aiba N, Tran HT, et al (2003) Simian TT virus (s-TTV) infection in patients with liver diseases. Hepatol Res 25:135–142

Johne R, Fernandez-de-Luco D, Hofle U, et al (2006) Genome of a novel circovirus of starlings, amplified by multiply primed rolling-circle amplification. J Gen Virol 87:1189–1195

Kamahora T, Hino S, Miyata H (2000) Three spliced mRNAs of TT virus transcribed from a plasmid containing the entire genome in COS1 cells. J Virol 74:9980–9986

Kekarainen T, Segales J (2008) Torque teno virus infection in the pig and its potential role as a model of human infection. Vet J Feb 22 [Epub ahead of print]

Kekarainen T, Sibila M, Segales J (2006) Prevalence of swine Torque teno virus in post-weaning multisystemic wasting syndrome (PMWS)-affected and non-PMWS-affected pigs in Spain. J Gen Virol 87:833–837

Khudyakov YE, Cong ME, Nichols B, et al (2000) Sequence heterogeneity of TT virus and closely related viruses. J Virol 74:2990–3000

Leary TP, Erker JC, Chalmers ML, et al (1999) Improved detection systems for TT virus reveal high prevalence in humans, non-human primates and farm animals. J Gen Virol 80:2115–2120

Lisitsyn N, Lisitsyn N, Wigler M (1993) Cloning the differences between two complex genomes. Science 259:946–951

Martelli F, Caprioli A, Di Bartolo I, et al (2006) Detection of swine torque teno virus in Italian pig herds. J Vet Med B Infect Dis Vet Public Health 53:234–238

Martin RD (1990) Are tree shrews primates? In: Martin RD (ed) Primate origins and evolution: a phylogenetic reconstruction. Chapman and Hall, London, pp 191–213

Martinez L, Kekarainen T, Sibila M, et al (2006) Torque teno virus (TTV) is highly prevalent in the European wild boar (*Sus scrofa*). Vet Microbiol 118:223–229

McKeown NE, Fenaux M, Halbur PG, et al (2004) Molecular characterization of porcine TT virus, an orphan virus, in pigs from six different countries. Vet Microbiol 104:113–117

Miyata H, Tsunoda H, Kazi A, et al (1999) Identification of a novel GC-rich 113-nucleotide region to complete the circular, single-stranded DNA genome of TT virus, the first human circovirus. J Virol 73:3582–3586

Muljono DH, Nishizawa T, Tsuda F, et al (2001) Molecular epidemiology of TT virus (TTV) and characterization of two novel TTV genotypes in Indonesia. Arch Virol 146:1249–1266

Mushahwar IK, Erker JC, Muerhoff AS, et al (1999) Molecular and biophysical characterization of TT virus: evidence for a new virus family infecting humans. Proc Natl Acad Sci U S A 96:3177–3182

Niel C, Saback FL, Lampe E (2000) Coinfection with multiple TT virus strains belonging to different genotypes is a common event in healthy Brazilian adults. J Clin Microbiol 38:1926–1930

Niel C, Diniz-Mendes L, Devalle S (2005) Rolling-circle amplification of Torque teno virus (TTV) complete genomes from human and swine sera and identification of a novel swine TTV genogroup. J Gen Virol 86:1343–1347

Ninomiya M, Nishizawa T, Takahashi M, et al (2007a) Identification and genomic characterization of a novel human torque teno virus of 3.2 kb. J Gen Virol 88:1939–1944

Ninomiya M, Takahashi M, Shimosegawa T, et al (2007b) Analysis of the entire genomes of fifteen torque teno midi virus variants classifiable into a third group of genus Anellovirus. Arch Virol 152:1961–1975

Ninomiya M, Takahashi M, Nishizawa T, et al (2008) Development of PCR assays with nested primers specific for differential detection of three human anelloviruses and early acquisition of dual or triple infection during infancy. J Clin Microbiol 46:507–514

Nishizawa T, Okamoto H, Konishi K, et al (1997) A novel DNA virus (TTV) associated with elevated transaminase levels in posttransfusion hepatitis of unknown etiology. Biochem Biophys Res Commun 241:92–97

Noppornpanth S, Chinchai T, Ratanakorn P, et al (2001) TT virus infection in gibbons. J Vet Med Sci 63:663–666

Okamoto H, Akahane Y, Ukita M, et al (1998a) Fecal excretion of a nonenveloped DNA virus (TTV) associated with posttransfusion non-A–G hepatitis. J Med Virol 56:128–132

Okamoto H, Nishizawa T, Kato N, et al (1998b) Molecular cloning and characterization of a novel DNA virus (TTV) associated with posttransfusion hepatitis of unknown etiology. Hepatol Res 10:1–16

Okamoto H, Kato N, Iizuka H, et al (1999a) Distinct genotypes of a nonenveloped DNA virus associated with posttransfusion non-A to G hepatitis (TT virus) in plasma and peripheral blood mononuclear cells. J Med Virol 57:252–258

Okamoto H, Nishizawa T, Ukita M (1999b) A novel unenveloped DNA virus (TT virus) associated with acute and chronic non-A to G hepatitis. Intervirology 42:196–204

Okamoto H, Nishizawa T, Ukita M, et al (1999c) The entire nucleotide sequence of a TT virus isolate from the United States (TUS01): comparison with reported isolates and phylogenetic analysis. Virology 259:437–448

Okamoto H, Takahashi M, Nishizawa T, et al (1999d) Marked genomic heterogeneity and frequent mixed infection of TT virus demonstrated by PCR with primers from coding and noncoding regions. Virology 259:428–436

Okamoto H, Fukuda M, Tawara A, et al (2000a) Species-specific TT viruses and cross-species infection in nonhuman primates. J Virol 74:1132–1139

Okamoto H, Nishizawa T, Tawara A, et al (2000b) Species-specific TT viruses in humans and nonhuman primates and their phylogenetic relatedness. Virology 277:368–378

Okamoto H, Nishizawa T, Tawara A, et al (2000c) TT virus mRNAs detected in the bone marrow cells from an infected individual. Biochem Biophys Res Commun 279:700–707

Okamoto H, Ukita M, Nishizawa T, et al (2000d) Circular double-stranded forms of TT virus DNA in the liver. J Virol 74:5161–5167

Okamoto H, Nishizawa T, Takahashi M, et al (2001a) Genomic and evolutionary characterization of TT virus (TTV) in tupaias and comparison with species-specific TTVs in humans and nonhuman primates. J Gen Virol 82:2041–2050

Okamoto H, Nishizawa T, Takahashi M, et al (2001b) Heterogeneous distribution of TT virus of distinct genotypes in multiple tissues from infected humans. Virology 288:358–368

Okamoto H, Takahashi M, Nishizawa T, et al (2002) Genomic characterization of TT viruses (TTVs) in pigs, cats and dogs and their relatedness with species-specific TTVs in primates and tupaias. J Gen Virol 83:1291–1297

Okamoto H, Nishizawa T, Takahashi M (2004) Torque teno virus (TTV): molecular virology and clinical implications. In: Mushahwar IK (ed) Viral hepatitis: molecular biology, diagnosis, epidemiology and control. Elsevier, Amsterdam

Peng YH, Nishizawa T, Takahashi M, et al (2002) Analysis of the entire genomes of thirteen TT virus variants classifiable into the fourth and fifth genetic groups, isolated from viremic infants. Arch Virol 147:21–41

Rector A, Tachezy R, Van Ranst M (2004) A sequence-independent strategy for detection and cloning of circular DNA virus genomes by using multiply primed rolling-circle amplification. J Virol 78:4993–4998

Reyes GR, Kim JP (1991) Sequence-independent, single-primer amplification (SISPA) of complex DNA populations. Mol Cell Probes 5:473–481

Romeo R, Hegerich P, Emerson SU, et al (2000) High prevalence of TT virus (TTV) in naive chimpanzees and in hepatitis C virus-infected humans: frequent mixed infections and identification of new TTV genotypes in chimpanzees. J Gen Virol 81:1001–1007

Saitou N, Nei M (1987) The neighbor-joining method: a new method for reconstructing phylogenetic trees. Mol Biol Evol 4:406–425

Takahashi K, Hijikata M, Samokhvalov EI, et al (2000a) Full or near full length nucleotide sequences of TT virus variants (Types SANBAN and YONBAN) and the TT virus-like mini virus. Intervirology 43:119–123

Takahashi K, Iwasa Y, Hijikata M, et al (2000b) Identification of a new human DNA virus (TTV-like mini virus, TLMV) intermediately related to TT virus and chicken anemia virus. Arch Virol 145:979–993

Takayama S, Yamazaki S, Matsuo S, et al (1999) Multiple infection of TT virus (TTV) with different genotypes in Japanese hemophiliacs. Biochem Biophys Res Commun 256:208–211

Tawara A, Akahane Y, Takahashi M, et al (2000) Transmission of human TT virus of genotype 1a to chimpanzees with fecal supernatant or serum from patients with acute TTV infection. Biochem Biophys Res Commun 278:470–476

Thom K, Morrison C, Lewis JC, et al (2003) Distribution of TT virus (TTV), TTV-like minivirus, and related viruses in humans and nonhuman primates. Virology 306:324–333

Todd D, Bendinelli M, Biagigi P et al (2005) Family Circoviridae. In: Fauquet CM, Mayo MA, Maniloff J, et al (eds) Virus taxonomy: classification and nomenclature of viruses, eight report of the international committee on taxonomy of viruses. Elsevier/Academic Press, London, pp 327–334

Tsuda F, Okamoto H, Ukita M, et al (1999) Determination of antibodies to TT virus (TTV) and application to blood donors and patients with post-transfusion non-A to G hepatitis in Japan. J Virol Methods 77:199–206

Ukita M, Okamoto H, Nishizawa T, et al (2000) The entire nucleotide sequences of two distinct TT virus (TTV) isolates (TJN01 and TJN02) remotely related to the original TTV isolates. Arch Virol 145:1543–1559

Verschoor EJ, Langenhuijzen S, Heeney JL (1999) TT viruses (TTV) of non-human primates and their relationship to the human TTV genotypes. J Gen Virol 80:2491–2499

# Replication of and Protein Synthesis by TT Viruses

L. Kakkola, K. Hedman, J. Qiu, D. Pintel, and M. Söderlund-Venermo(✉)

**Contents**



**Abstract** The host cells and the events in the cells during Torque teno (TT) virus infection are at present unknown. Replicating TT virus DNA has been detected in liver, in peripheral blood mononuclear cells (PBMC), and in bone marrow. By alternative splicing this small virus generates three mRNA species, from which by alternative translation initiation at least six proteins are produced. The functions of the proteins are not yet fully understood. However, functions associated with, e.g., DNA replication, immunomodulation, and apoptosis have been suggested to reside in the multifunctional proteins of anelloviruses.

## Replication of TTV

Torque teno virus (TTV) DNA is present in several organs and tissues, reflecting a potentially wide host cell tropism for TTV. Identification of a host cell has been attempted by detecting the presence of TTV DNA (*in situ* methods or DNA quantification) in certain cells, or by detecting replicating TTV DNA [gel separation methods, strand-specific primer extension, or full-length polymerase chain reaction (PCR)] in tissues. With the latter methods, double-stranded, circular, and potentially replicating TTV DNA has been detected in a variety of tissues: e.g., lung,

M. Söderlund-Venermo
Department of Virology, Haartman Institute, University of Helsinki and Helsinki University Central Hospital Laboratory Finland
maria.soderlund-venermo@helsinki.fi

E.-M. de Villiers, H. zur Hausen (eds.), *TT Viruses: The Still Elusive Human Pathogens,* 

stimulated peripheral blood mononuclear cells (PBMC), bone marrow, liver, lymph node, thyroid gland, spleen, pancreas, and kidney (Bando et al. 2001; Mariscal et al. 2002; Okamoto et al. 2000b; Okamoto et al. 2000c; Okamoto et al. 2001). In experimental TTV infections of Rhesus monkeys, viral DNA was detected in various organs, whereas replicating TTV DNA was detected only in liver, bone marrow, and the small intestine (Luo et al. 2000; Xiao et al. 2002). Could all these cell types be the hosts for TTVs? As noted by Takahashi and co-workers, since TTV is present in the cells of the immune system that invade multiple tissues in inflammation, it should be carefully determined whether TTV really harbors in the cells inherent in each of the tissues (Takahashi et al. 2002).

Cells circulating in blood and originating from the hematopoietic compartment have been strong candidates for TTV host cells, and the presence of TTV in PBMC was confirmed by fluorescent *in situ* hybridization (Lopez-Alcorocho et al. 2000; Mariscal et al. 2002; Zhong et al. 2002). Furthermore, PBMC can be infected with TTV, and when stimulated, the cells produce infectious TTVs (Maggi et al. 2001a; Mariscal et al. 2002). It has been suggested that some genotypes could have preference for PBMC (Okamoto et al. 1999) and that PBMC could serve as a reservoir for TTV (Chan et al. 2001a; Garbuglia et al. 2003; Maggi et al. 2001b). The precursors for hematopoietic cells reside in bone marrow. The observation that during myelosuppression in bone marrow-transplant recipients the levels of TTV DNA are decreased, led to a suggestion that hematopoietic cells could sustain TTV replication (Kanda et al. 1999). Further evidence showing replicating TTV DNA in bone marrow (Okamoto et al. 2000b; Yu et al. 2002; Zhong et al. 2002) and clearance of TTV after bone marrow transplantation (Chan et al. 2001b) support the hypothesis that the hematological compartment could be the site of replication and/or sustained persistence of TTV.

In addition to the hematopoietic cells, the presence of high TTV DNA levels in saliva (Deng et al. 2000; Gallian et al. 2000) has led to investigations of oropharyngeal tissue as a putative target/host tissue for TTV infection. TTV DNA has been detected by *in situ* hybridization in the cytoplasm of oral epithelial cells (Rodriguez-Inigo et al. 2001), and TTV DNA loads in the nasal cavity exceed those in serum, suggesting that the nasal cavity could be the primary site of TTV infection (Maggi et al. 2003).

Due to the history of TTV as a potential hepatitis virus, liver as a main target organ for TTV infection has also been studied extensively. TTV is found by *in situ* methods in the nucleus and/or the cytoplasm of hepatocytes in patients with liver damage (Cheng et al. 2000; Comar et al. 2002; Jiang et al. 2000) without any cytopathological changes (Ohbayashi et al. 2001; Rodriguez-Inigo et al. 2000).

Infection and replication of TTV has been examined also in various cell lines. Desai and co-workers infected a liver cell line, Chang, a B lymphoblast cell line, Raji, and stimulated PBMCs with TTV genotype 1 (genogroup 1)-positive sera. The infected Chang cells produced constant, low amounts of TTV, whereas in PBMC and Raji cells the release of viruses was transient (Desai et al. 2005). Leppik and co-workers transfected a Hodgkin's lymphoma-derived cell line (L428) with a full-length TTV clone of tth8-isolate (genogroup 5), whereas Kakkola and co-workers transfected monkey kidney-, human erythroid-, liver-, and kidney-derived cell lines (Cos-1, KU812Ep6, Huh7, Chang liver, UT/Epo-S1, 293T, and 293) with a full-length TTV clone of isolate HEL32 of genotype 6, a member of genogroup 1 (Kakkola et al. 2007; Leppik et al. 2007). Both studies showed, by *Dpn*I restriction enzyme and Southern analysis, the recirculation and replication of TTV DNA in human-derived cell types.

In addition to the presence of TTV DNA in certain cell types, the enhancer activity of the TTV promoter of isolate VT416 (genotype 1) has been investigated in various cell lines. The activity was highest in K562 human erythroleukemia cells, and also in HepG2 human hepatocellular carcinoma cells (Kamada et al. 2004), which further support that these two cell types are targets of TTV infection. Taken together, the best candidates at the moment for TTV host cells are cells of hepatic and erythroid origin, as well as the epithelium of the respiratory tract. However, it is also possible that only a specific subset of those cells or yet unidentified cell types could be the main targets for TTV infection and replication. In addition, the various TTV genotypes can differ in host cell tropism thereby further complicating the picture.

## Replication Mechanism

The majority of small DNA viruses depend on the host cell replication machinery for virus replication. Viruses either infect actively dividing cells or induce their host cells to enter S-phase, and subsequently prevent apoptosis. Based on sequence analysis TTV is not assumed to encode a DNA polymerase for replication, but instead it would use cellular polymerases. Indeed, in the presence of aphidicolin, a drug blocking the cellular DNA polymerase, DNA replication of HEL32 TTV (genotype 6) did not occur, indicating that TTV utilizes for DNA replication the cellular replication machinery (Kakkola et al. 2007).

The exact mechanism of TTV replication is unknown. It is assumed, based on similarities with other circular single-stranded (ss)DNA viruses, that TTV could use the rolling circle mechanism (Mushahwar et al. 1999). Probably, as has been shown with the TTV-related circoviruses, viral proteins are needed to interact with cellular proteins for the initiation of replication. For this task circoviruses encode replication-associated proteins with specific Rep-motifs that bind to the replication initiation site (Mankertz et al. 1998; Niagro et al. 1998). Based on the amino acid sequence, TTV open reading frame (ORF)1 seems to contain similar Rep-motifs (Erker et al. 1999; Mushahwar et al. 1999; Tanaka et al. 2001). In addition to replication-associated proteins, animal circoviruses have conserved sequences and genomic structures that are involved in replication (Mankertz et al. 2004; Niagro et al. 1998; Todd et al. 2004). The untranslated region (UTR) of TTV also contains sequences that could form similar structures (Hijikata et al. 1999; Mushahwar et al. 1999; Peng et al. 2002). Whether these structures and proteins of TTV are used in viral replication is currently unknown.

## RNA Processing of TTV

Due to the lack of an efficient culture system to support TTV replication, the transcription profile of TTV has been largely gained from cells transfected with TTV plasmids. Kamahora and co-workers analyzed in transfected Cos-1 cells the TTV mRNAs transcribed from a construct that contained a linearized TTV genome (VT416, genotype 1) (Kamahora et al. 2000). Subsequently, Okamoto and co-workers detected TTV mRNAs (of TYM9-isolate; genogroup 3) in human bone marrow cells from a TTV-infected patient with acute myeloblastic leukemia, in which TTV was actively replicating (Okamoto et al. 2000a). More recently, the detailed TTV transcription profile was characterized by transfection of a clone of the TTV genotype 6, HEL32 isolate in human 293 cells (Qiu et al. 2005). In addition, Leppik and co-workers showed the presence of additional splice variants in L428 Hodgkin's lymphoma cells transfected with tth7 and tth8 isolates (genogroup 5) (Leppik et al. 2007).

Three species of TTV mRNAs (2.8–3.0 kb, 1.2 kb, and 1.0 kb) have been detected in transfected Cos-1, 293 and L428 cells, as well as in infected bone marrow cells (Kamahora et al. 2000; Leppik et al. 2007; Okamoto et al. 2000a; Qiu et al. 2005). The genetic map of TTV generated by transfection of a clone of TTV genotype 6 is shown in Fig. 1 (Qiu et al. 2005), with the nucleotide numbers referring to GenBank accession number AY666122 (Kakkola et al. 2002). All the three TTV mRNAs are transcribed from a single promoter, which is located in the region −154/−76 (the RNA initiation site is denoted as position +1) (Kamada et al. 2004), and are polyadenylated at a single site at nt 2978 in HEL32. Therefore, all three mRNAs have the same 5′- and 3′-ends in common. All the mRNAs splice out a small intron (~100 nt) located approx. 70 nt from the RNA initiation site. The two short mRNAs (1.2 kb and 1.0 kb) have further been spliced approx. 400 nt downstream of the small intron, excising another larger intron with alternative 3′ splice sites 2A1 (nt 2315) and 2A2 (nt 2505) (Qiu et al. 2005). The large 2.8-kb mRNA is unspliced in the region of the large intron. All the splice junctions detected in transfected cells or infected bone marrow cells are at their corresponding locations in all TTV genomes tested (Kamahora et al. 2000; Leppik et al. 2007; Okamoto et al. 2000a; Qiu et al. 2005).

Approximately half of the TTV mRNAs of HEL32 (genotype 6) are unspliced at the second intron, which generates the abundant 2.8-kb mRNA. The ratio of mRNAs spliced at 2A1 relative to mRNAs spliced at 2A2 is approx. 1:10. The 2.8-, 1.2-, and 1.0-kb mRNAs comprise approx. 60%, 5%, and 35% of the total TTV RNA, respectively (Qiu et al. 2005). Consistent with this observation, it has been shown that during TTV genotype-1 infection of bone marrow cells, the large mRNA was the predominant mRNA (Okamoto et al. 2000a).

The polyadenylation signal (AUUAAA) at nt 2978 of the HEL32 TTV isolate is nonconsensus but functional (Qiu et al. 2005), while the isolates VT416 and TYM9 use a canonical polyadenylation site (Kamahora et al. 2000; Okamoto

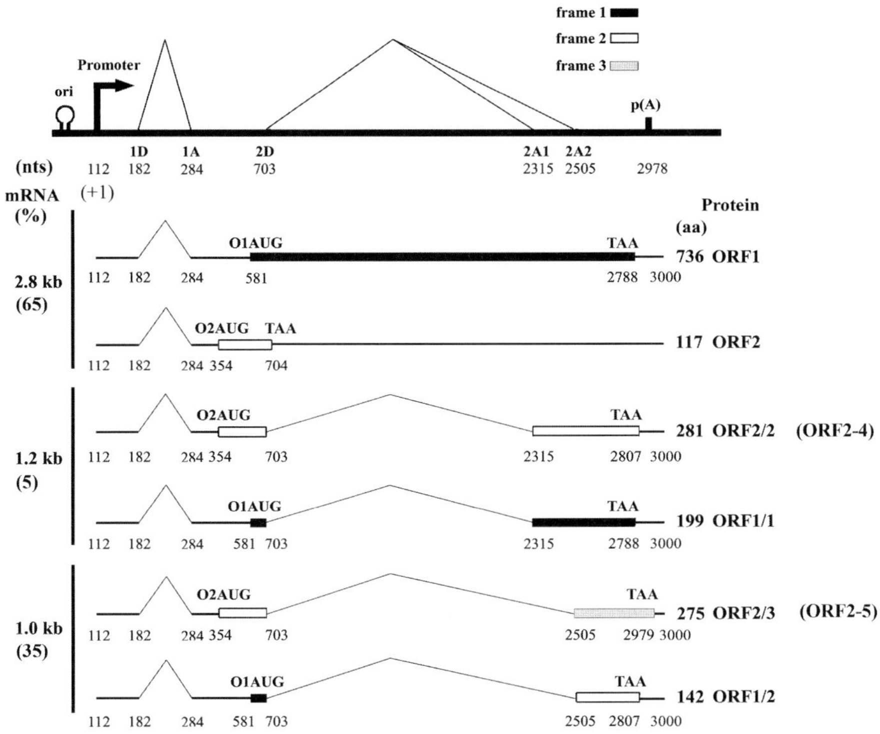


**Fig. 1** The genetic map of TTV genotype 6. The RNA initiation site (+*1*), splice donors (*D*), and acceptors (*A*), and the polyadenylation signal, *p(A)*, are shown with nucleotide numbers (GenBank accession number AY666122). The three species of TTV mRNA are shown with their relative sizes and abundance (% of total) on the *left*. Six ORF expression strategies are diagrammed in the map. The ORF1 and ORF2 proteins are encoded from the singly spliced 2.8-kb mRNA by alternative translation, using O1AUG at nt 581 and O2AUG at nt 354, respectively. The ORF2/2 and ORF2/3 proteins (referred to as *ORF2–4* and *ORF2–5* proteins, respectively, by Kamahora et al. 2000) are initiated at the O2AUG, and are translated from the doubly spliced 1.2- and 1.0-kb mRNAs, respectively. ORF1/1 and ORF1/2 proteins are encoded by the 1.2- and 1.0-kb mRNAs, respectively, using the O1AUG at nt 581. The different reading frames are indicated by different *fill patterns*. (Adapted from Qiu et al. 2005)

et al. 2000a). All the TTV RNAs are cleaved at the same site at approx. nt 3000 (Qiu et al. 2005).

The transcription map of the TTV genotype 1 was determined by transfection of a clone of VT416, and is shown in Fig. 2. Analysis of the sequences of the donor and acceptor sites in 16 TTVs from various genotypes retrieved from the database showed that all the splice sites use the conserved GT–AG donor and acceptor sequences (Kamahora et al. 2000).

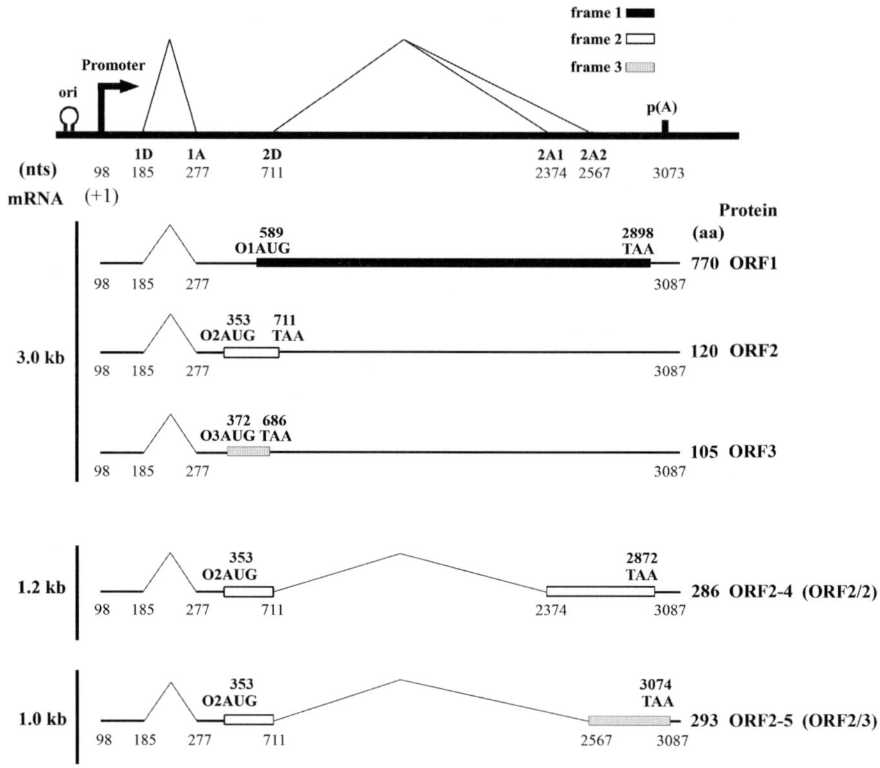


**Fig. 2** The genetic map of TTV genotype 1. The RNA initiation site, splice junctions, and the polyadenylation signal are shown with nucleotide numbers (GenBank accession number NC_002076). The three species of TTV mRNAs with the relative sizes are shown on the *left*. Five predicted ORF expression strategies are diagrammed in the map. The ORF1, ORF2, and ORF3 proteins are predicted to be encoded from the singly spliced 3.0-kb mRNA by alternative translation, using O1AUG at nt 589, O2AUG at nt 353 and O3AUG at nt 372, respectively. The *ORF2–4* and *ORF2–5* proteins are equivalent to the *ORF2/2* and *ORF2/3* proteins, respectively, of genotype 6 as diagrammed in Fig. 1. The proteins are predicted from the coding capabilities of the genomes of TTVs (TA278 and VT416, genotype 1). (Adapted from Kamahora et al. 2000)

Analysis of RNA transcripts produced in the L428 Hodgkin's lymphoma cells transfected with isolates tth7 and tth8 (genogroup 5) confirmed the presence of the above-described TTV mRNAs (Leppik et al. 2007). However, 12 additional transcripts were also identified by RT-PCR. These transcripts were suggested to have resulted from additional splicing events and also from intragenomic rearrangements (leading to subgenomic isolates) within the TTV genome. Transfection of T cell leukemia-, Burkitt's lymphoma-, kidney-, and liver-derived cell lines (Jurkat, HSB2, BJAB, Raji, 293, HepG2) resulted in similar transcription profiles with both tth7 and tth8 isolates. Interestingly, differences between cell lines were observed; the T cell leukemia cell line HSB2 showed no TTV transcription, and 2/12 transcripts were produced specifically only in L428 cells (Leppik et al. 2007). This variation in transcription, possibly resulting in translation of a variety of proteins from different TTV strains with different cell types, could have implications on the biological and/or pathogenetic differences of the TTV strains.

## Protein Expression of TTV

The genome of TTV is a circular single-stranded negative-sense DNA of approx. 3,800 nt in length. Thus expression of TTV mRNA must employ efficient strategies, including both alternative splicing and alternative translation initiation. All three TTV mRNAs use alternative AUGs for translation, and therefore at least six TTV proteins are expressed in genotype 6 (Qiu et al. 2005). Antibodies in TTV-infected individuals have been detected toward all six proteins of the HEL32 isolate of genotype 6 (Kakkola et al. 2002; Kakkola et al. in press), and toward the N- and C-termini of the ORF1-encoded protein of genotype-1 isolates (Handa et al. 2000; Ott et al. 2000).

The abundant 2.8-kb mRNA of HEL32 expresses the ORF1 and ORF2 proteins by alternative AUG usage. The TTV ORF1 protein (~81 kDa) is the largest encoded by TTV mRNA (Fig. 1), and (for genotype 6) is predicted to be 736 amino acids (aa) in length, initiating at the AUG at nt 581 (O1AUG). The ORF2 protein (~13 kDa; Fig. 1) is encoded by the 2.8-kb mRNA in the second ORF, initiating at the AUG at nt 354 (O2AUG) and extending for 117aa (Qiu et al. 2005). Although the Kozak translation signal of the O1AUG (ORF1) in most of the TTV isolates represents more of a consensus sequence (a/gcc AUG g) than that of the O2AUG (Kamahora et al. 2000), the relative usage of these two AUGs has not been conclusively determined.

The small 1.2-kb mRNA of HEL32 expresses TTV proteins ORF2/2 and ORF1/1 with sizes of 31 and 22 kDa, respectively; the smallest 1.0-kb mRNA expresses TTV proteins ORF2/3 and ORF1/2 with sizes of 30 and 16 kDa respectively. Expressions of these proteins were confirmed by transfection of the TTV clone in 293 cells (Qiu et al. 2005). The presence of ORF2 (nt 2315–2807) and ORF3 (nt 2502–2979) at the 3′ termini of both the 1.2- and 1.0-kb mRNAs in virtually all TTV sequences in the database supports their usage during the viral life cycle. The 1.0-kb spliced mRNAs of HEL32, which use the 2A2 splice acceptor and encode the ORF2/3 and ORF1/2 proteins, are approx. 7 times more abundant than the 1.2-kb spliced mRNAs that use the 2A1 acceptor and encode the ORF2/2 and ORF1/1 proteins (Fig. 1). Whether these differences can be seen also in the corresponding protein abundances is unknown. The transcription maps of the three different isolates, VT416 of genotype 1, HEL32 of genotype 6 (both belonging to genogroup 1), and isolate TYM9 (belonging to genogroup 3) are similar (Kamahora et al. 2000; Okamoto et al. 2000a; Qiu et al. 2005), and this is likely to hold true also for the other TTV genotypes. Indeed, similarities have been shown also in representatives of genogroup 5 (isolates tth7 and tth8) in which, however, additional transcripts—produced by variant splice sites and by intragenomic rearrangements—have been detected (Leppik et al. 2007).

## Functions of TTV Proteins

The functions of the TTV proteins are poorly understood since neither a virus culture system nor an animal model is available. Furthermore, there are as yet no reports of TTV purification, whereby the structural proteins forming the virus capsid have not been identified. An animal circovirus, chicken anemia virus (CAV), which causes severe immunosuppression, anemia, and thrombocytopenia (Noteborn et al. 1991), has one structural protein, VP1. The TTV ORF1 and ORF2 encoded proteins have been shown (for the HEL32 isolate) to be predominantly localized in the cytoplasm (Qiu et al. 2005), and the former is assumed to be a structural and replication-associated protein based on similarities with CAV VP1, i.e., possessing an arginine-rich N-terminus and Rep-motifs (Erker et al. 1999; Mushahwar et al. 1999; Tanaka et al. 2001). The expression level of ORF1 was, for an unknown reason, extremely low following transfection of the full-length clone of genotype 6 into 293 cells (Qiu et al. 2005). In HEL32 the two ORF1-related proteins, ORF1/1 and ORF1/2, are distributed evenly both in the cytoplasm and the nucleus, and when compared to the ORF2-containing genes (described below), are transcribed at a low level (Qiu et al. 2005). Interestingly, TTV (isolate TRM1, genotype 1) *ORF1*-transgenic mice have morphological changes in renal epithelial cells. The expression of the ORF1 gene in mice affected the maturation of renal epithelial cells in a dose-dependent manner. However, it was not determined which of the three ORF1-containing proteins was responsible for the changes, as small spliced mRNAs with a potential to express ORF1/1 or ORF1/2 proteins were also detected (Yokoyama et al. 2002).

The TTV ORF2 protein (SANBAN of genogroup 3) has been studied in cells transfected with ORF2 complementary DNA (cDNA) (Zheng et al. 2007). Following transfection into human cell lines the TTV ORF2 protein was shown to suppress both the canonical and the noncanonical nuclear factor (NF)-κB pathways by interacting with the catalytic subunits (inhibitor of κB[1κB] kinase) IKKα and IKKβ of the IκBα kinase complex, thus inhibiting its degradation. By hindering NF-κB from reaching the nucleus the TTV ORF2 protein also indirectly decreased the expression of the inflammation factors interleukin (IL)-6, IL-8, and cyclo-oxygenase (COX)-2. NF-κB activation is a protective response of the host to viral pathogens. This suggests that the TTV ORF2 protein may be involved in the regulation of the host innate and adaptive immunity. Thus, the role of the ORF2 protein in immune evasion may be an important aspect of TTV biology.

CAV VP2 possesses a functional protein-tyrosine phosphatase (PTPase)-like domain (Peters et al. 2002; Takahashi et al. 2000). The ORF2 proteins of TTV, SAV/TT midi virus (TTMDV) and Torque teno minivirus (TTMV) also share a high similarity with the CAV VP2 (Andreoli et al. 2006; Biagini et al. 2001; Takahashi et al. 2000), of which the latter has been shown to exert phosphatase activity (Peters et al. 2002). Protein phosphatases are known to have important functions in the regulation of gene transcription, signal transduction, and mitogenesis, and in cytokine responses of lymphocytes (Ong et al. 1997; Schievella et al. 1993). The CAV VP2 PTPase activity may regulate these events in infected lymphocytes, which could lead to profound immunosuppression and anemia in chickens

(Noteborn et al. 1991; Peters et al. 2002). The five TTV ORF2 coding sequences of the TTV SANBAN group (genogroup 3) have a high amino acid similarity (53%–55%), with an identity of 22%–26%, with the CAV VP2 (Peters et al. 2002). However, the sequence variation of TTV ORF2 may interfere with the PTPase function, which could contribute to the pathogenesis of different anelloviruses.

The ORF2/2 and ORF2/3 proteins of HEL32, genotype 6 [corresponding to the respective ORF2-ORF4 and ORF2-ORF5 proteins of genotype 1 (Kamahora et al. 2000)] are exclusively localized in discrete foci in the nucleus (Qiu et al. 2005) implying roles in genome expression and replication. Both contain the ORF2 protein at the N-termini, with the ORF2/3 transcript being the most abundant (Qiu et al. 2005). Considering the high expression level of the ORF2-containing TTV transcripts of HEL32 (ORF2, ORF2/2 and ORF2/3), it will be interesting to determine whether all of these corresponding proteins have PTPase activities and to understand their roles in TTV pathogenesis. The ORF2/2 protein not only shares the N-terminus with the ORF2 protein (Kamahora et al. 2000; Qiu et al. 2005), but also has a serine-rich domain at the C-terminus (Asabe et al. 2001), capable of generating different phosphorylation sites, similar to hepatitis C virus (HCV) NS5A (Tellinghuisen et al. 2008). This suggests that a phosphorylated version of the ORF2/2 protein may play a similar role in binding the DNA template as does the NS5A of HCV in viral replication.

In TTV genotype 1, isolate TA278, an additional protein at nt 372–686 in reading frame 3 was proposed to be encoded from the N-terminus of the 3.0-kb mRNA (Fig. 2). Due to the similarity to apoptin, the main apoptosis-inducing agent of CAV (Kooistra et al. 2004; Miyata et al. 1999), this putative 105-aa protein (~12 kDa) was termed TTV-derived apoptosis-inducing protein (TAIP). Both TAIP and apoptin induced p53-independent apoptosis in human hepatocellular carcinoma (HCC) cell lines; however, unlike apoptin, TAIP was only weakly apoptotic in other human cancer cell lines (Kooistra et al. 2004). Interestingly, the 5′-end of the ORF3-encoding region in e.g., HEL32 (genotype 6) is interrupted with stop codons, whereby the TAIP-related ORF3 protein is not expressed from TTV genotype 6. The activity of TAIP, coupled with the heterogeneity of TTV isolates, could contribute to some extent to the putatively variable pathogenesis of TTV.

Overall, our knowledge of the function of TTV proteins is to a large extent limited by the lack of an efficient cell culture system and by restricted availability of the virus itself. However, it is possible that, for example, the multifunctional nature of TTV ORF2-containing proteins could confer pathogenetic diversity to this large genus of anelloviruses.

## References

Andreoli E, Maggi F, Pistello M, et al (2006) Small anellovirus in hepatitis C patients and healthy controls. Emerg Infect Dis 12:1175–1176

Asabe S, Nishizawa T, Iwanari H, Okamoto H (2001) Phosphorylation of serine-rich protein encoded by open reading frame 3 of the TT virus genome. Biochem Biophys Res Commun 286:298–304

Bando M, Ohno S, Oshikawa K, et al (2001) Infection of TT virus in patients with idiopathic pulmonary fibrosis. Respir Med 95:935–942
Biagini P, Gallian P, Attoui H, et al (2001) Genetic analysis of full-length genomes and subgenomic sequences of TT virus-like mini virus human isolates. J Gen Virol 82:379–383
Chan PK, Chik KW, Li CK, et al (2001a) Prevalence and genotype distribution of TT virus in various specimen types from thalassaemic patients. J Viral Hepat 8:304–309
Chan PK, Chik KW, To KF, et al (2001b) Clearance of TT virus after allogeneic bone marrow transplantation. J Pediatr Hematol Oncol 23:57–58
Cheng J, Hada T, Liu W, et al (2000) Investigation of TTV by in situ hybridization in patients with chronic hepatitis. Hepatol Res 18:43–53
Comar M, Ansaldi F, Morandi L, et al (2002) In situ polymerase chain reaction detection of transfusion-transmitted virus in liver biopsy. J Viral Hepat 9:123–127
Deng X, Terunuma H, Handema R, et al (2000) Higher prevalence and viral load of TT virus in saliva than in the corresponding serum: another possible transmission route and replication site of TT virus. J Med Virol 62:531–537
Desai M, Pal R, Deshmukh R, Banker D (2005) Replication of TT virus in hepatocyte and leucocyte cell lines. J Med Virol 77:136–143
Erker JC, Leary TP, Desai SM, et al (1999) Analyses of TT virus full-length genomic sequences. J Gen Virol 80:1743–1750
Gallian P, Biagini P, Zhong S, et al (2000) TT virus: a study of molecular epidemiology and transmission of genotypes 1, 2 and 3. J Clin Virol 17:43–49
Garbuglia AR, Iezzi T, Capobianchi MR, et al (2003) Detection of TT virus in lymph node biopsies of B-cell lymphoma and Hodgkin's disease, and its association with EBV infection. Int J Immunopathol Pharmacol 16:109–118
Handa A, Dickstein B, Young NS, Brown KE (2000) Prevalence of the newly described human circovirus, TTV, in United States blood donors. Transfusion 40:245–251
Hijikata M, Takahashi K, Mishiro S (1999) Complete circular DNA genome of a TT virus variant (isolate name SANBAN) and 44 partial ORF2 sequences implicating a great degree of diversity beyond genotypes. Virology 260:17–22
Jiang XJ, Luo KX, He HT (2000) Intrahepatic transfusion-transmitted virus detected by in situ hybridization in patients with liver diseases. J Viral Hepat 7:292–296
Kakkola L, Hedman K, Vanrobaeys H, et al (2002) Cloning and sequencing of TT virus genotype 6 and expression of antigenic open reading frame 2 proteins. J Gen Virol 83:979–990
Kakkola L, Tommiska J, Boele LC, et al (2007) Construction and biological activity of a full-length molecular clone of human Torque teno virus (TTV) genotype 6. FEBS J 274:4719–4730
Kakkola L, Bondén H, Hedman L, et al (in press) Expression of all six human Torque teno virus (TTV) proteins in bacteria and in insect cells, and analysis of their ig6 responses. Virology.
Kamada K, Kamahora T, Kabat P, Hino S (2004) Transcriptional regulation of TT virus: promoter and enhancer regions in the 1.2-kb noncoding region. Virology 321:341–348
Kamahora T, Hino S, Miyata H (2000) Three spliced mRNAs of TT virus transcribed from a plasmid containing the entire genome in COS1 cells. J Virol 74:9980–9986
Kanda Y, Tanaka Y, Kami M, et al (1999) TT virus in bone marrow transplant recipients. Blood 93:2485–2490
Kooistra K, Zhang YH, Henriquez NV, et al (2004) TT virus-derived apoptosis-inducing protein induces apoptosis preferentially in hepatocellular carcinoma-derived cells. J Gen Virol 85:1445–1450
Leppik L, Gunst K, Lehtinen M, et al (2007) In vivo and in vitro intragenomic rearrangement of TT viruses. J Virol 81:9346–9356
Lopez-Alcorocho JM, Mariscal LF, de Lucas S, et al (2000) Presence of TTV DNA in serum, liver and peripheral blood mononuclear cells from patients with chronic hepatitis. J Viral Hepat 7:440–447
Luo K, Liang W, He H, et al (2000) Experimental infection of nonenveloped DNA virus (TTV) in rhesus monkey. J Med Virol 61:159–164

Maggi F, Fornai C, Zaccaro L, et al (2001a) TT virus (TTV) loads associated with different peripheral blood cell types and evidence for TTV replication in activated mononuclear cells. J Med Virol 64:190–194
Maggi F, Pistello M, Vatteroni M, et al (2001b) Dynamics of persistent TT virus infection, as determined in patients treated with alpha interferon for concomitant hepatitis C virus infection. J Virol 75:11999–12004
Maggi F, Pifferi M, Fornai C, et al (2003) TT virus in the nasal secretions of children with acute respiratory diseases: relations to viremia and disease severity. J Virol 77:2418–2425
Mankertz A, Mankertz J, Wolf K, Buhk HJ (1998) Identification of a protein essential for replication of porcine circovirus. J Gen Virol 79:381–384
Mankertz A, Caliskan R, Hattermann K, et al (2004) Molecular biology of Porcine circovirus: analyses of gene expression and viral replication. Vet Microbiol 98:81–88
Mariscal LF, Lopez-Alcorocho JM, Rodriguez-Inigo E, et al (2002) TT virus replicates in stimulated but not in nonstimulated peripheral blood mononuclear cells. Virology 301:121–129
Miyata H, Tsunoda H, Kazi A, et al (1999) Identification of a novel GC-rich 113-nucleotide region to complete the circular, single-stranded DNA genome of TT virus, the first human circovirus. J Virol 73:3582–3586
Mushahwar IK, Erker JC, Muerhoff AS, et al (1999) Molecular and biophysical characterization of TT virus: evidence for a new virus family infecting humans. Proc Natl Acad Sci U S A 96:3177–3182
Niagro FD, Forsthoefel AN, Lawther RP, et al (1998) Beak and feather disease virus and porcine circovirus genomes: intermediates between the geminiviruses and plant circoviruses. Arch Virol 143:1723–1744
Noteborn MH, de Boer GF, van Roozelaar DJ, et al (1991) Characterization of cloned chicken anemia virus DNA that contains all elements for the infectious replication cycle. J Virol 65:3131–3139
Ohbayashi H, Tanaka Y, Ohoka S, et al (2001) TT virus is shown in the liver by in situ hybridization with a PCR-generated probe from the serum TTV-DNA. J Gastroenterol Hepatol 16:424–428
Okamoto H, Kato N, Iizuka H, et al (1999) Distinct genotypes of a nonenveloped DNA virus associated with posttransfusion non-A to G hepatitis (TT virus) in plasma and peripheral blood mononuclear cells. J Med Virol 57:252–258
Okamoto H, Nishizawa T, Tawara A, et al (2000a) TT virus mRNAs detected in the bone marrow cells from an infected individual. Biochem Biophys Res Commun 279:700–707
Okamoto H, Takahashi M, Nishizawa T, et al (2000b) Replicative forms of TT virus DNA in bone marrow cells. Biochem Biophys Res Commun 270:657–662
Okamoto H, Ukita M, Nishizawa T, et al (2000c) Circular double-stranded forms of TT virus DNA in the liver. J Virol 74:5161–5167
Okamoto H, Nishizawa T, Takahashi M, et al (2001) Heterogeneous distribution of TT virus of distinct genotypes in multiple tissues from infected humans. Virology 288:358–368
Ong CJ, Dutz JP, Chui D, et al (1997) CD45 enhances positive selection and is expressed at a high level in large, cycling, positively selected CD4+CD8+ thymocytes. Immunology 91:95–103
Ott C, Duret L, Chemin I, et al (2000) Use of a TT virus ORF1 recombinant protein to detect anti-TT virus antibodies in human sera. J Gen Virol 81:2949–2958
Peng YH, Nishizawa T, Takahashi M, et al (2002) Analysis of the entire genomes of thirteen TT virus variants classifiable into the fourth and fifth genetic groups, isolated from viremic infants. Arch Virol 147:21–41
Peters MA, Jackson DC, Crabb BS, Browning GF (2002) Chicken anemia virus VP2 is a novel dual specificity protein phosphatase. J Biol Chem 277:39566–39573
Qiu J, Kakkola L, Cheng F, et al (2005) Human circovirus TT virus genotype 6 expresses six proteins following transfection of a full-length clone. J Virol 79:6505–6510
Rodriguez-Inigo E, Casqueiro M, Bartolome J, et al (2000) Detection of TT virus DNA in liver biopsies by in situ hybridization. Am J Pathol 156:1227–1234

Rodriguez-Inigo E, Arrieta JJ, Casqueiro M, et al (2001) TT virus detection in oral lichen planus lesions. J Med Virol 64:183–189

Schievella AR, Paige LA, Johnson KA, et al (1993) Protein tyrosine phosphatase 1B undergoes mitosis-specific phosphorylation on serine. Cell Growth Differ 4:239–246

Takahashi K, Hijikata M, Samokhvalov EI, Mishiro S (2000) Full or near full length nucleotide sequences of TT virus variants (Types SANBAN and YONBAN) and the TT virus-like mini virus. Intervirology 43:119–123

Takahashi M, Asabe S, Gotanda Y, et al (2002) TT virus is distributed in various leukocyte subpopulations at distinct levels, with the highest viral load in granulocytes. Biochem Biophys Res Commun 290:242–248

Tanaka Y, Primi D, Wang RY, et al (2001) Genomic and molecular evolutionary analysis of a newly identified infectious agent (SEN virus) and its relationship to the TT virus family. J Infect Dis 183:359–367

Tellinghuisen TL, Foss KL, Treadaway J (2008) Regulation of hepatitis C virion production via phosphorylation of the NS5A protein. PLoS Pathog 4:e1000032

Todd D, Bendinelli M, Biagini P, et al (2004) Circoviridae. In: Fauquet CM, Mayo MA, Maniloff J, Desselberger U, Ball LA (eds) Virus taxonomy: eighth report of the International Committee on Taxonomy of Viruses. Elsevier/Academic Press, London

Xiao H, Luo K, Yang S, et al (2002) Tissue tropism of the TTV in experimentally infected rhesus monkeys. Chin Med J (Engl) 115:1088–1090

Yokoyama H, Yasuda J, Okamoto H, Iwakura Y (2002) Pathological changes of renal epithelial cells in mice transgenic for the TT virus ORF1 gene. J Gen Virol 83:141–150

Yu Q, Shiramizu B, Nerurkar VR, et al (2002) TT virus: preferential distribution in CD19(+) peripheral blood mononuclear cells and lack of viral integration. J Med Virol 66:276–284

Zheng H, Ye L, Fang X, et al (2007) Torque teno virus (SANBAN isolate) ORF2 protein suppresses NF-kappaB pathways via interaction with IkappaB kinases. J Virol 81:11917–11924

Zhong S, Yeo W, Tang M, et al (2002) Frequent detection of the replicative form of TT virus DNA in peripheral blood mononuclear cells and bone marrow cells in cancer patients. J Med Virol 66:428–434

# Immunobiology of the Torque Teno Viruses and Other Anelloviruses

**F. Maggi(✉) and M. Bendinelli**

**Contents**



**Abstract** Many features of the Torque teno virus and the other anelloviruses (AVs) that have been identified after this virus was discovered in 1997 remain elusive. The immunobiology of the AVs is no exception. However, evidence is progressively accumulating that at least some AVs have an interesting interplay with cells and soluble factors known to contribute to the homeostasis of innate and adaptive immunity. Evidence is also accumulating that this interplay can have a significant impact on how effectively an infected host can deal with superimposed infectious and non-infectious noxae. This review article discusses the scanty information available on these aspects and highlights the ones that would be more urgent to precisely understand in order to get an adequate assessment of how important for human health these extremely ubiquitous and pervasive viruses really are.

F. Maggi
Virology Unit, Pisa University Hospital, 35–37, Via San Zeno, I-56127 Pisa, Italy
maggif@biomed.unipi.it

E.-M. de Villiers, H. zur Hausen (eds.), *TT Viruses: The Still Elusive Human Pathogens*, 

**Abbreviations** ADCC: Antibody-dependent cellular cytotoxicity; AVs: Anelloviruses; CNS:Central nervous system; CTL: Cytotoxic T lymphocyte; ECP: Eosinophil cationic protein; eNO: Exhaled nitric oxide; HCV:Hepatitis C virus; HIV: Human immunodeficiency virus: ICs: Immunocomplexes; IFN: Interferon; IL: Interleukin; NF-κB: Nuclear factor kappa B; ORF: Open reading frame: PAMPs: Pathogen-associated molecular patterns; PBMC: Peripheral blood mononuclear cell; SLE: Systemic lupus erythematosus; TLR: Toll-like receptor; TTV: Torque teno virus; TTMDV: Torque teno midi virus; TTMV: Torque teno mini virus; WBC: White blood cell

## Introduction

Torque teno virus (TTV), first identified using molecular techniques in the blood of a patient who post-transfusion presented with abnormal liver enzymes levels but no classic hepatitis viruses a little over one decade ago (Nishizawa et al. 1997), is now known to be the prototype of a vast group of related yet distinct viruses that infect humans and various animal species (Biagini 2004; Biagini et al. 2004; Hino and Miyata 2007).

Among the anelloviruses (AVs), those of humans are the best characterised. They possess several remarkable properties, including: (1) a particularly small (2.2–3.7 kb) single-stranded circular DNA genome that makes them the genetically simplest of all known replication-competent viruses hitherto detected in humans (Miyata et al. 1999; Takahashi et al. 2000; Jones et al. 2005; Ninomiya et al. 2007); (2) an extremely high degree of genetic heterogeneity which has brought us to recognise at least three major viral species [TTV; Torque teno mini virus (TTMV); and Torque teno midi virus (TTMDV)], each consisting of numerous genogroups, genotypes (over 80 already identified), and strains (Bendinelli and Maggi 2005); (3) a remarkable ability to produce persistent, possibly life-long infections with variably elevated levels of plasma viraemia with no clearly associated clinical manifestations (Bendinelli et al. 2001); (4) a general ubiquity in the body, as revealed by the facility with which they are detected in all the tissues and organs, except possibly the uninjured central nervous system (CNS) (Maggi et al. 2001a; Okamoto et al. 2001); (5) a highly unequal prevalence in the populations examined of the various viruses in the group, with some acquired very early in life or even prenatally and found in large proportions of individuals regardless of age, socio-economical standing and health conditions, and others detected only very occasionally (Maggi et al. 2003a; 2005a; 2006; Ninomiya et al. 2008).

From the above properties and all the rest that is currently known, it is clear that the AVs have established a highly successful interaction with their hosts and that has extensively helped them to evolve and diversify (Simmonds 2002). However, the lack of important investigational tools, including sufficiently sensitive in vitro culture systems and easy-to-handle experimental animals (the only animal shown to be susceptible to at least some human AVs is the chimpanzee, a species that possesses its own AVs which may influence the outcome of experimental infections Inami et al. 2000; Tawara et al. 2000), has so far considerably hindered a thorough analysis of many aspects of their biology and life cycle. Of outmost importance, among the aspects that are still poorly understood, are the immunological properties that make the AVs so successful at evading control by the host's innate and adaptive defences. Also essentially unexplored is how extensively and in what direction these defences are affected by continued exposure to the variable but generally sustained loads of AVs that are found in most, if not all, individuals (Pistello et al. 2001). In this chapter we review the studies that have dealt with this most stimulating area of research, warning the reader that investigations to date are not only limited in number but have also almost exclusively dealt with the prototype member of the group, TTV, and, more precisely, to some genetic forms of this viral species. Whether or not what has been learned for TTV also holds true for the other AVs is still almost completely unknown. As a matter of fact, the entire field of the interactions these viruses establish with their hosts and of how such interactions impact on the host's innate and adaptive mechanisms of defence is still essentially unploughed.

## Innate Immunity

Although the AVs have not been linked directly to any overt disease, they have been seen to worsen the impact of several other clinically relevant infections (see the following sections). Furthermore, they have been shown to be associated with an

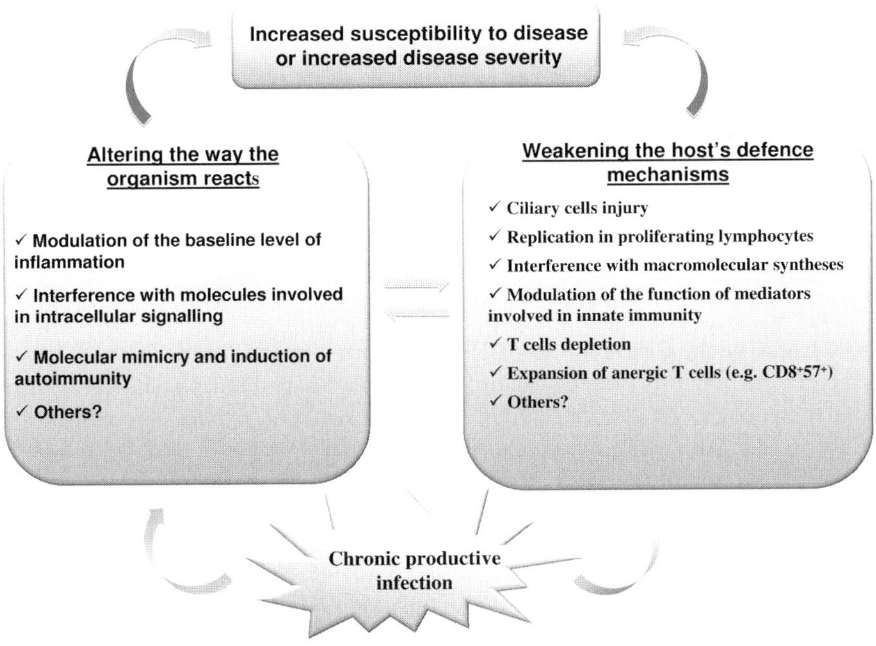


**Fig. 1** Possible mechanisms whereby chronic anellovirus infections might contribute to disease induction by infectious (*right panel*) and non-infectious noxae (*left panel*)

aggravated course of multifactorial diseases having in common the existence of a large inflammatory component in the absence of a clearly recognised causing agent, such as asthma and selected forms of arthritis (Bendinelli and Maggi 2005; Pifferi et al. 2005, 2006; Maggi et al. 2007b). This has led investigators to suggest that the AVs weaken the host's defence mechanisms against other noxae and/or alter the way the organism reacts to them, for example by keeping baseline levels of inflammation elevated (Fig. 1), so that the superimposed noxae deal with an already compromised, although subclinical, state of affairs (Bendinelli et al. 2001; Maggi et al. 2003a, 2004; Pifferi et al. 2005).

A possible example of the first possibility is provided by a recent investigation of the ciliated cells in the respiratory epithelium, cells that exert the important non-specific defence function of continuously sweeping outwards inhaled particles in collaboration with mucous secretions. Previous studies have shown that the respiratory tract, at least its upper portion, is continuously exposed to abundant TTV and represents not only a major route through which this virus can enter the body and is shed into the environment but also a site of continued TTV replication (Maggi et al. 2003a). The type(s) of cells supporting TTV replication in the respiratory tract is currently unknown, but recent evidence from our group indicates that ciliated cells are a likely possibility. Laser micro-dissected epithelial cells obtained from nasal brushings of children with persistent recurrent bronchopneumonia were found to contain abundant TTV DNA (Pifferi et al. 2008). Furthermore, in the same study a significant correlation was observed between the size of TTV load in nasal fluids and reduced functional and anatomical integrity of the ciliary cells. Collectively these findings seem to indicate that TTV may disturb the "mucociliary escalator", thus exerting effects similar to influenza and other common respiratory viruses.

On the other hand, such a modulation by the AVs of the level of inflammation in infected tissues is also a clear possibility as suggested by the lines of evidence discussed in the following sections.

## *TTV and NF-κB*

Information on the consequences of TTV replication on infected cells is still scanty. There are, for example, essentially no data on how biosynthetic machinery, expression of receptors on the cell surface, and intracellular signalling pathways are affected. Recently, however, Zheng et al. (2007) have demonstrated that the ORF2 protein of a genogroup-3 TTV, the so-called SANBAN strain, has the potential to interfere with the activity of NF-κB, a well-characterised intracellular signal transcription factor known to play a myriad of roles in inflammation, immunomodulation, lymphocyte development, cell growth, and cancer. These investigators have shown that recombinant ORF2 protein, which probably also possesses phosphatase activity (Peters et al. 2002), has the ability to suppress NF-κB activity in vitro in a dose-dependent manner, affecting translocation of the factor to the cell nucleus and, consequently, inhibiting the transcription of downstream genes such as interleukin (IL)-6, IL-8, and cyclo-oxygenase-2. Thus, since NF-κB promotes the expression of over 100 genes including pro-inflammatory cytokines, chemokines, adhesion molecules, inducible enzymes, and immune receptors (Hiscott 2007), TTV, or at least some isolates within this viral species, is potentially capable of impacting significantly, by means of the ORF2 protein, many factors that concur at effecting or regulating the inflammatory response (Maggi et al. 2004; Zheng et al. 2007).

## *TTV and Toll-Like Receptor-9*

NF-κB is also a key controlling element that participates in the synthesis of pro-inflammatory cytokines triggered by the stimulation of TLRs. These type I trans-membrane receptors sense specific patterns on the surface of exogenous or endogenous abnormal molecules or molecular complexes [pathogen-associated molecular patterns (PAMPs)] and, when properly stimulated, induce the activation of inflammatory and immune cell responses directed against such structures. Thus, the TLRs are involved in many facets of the anti-microbial defence system and, in vertebrates, also provide important connections between innate and adaptive immune responses.

Among the 13 TLRs so far described in humans, TLR-9 recognises intracellular unmethylated, and therefore mainly microbial, DNA through the recognition of specific short deoxyribonucleotide sequences formed by heterodimers of guanosine and cytosine (CpGs). The results of such recognition are not uniform but vary depending on the types of nucleotides present around the CpGs in the unmethylated DNA: thus, while some CpGs stimulate TLR-9 to trigger a cascade of events culminating in the induction of several pro-inflammatory cytokines, including interferon (IFN)-γ, IL-6, and IL-12, other CpGs are inhibitory and block (neutralise) the activity of the stimulatory ones. Prompted by the observation that both stimulatory and neutralising CpGs are variously abundant in the genome of DNA viruses (Lundberg et al. 2003), our group has recently observed that, relative to most other viral DNAs, the genomes of the AVs, as well as the replicative molecular intermediates these form in infected cells, are unusually rich in CpG sequences (Maggi et al. 2007a). Moreover, the DNA of one genogroup-4 TTV isolate (ViPiSAL strain) was found to provoke a robust activation of TLR-9 in ex vivo grown mouse spleen cells, as revealed by the production of pro-inflammatory cytokines, and to amplify the consequences of exposing these cells to independent DNA preparations known to be stimulatory for TLR-9. In addition, when the genomes of different TTVs were compared in silico for potential overall TLR-9 activity, it was found that different TTVs may differ considerably in this respect (J. Rocchi, V. Ricci, M. Bendinelli, F. Maggi, manuscript in preparation). Collectively, these findings point to TLR-9 as one of the cell sensors through which the AVs may impact the inflammatory response and suggest that the resulting effect of infection on the body's inflammatory status may vary greatly depending on the AV or cocktail of AVs replicating in the host. This would appear to be in line with findings showing an association between certain chronic diseases of undefined aetiology and particularly florid infection with specific AV genogroups (Maggi et al. 2003a, 2007b). Moreover, the fact that the AVs are characterised by a unique level of ubiquity and persistence throughout the body makes them theoretically very suitable at interacting in a continued and systematic fashion with many cellular sensors, including the TLRs.

## *TTV, Eosinophil Cationic Protein and Exhaled Nitric Oxide*

That TTV may represent an inducer or, more likely, a facilitator of the biological processes which lead to inflammation is also suggested by findings showing the existence of a positive correlation between loads of TTV and concentration of ECP circulating in the peripheral blood of children with acute respiratory diseases (Maggi et al. 2004) and between TTV loads in nasal fluid and concentration of eNO in the expired air of asthmatic children (F. Maggi, M. Pifferi, V. Ricci, J. Rocchi, M. Bendinelli, manuscript in preparation). ECP and eNO have complex and diverse functions in the pathophysiology of various respiratory diseases, are critically involved in innate and immunological defences, and, most importantly, exert key roles in airway inflammation. In particular, ECP is widely used in clinical medicine as a reliable indicator of the extent of inflammatory activity on-going in the respiratory tract. Whether these effects are a direct consequence of TTV replication or are mediated by the abundant immune complexes TTV forms in the blood (see the following sections), and presumably also in solid tissues, remains to be established. In any case, although work on these aspects is still in its infancy, there is suggestive correlative evidence that an active AV replication can be associated with an increased inflammatory status in the body or in specific districts within it and possibly augment the inflammatory response to superimposed stimuli, thus impacting significantly the resulting pathologic consequences, when these are present.

## Adaptive Immune Responses

Since the AVs are so pervasive, it seems logical to expect that antibodies and other immune effectors specific for one or more of these viruses are also quite common in the human population. The contrary would imply that the AVs (1) lack immunogenicity, which would be unprecedented in true viruses, (2) induce a complete immune tolerance, which would appear more feasible given that intrauterine anellovirus infections are a frequent occurrence (Bendinelli et al. 2001)—although this is still unlikely in the light of what happens in other congenital viral infections—or (3) elicit immune responses that wane much faster than those observed in any other viral infection of humans. Indeed, generally it is the viruses that cause chronic systemic infections, as many if not all the AVs do, that give rise to the most robust and long-lasting specific immune responses observable in humans.

## *Antigens of the AVs*

Among the few species of proteins encoded by the genome of the AVs (Qiu et al. 2005), the most interesting from an immunological standpoint is certainly the product of ORF1, which is by far the largest coding sequence of the AVs, consisting of between 628 and 783 amino acids in different members of the genus, and represents the putative nucleocapsid protein. In fact, at least in TTV this molecule, (1) when analysed for hydrophilicity profile, it has revealed the presence of major hydrophilic regions predicted to have potential antigenic properties (Okamoto et al. 1999b; Luo et al. 2002); (2) in its central portion it contains highly hypervariable regions that resemble similar regions that in other viruses are known to be modelled by the host's immune responses and to help dodge their anti-viral action (Nishizawa et al. 1999; Ukita et al. 2000; Yusufu et al. 2001; Jelcic et al. 2004); (3) contains potential glycosylation sites, again in its central region, which might provide further mechanisms of immune escape (Takahashi et al. 1998; Hijikata et al. 1999); (4) may contain epitopes that, being exposed on the outer surface of the viral particle, are likely to be involved in virus neutralisation and other immune mechanisms implicated in infection control (Nishizawa et al. 1999; Bendinelli et al. 2001); (5) shares substantial structural homologies (a highly arginine-rich domain at its N-terminus and a conserved Rep protein motif; Takahashi et al. 1998; Hijikata et al. 1999; Mushahwar et al. 1999) with the capsid protein of chicken anaemia virus (CAV), known not only to be immunogenic but also to contain important neutralisation epitopes (Koch et al. 1995; Noteborn et al. 1998).

In the light of the such circumstances, it is unsurprising that the serosurveys performed for the purpose of assessing the prevalence of TTV antibodies in the human population (discussed in Sect. 3.3) as test antigens have either used partially purified preparations of whole virions or recombinant products reproducing parts of the ORF1 protein. Importantly, these surveys have provided clear direct evidence that the ORF1 protein of TTV is indeed immunoreactive and that antibodies to it persist for long periods of time regardless of whether TTV DNA is detectable or not in blood (Handa et al. 2000; Ott et al. 2000; Tsuda et al. 2001).

More recent data supporting the notion that the ORF1 protein of TTV is immunogenic have been generated by Gergely et al. (2005) who probed selected human sera against several synthetic 15-mer peptides deduced from TTV DNA. In this study, the specimens tested were a subset of sera from systemic lupus erythematosus (SLE) patients selected because they contained binding activity for the nuclear autoantigen HRES-1/p28; the TTV oligopeptides were deduced from conserved regions of TTV, some of which exhibited amino acid sequence homology to HRES-1/p28; and the assay used was an immunoblotting in which the oligopeptides were attached to strips of cellulose, and antibody detection was performed with goat anti-human IgG using a chemiluminescent substrate. Oligopeptides derived from other common viruses were screened in parallel. All the sera tested recognised at least one TTV oligopeptide, and among those having no homology to HRES-1/p28 the most often recognised reproduced the N-terminal portion (amino acids 181–195) of the ORF1 of a genotype 1a TTV (TA278 strain), which was seen by 9 out of 16 sera. Interestingly, in the same study the prevalence of circulating TTV DNA was found to be greater in the SLE patients than in controls (healthy relatives of the SLE patients), and this was especially evident in those patients who possessed antibodies for the HRES-1/p28 autoantigen. Based on these findings, Gergely and co-workers proposed the interesting hypothesis that TTV infection, possibly in collaboration with other similarly widespread viruses, might influence the auto-antigenicity of HRES-1/p28. However, molecular mimicry could not be invoked as a factor triggering such process, since no amino acid sequence homologies existed between HRES-1/p28 and the TTV peptide most frequently recognised. Evidence that ORF1, in addition to several B cell epitopes, also encodes at least one T cell epitope is discussed in Sect. 3.6.

The other proteins believed to be expressed by TTV are still essentially uncharacterised both structurally and functionally. With regard to their immunological properties, only the product of ORF2 has been investigated to some extent. In different TTV genotypes, this ORF may represent a single coding unit or be divided into two separate ones, designated ORF2a and 2b, the first of which is well conserved among different TTVs, while the second is rather variable. Recently, Kakkola et al. (2002) have investigated the serological reactivity of the ORF2 of the HEL32 strain of TTV, belonging to genotype 6 and having this coding region bipartite. By testing the sera of 87 healthy donors in immunoblotting assays against the two small ORF2 proteins of this viral strain expressed in prokaryotes, these workers detected IgM antibody reactivity to one or both proteins in a surprisingly high proportion of subjects (9%) and, also quite surprisingly, identified some subjects in which this class of ORF2 antibody persisted for several years. Another unexpected finding of this study was that the overall prevalence rate of IgG specific for the two ORF2 proteins was in the same range (10%). Furthermore, the antibodies to the two ORF2 proteins were equally represented, despite the already mentioned difference in conservation of these molecules. We share the view of Kakkola et al. (2002) that currently these findings are hard to interpret. A possibility put forward by the authors of the study was that the ORF2 products are weakly immunogenic and unable to trigger the IgM-to-IgG class shift. On the other hand, in their study of SLE patients discussed above, Gergely et al. (2005) detected strong IgG reactivity to a synthetic oligopeptide derived from the ORF2 of a genotype-1a TTV (09Suz/2-3G strain) in nearly 90% of the sera tested. This raises the intriguing possibility that antibodies against this TTV protein are produced at low levels in healthy individuals but more frequently so in patients with autoimmune pathologies. Other, more trivial explanations are also possible, however. For example, it cannot be excluded that in the whole ORF2a protein used by Kakkola and co-workers the segment reproduced by the oligopeptide which was used by Gergely et al. is buried inside the molecule and therefore inaccessible to antibodies.

To date, the highest rate of TTV antibody detection in the population was obtained by Ott et al. (2000), who in a study discussed in more detail in the following section probed their test sera against the C-terminal portion of the ORF1 protein of one genotype-1a virus (X94-TTV strain). This might indicate that this region of the ORF1 protein contains a widely conserved immunodominant epitope(s) but might also reflect a particularly high prevalence in the population tested of the genotype-1a TTV from which the protein was obtained. Indeed, regrettably at the time of writing, the issue of antigenic cross-reactivity among the AVs remains completely unaddressed. In particular, it must be ascertained whether the AVs possess shared B epitopes common to all the members of the genus, or whether the different viral species, genogroups and genotypes within it are antigenically fully distinct. Given the extensive sequence heterogeneity exhibited by the coding regions of the AVs, it is reasonable to expect that, when this pivotal aspect will be investigated, a substantial diversity in the antigens expressed will be encountered. Thus, even the apparent contrast between the results of Gergely et al. (2005) with those of Kakkola et al. (2002) might find another explanation in the antigenic heterogeneity of the AVs, since the latter group obtained their antigens from a TTV of genotype 6 which is relatively uncommon in the human population.

## *Methods for Demonstrating Anti-AV Antibodies*

To date, the types of serological assays that have been used for the detection of AV-specific antibodies are few. The ones used for IgG antibody have in fact been limited to the immunoprecipitation of intact virions or to immunoblots in which recombinant full-length or truncated viral proteins or synthetic oligopeptides were blotted on strips and then reacted with the test sera followed by variously labelled anti-human IgG sera (Handa et al. 2000; Ott et al. 2000; Gergely et al. 2005). On the other hand, IgM antibodies have been demonstrated by capturing the total IgM present in the test sera on a solid support by means of an anti-IgM and then analysing whether the captured IgM have the ability to bind virions, as determined by measuring with PCR the numbers of viral DNA copies remaining bound to the support (Tsuda et al. 2001; Tawara et al. 2000). Some of these methods, and especially the immunoblots, are known to offer good to excellent specificity, but all are rather insensitive and difficult to standardise and provide little or no clues on whether the antibodies detected are potentially capable of inhibiting viral infectivity. To obtain information on the latter aspect, the availability of in vitro virus neutralisation assays would be indispensable, but the development of these assays has been prevented by the lack of sufficiently dependable systems for propagating the AVs in culture. On the other hand, there are also no reported attempts to develop specific enzyme-linked immunosorbent assays (ELISA) for any of the AVs. ELISA assays offer exquisite sensitivity and good specificity and currently represent the serological tests of choice for many human and non-human viruses.

In essence, at this time there are no serological methods that have been proposed, let alone validated, that can be considered suitable for making, or helping in the making, of a diagnosis of AV infection. Unfortunately, the methods rely on the detection of viral DNA in plasma or other clinical specimens, which is the only diagnostic approach currently available for the AVs, and are not exempt from limitations. These, for example, include the need to frequently update primers and probes to take into account the novel anellovirus sequences that are continually being described, and the fact that the breadth of viral variants being recognised is either too comprehensive (as in the so-called *universal* PCR assays) or too narrow (as in the *genogroup-specific* PCR assays) to permit a straightforward and reliable identification of the infecting AV(s). An even greater limitation of molecular methods, especially in the context of this article, is that they do not allow us to distinguish between a recently and a remotely acquired infection and to pinpoint past infections that have completely resolved but have left their antibody imprints, as adequate serological methods would permit us to do.

## Prevalence of Anti-viral Antibodies in the Population

Assessing the precise prevalence of antibodies specific for the different AVs would represent an important step forward in our understanding of many features of the epidemiology of these viruses and would also provide key clues about important facets of their life cycle that are still poorly defined, such as for example the relative proportions of the infections that resolve versus those that become persistent. However, even this task, which for most other human viruses has generally proved simple to achieve, in the instance of the AVs is proving difficult.

Indeed, at the time of writing, there are only a few reported studies investigating the prevalence of antibodies to the AVs, and these must be considered just preliminary, having used antibody detection methods with the limitations discussed above and having examined relatively small groups of viraemia-positive and viraemia-negative subjects. Furthermore, the data generated are quite diverse. In an early study, Tsuda et al. (1999) immunoprecipitated whole TTV virions concentrated from the faeces of an infected individual with the sera of healthy Japanese blood donors, assayed the immunoprecipitates and supernatants thus obtained by a TTV-specific PCR, and finally measured the optical density of the signals produced in comparison with a negative control serum. The latter precaution was considered necessary because any IgG–anti-IgG precipitate was likely to non-specifically co-precipitate at least some viral particles. By this complex procedure, anti-TTV antibodies were detected in 1 of 6 TTV DNA-positive subjects and in 11 of 38 TTV DNA-negative sera (overall prevalence, 27%). Similarly, Handa et al. (2000), using immunoblotting and a polyhistidine-tagged, bacterially expressed truncated ORF1 fusion protein spanning amino acids 1–411 of one genotype 1b TTV isolate (unspecified strain) as test antigen, detected TTV antibodies in 38 of 100 American blood donors, and 31 (84%) of the antibody-positive sera were also TTV DNA-positive. However, subsequent studies have indicated that most probably these were huge underestimates of both the proportion of individuals who possess TTV antibodies and of those who co-circulate TTV virions and TTV antibodies. Indeed, Ott et al. (2000), using as antigen a bacterially expressed recombinant peptide representing the C-terminal portion of the ORF1 protein (amino acids 504–752) of one genotype-1a TTV (X94-TTV strain) and polyspecific goat anti-human Ig serum as the developing agent, found anti-TTV reactivity in the serum of all but one of 70 French subjects examined (30 blood donors, 30 patients with cryptogenetic hepatitis, and 10 healthy children), approximately 1/4 of whom was TTV DNA-negative. None of the subjects tested positive when the blots were developed instead with an anti-human IgM monoclonal antibody. Another interesting observation in this study was that all the children aged less than 5 years that were tested reacted antibody-positive, thus confirming findings obtained with molecular methods that TTV infection is acquired very early in life and emphasising the role children may play in the spread of TTV infection (Biagini et al. 2003; Maggi et al. 2003a).

In conclusion, what is reviewed above can hardly be considered a satisfactory assessment of the serological status of people versus the AVs. Clearly, a better understanding of the antigenic diversity of these viruses and the consequent development of a panel of serological assays, each one with well-characterised specificity and sensitivity, is strongly needed. Unfortunately, as discussed in the two previous sections, the practice in these two pivotal areas is still rather primitive, and the hurdles that have slowed progress in the area are proving hard to remove.

## *Kinetics of Antibody Production During Primary Infections*

In spite of the many limitations outlined above, there have been a few attempts to determine the kinetics of TTV antibody production in individuals experiencing putative primary infections. Sequential examination of patients accidentally infected through contaminated blood and of chimpanzees experimentally infected with human viruses has shown that this response mounts rather slowly (Tsuda et al. 1999, 2001; Tawara et al. 2000; Ott et al. 2000), resembling what occurs with hepatitis C virus (HCV), which shares with the AVs a strong tendency to produce chronic productive infections. TTV-specific IgM were first detected in serum 10–21 weeks after the viral inoculum, and this generally happened at least 2–7 weeks after TTV DNA had become detectable in blood, although in one patient IgM appeared before TTV DNA. Subsequently, this antibody class declined in titre to disappear in 5–11 weeks, although cases of longer persistence were also reported (Tsuda et al. 2001; Kakkola et al. 2002). IgG antibodies generally developed with a lag of several weeks, increased in titre while the IgM faded, and then lasted with little variations in titre throughout the observation periods (Fig. 2). In one study, anti-TTV IgG could be detected in the serum of a transfusion-infected patient up to 4.4 years after infection (Tsuda et al. 2001), but it is likely that the antibodies last longer or even indefinitely. To our knowledge, AV-specific IgA responses in serum or mucosae have never been investigated.

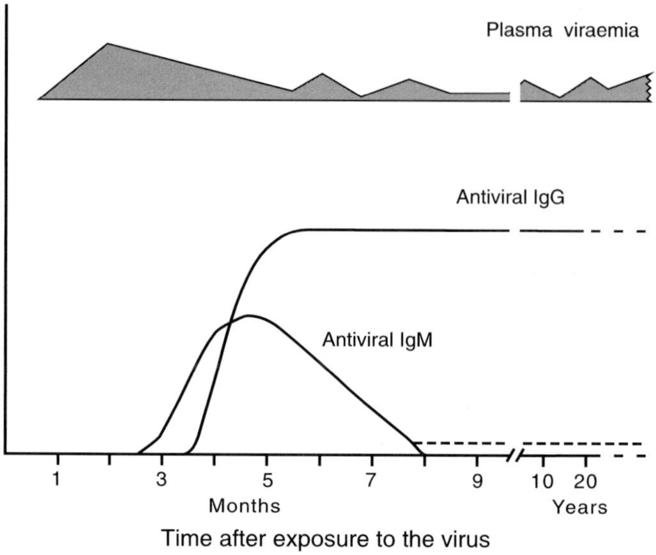


**Fig. 2** Time course of the antibody response in primary anellovirus infections as inferred from experimental studies in chimpanzees (Tawara et al. 2000) and accidentally infected individuals (Tsuda et al. 1999, 2001; Ott et al. 2000)

## *Immunocomplexed AVs in the Blood of Chronically Infected Patients*

In their seroprevalence study discussed above, Tsuda et al. (1999) also observed that a variable fraction of the TTV detected by molecular methods in the serum of some infected individuals could be immunoprecipitated with anti-human IgG sera, thus indicating that—similar to what is already known for other chronic viral infections such HCV and HIV—TTV can circulate in the blood bound to Ig. Observations in the same direction were done by Nishizawa et al. (1999), who used a similar method and found that most if not all the TTV found in the blood of chronically infected subjects is immunocomplexed. By contrast, in the blood of individuals with putative acute infections, free TTV particles were well represented and could exceed those of IgG-complexed particles.

A quantitative analysis of immunocomplexed virus done by Itoh et al. (2000) in symptomatic and asymptomatic HIV-1-positive patients confirmed that immunocomplexed virus represents a large fraction of the TTV found in serum, with moderate individual variations (from 80% to 95% of total TTV). By contrast, most of the TTV found in faecal extracts was not precipitated by polyspecific anti-human Ig sera, showing that the virus excreted with the faeces is essentially free from bound IgG antibody (Itoh et al. 2000; Tawara et al. 2000). Of note, Itoh et al. exploited this approach for the first visualisation of TTV virions using immunogold electron microscopy. More recently, our group could demonstrate that an apparently self-limited superinfection by a genogroup-2 TTV (ViPi04 strain), observed in a subject who already carried three different types of the virus, was accompanied by a rapid increase of total plasma viraemia and by a parallel marked decline in the proportion of immunocomplexed circulating virus. These two changes in viral parameters were interpreted as likely consequences of the florid replication undertaken by the superimposed TTV during the time interval when antibodies that could specifically bind to it were still absent or very scarce (Maggi et al. 2006).

## Cell-Mediated Immunity

Based on common sense, it is clearly to be expected that the AVs will one day be found to trigger cell-mediated immunity as effectively as other chronic viraemia-inducing viruses, such as HIV-1, hepatitis B virus and HCV. At the present there is, however, a single report in the literature dealing with this matter. In this study, Sospedra et al. (2005) isolated and propagated in vitro clones of $CD4^+$ T cells obtained from the cerebrospinal fluid of a multiple sclerosis patient during an exacerbation of the disease, by using phytohaemagglutinin as an unbiased stimulus to expand the cells. Interestingly, the $CD4^+$ T cell clones thus obtained were found to recognise a large number of oligopeptides predicted from TTV and TTMV sequences. Furthermore, apart from a few exceptions, the oligopeptides recognised were from a short poly-arginine sequence (74 amino acids) present at the N-terminal region of the viral ORF1. This study might have significant implications for the pathogenesis of multiple sclerosis in that, as the authors suggest, repeated active infections and/or chronic infections with different AVs might result in consecutive cycles of stimulation and expansion of $CD4^+$ T cells specific for arginine-rich motifs, motifs that are present in molecules of the CNS involved in the disease as well as in the AVs. In any case, the study represents an indication that a systematic investigation of cell-mediated immune responses specific for the AVs in healthy and diseased people might permit the harvest of some succulent fruits.

## Role of Adaptive Immune Responses in the Control of AVs

A well-documented feature of the AVs is that, once they have gained a foothold in the body, they are not easily eradicated by the host (Lefrere et al. 1999; Bendinelli et al. 2001; Bendinelli and Maggi 2005). Indeed, judging from the proportion of people who carry chronic productive infections by one or, most often, several distinct AVs regardless of any population variable investigated (Maggi et al. 2005a; Devalle and Niel 2004; Ninomiya et al. 2008), one might suspect that self-limited anellovirus infections do not exist or are very rare. However, one should also reason that, due to the large doses of AVs shed with the faeces and body fluids by those acutely or chronically infected (Bendinelli et al. 2001; Bendinelli and Maggi 2005) and to the high resistance of these viruses in the environment (Maggi et al. 2001b; Welch et al. 2006), the occasions of contagion must be quite frequent. Therefore, even though serology to date has helped little in shedding light on this matter (Sect. 3.3), it is plausible that self-limited anellovirus infections are a frequent occurrence as well. A few putative self-limited primary TTV infections, in which the infecting viruses were no longer detected in the circulation after variably long time intervals during which they had instead been readily detected, have been described both in humans and experimentally infected chimpanzees (Lefrere et al. 1999; Tawara et al. 2000; Bendinelli et al. 2001), although the doubt may exist that in these subjects the virus could have just dropped in titre below the lower limit of sensitivity of the detection methods used and/or was still persisting in body tissues other that peripheral blood. AVs have repeatedly been detected in extracts prepared from the tissues of individuals who failed to show the corresponding viruses in plasma, suggesting that sanctuaries of viral latency may exist for these as well as for other viruses (Okamoto et al. 2000b; Okamoto and Mayumi 2001).

It is also well established that in people chronically infected with TTV the titres of plasma viraemia generally remain rather stable over protracted periods of time (Maggi et al. 2005b), but the level at which viraemia stabilises in individual subjects can vary greatly [from $10^2$ to $10^8$ copies of viral DNA per millilitre of plasma in a study by Pistello et al. (2001)]. Studies have attributed at least part of these large variations to the variable numbers of different AVs people may carry (Maggi et al. 2005a); however, there are also indications that they may reflect differences in the rates the infecting virus(es) is shed into and/or removed from the circulation. On the basis of the rate with which TTV disappeared from the plasma of patients treated with IFN-$\alpha$ therapy for concomitant hepatitis C treatment, our group could calculate that the numbers of TTV virions generated per day by chronically infected individuals are generally very high ($3.8\times10^{10}$ minimum) and that approximately 90% of the virions in plasma are cleared and replenished daily, but individual patients exhibited variations of these viral parameters (Maggi et al. 2001c). In any case, the factors that may impact on the size of a chronic plasma viraemia are quite numerous, and only some have been investigated while studying the AVs (Table 1).

Based on what is known about human viruses in general, it seems safe to assume that the adaptive immune responses mounted by the infected hosts play a key role both in determining whether a primary anellovirus infection will resolve and in setting the extent at which the AVs circulate in the peripheral blood of those who fail to clear them. Direct evidence supporting this view comes from two chimpanzees that were inoculated with the same amount of TTV-containing human serum: one chimp developed IgM and IgG antibodies to the virus and resolved the infection within a few months, whereas the other developed no detectable antibody and became persistently infected (Tawara et al. 2000).

Strong albeit circumstantial evidence arguing for an important role of adaptive immune responses in the equilibrium that AVs establish with their hosts also includes data showing that HIV-1-infected individuals and other immunocompromised people present higher prevalence rates and/or higher concentrations in blood of these viruses than healthy controls. While studies where HIV patients were simply investigated qualitatively for the prevalence of AVs showed no major differences with the controls (Puig-Basagoiti et al. 2000; Sagir et al. 2005b), those in

**Table 1** Some factors that may impact on the size of plasma viraemia in chronic anellovirus infections

| Factor type | Factor | Evidence[a] |
|---|---|---|
| Virus-related factors | Number and spectrum of cells replicating the infecting AV(s) | H |
| | Turnover of virus-producing cells | H |
| | Number of virions produced per cell | H |
| | Total daily production of virions | H |
| | Rate of virions release into the circulation | H |
| | Rate of virions clearance from the circulation | H |
| | Proportion of immunocomplexed virions | H |
| | Number of different AVs harboured | EA |
| | Synergy or interference between the AVs carried | EA |
| | Acute intercurrent superinfection by a different AV | EA |
| Immunological factors | Depressed humoral and/or cell-mediated responses | EA |
| | Changes in the functional integrity and relative proportion of cells participating in such responses | H |
| | Counts of circulating lymphocytes | EA |
| | Immune activation by superimposed exogenous immunogens | EA |
| Other host-related factors | Concomitant infections by other pathogens | H |
| | Presence of concomitant non-infectious pathologies (tumours, etc.) | EA |
| | Immunosuppressive and cytotoxic therapies | EA |
| | Local accumulation of proliferating lymphoid cells | H |
| | Regeneration rate of susceptible cells | H |

[a]EA, at least some direct evidence available; H, hypothetical

which the patients' loads of AVs were also measured, with few exceptions (Moen et al. 2002), have produced evidence arguing for the existence of a correlation between severity of the patients' immunosuppression and burdens of AVs carried.

In a study involving 185 Danish HIV patients, Christensen et al. (2000) reported that 72 (51%) of 140 HIV-1-infected patients who were also TTV-positive had TTV loads at or exceeding fivefold greater than the maximum level that was detected among the control blood donors that were examined in parallel. These workers also found significantly higher CD4 T cell counts in the patients with low or undetectable TTV loads in plasma (68 and 45 patients, respectively) than in those with high TTV loads (72 patients), and multivariate regression analysis showed an association between high TTV loads (>$10^5$ copies/ml) and decreased survival rates. Similarly, Touinssi et al. (2001), examining 75 HIV-1-infected patients with CD4 cell counts of less than 200/mm$^3$, found that they had plasma TTV loads significantly greater than healthy controls. In a study of 144 HIV-infected Japanese patients using two PCR assays amplifying different regions of the viral genome, Shibayama et al. (2001) found that TTV prevalence and load in their plasma were significantly higher than in blood donors and detected an association between high TTV loads, on the one hand, and low CD4 T cell counts, high HIV viral loads and overt AIDS on the other. Sherman et al. (2001) performed a cross-sectional analysis of 86 HIV-positive subjects and 118 HIV-negative controls in the United States and characterised the TTVs detected in terms of genotypic variability by sequence and/or restriction fragment length polymorphism analysis and found that mixed genotypes were more common in those with HIV. More recently, Thom and Petrik (2007) compared the levels of TTV and TTMV in bone marrow and spleen specimens collected at autopsy from 13 patients with a clinical diagnosis of AIDS, 6 HIV-infected patients who had not progressed to AIDS, and 7 HIV-negative subjects and found the highest titres of both AVs in the tissues of those with AIDS. Also, the viral titres were inversely correlated with the CD4 T cell counts obtained from the subjects' medical records.

Clearly, these taken together, along with additional studies (Madsen et al. 2002; Sagir et al. 2005a, b), have corroborated the idea that the poorly functional immune system of HIV-1-infected patients permits the AVs to replicate in the host more freely than would occur if the immune system were functionally intact.

A major problem with the above findings and with similar observations obtained in transplant patients and in patients with tumours (Shang et al. 2000; Zhong et al. 2001; Tokita et al. 2002; Sagir et al. 2004) is that they can be interpreted in the opposite way, that is that the co-infecting AVs contributed to overall immunodepression and CD4 T cell depletion of the patients (see Sect. 3.8). Thus, the study by Moen et al. (2003) appears to be especially informative. These workers followed 10 kidney transplant patients receiving high doses of cyclosporine and corticosteroids and found that the TTVs and TTMVs they carried in blood underwent a marked increase in titre as a result of this drastic immunosuppressive treatment. In contrast, in 25 military recruits who had gone through an intensive training programme the titres of both such viruses in blood increased slightly but not significantly, suggesting that low levels of physiologically induced immunosuppression are not sufficient for affecting AV replication.

The two chimps inoculated with a human TTV by Tawara et al. (2000) in the study discussed already showed peaks of viraemia that were surprisingly low ($10^3$ to $10^4$ log DNA copies/ml, respectively), late and short-lived (lasting from week 12 to 13 and week 14 to 16, respectively), possibly due to the fact that the infecting virus derived from a different species. On the other hand, in most primary TTV infections described in humans the replicating viruses were studied only qualitatively. Thus, we have essentially no clues to the extent to which the host's immune response can curb the size of viraemia in the course of a primary infection, which could be inferred from the kinetics of viraemia during and after the acute phase of infection, as has been possible in other viral infections of humans that evolve to chronicity. The only quantitative study of viral load performed in humans with a putative self-limited AV infection reported so far was conducted by our group in a woman who was already chronically infected with three different genotypes of TTV. Along with the development of mild flu-like symptoms, this woman underwent an abrupt increase of pre-existing plasma viraemia from around 5.5 log to over 9 log, which corresponded temporally with the appearance in her blood of an additional TTV genotype which had not been detected in any of several earlier samplings. This peak lasted only briefly, since viraemia returned to her baseline values in a few weeks, in coincidence with the disappearance of the superimposed

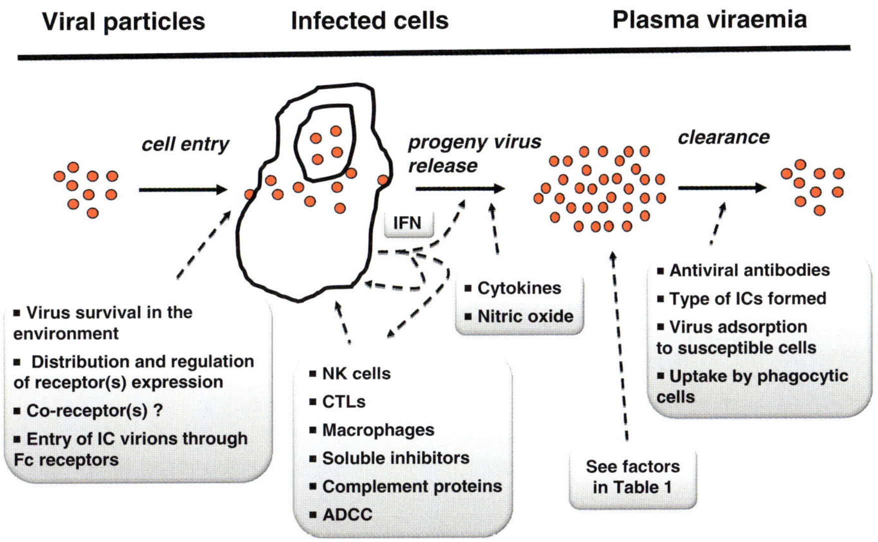


**Fig. 3** Some immunological factors that may play a role in the control of anellovirus infections, tabulated according to main site of action

virus. It could not, however, be established whether the relative promptness with which the superinfection was controlled was somehow due to the immunity the woman already had developed against the pre-existing AVs, immunity revealed by the presence of abundant IgG-bound virus in the circulation prior to the superinfection (Maggi et al. 2006). On the other hand, it cannot be excluded that different AVs are inherently diverse in their ability to persist, which might contribute to explaining why there are such large variations in the frequency with which the various genotypes are detected in the population.

What effectors of the adaptive immune responses may concur at keeping the AVs under partial or total check is still a matter of speculation (Fig.3). The chimp study by Tawara et al. discussed above has shown a clear albeit anecdotal correlation between the evolution TTV infection to chronicity and the failure to develop TTV antibodies. On the other hand, the well-established fact that a large fraction of the TTV found in chronically infected patients (Sect. 3.5) is bound to antibodies implies that antibodies are unable to eradicate an infection that has already established itself in the host, but does not exclude that antibodies are capable of preventing or curing a primary infection and of containing viral replication in chronic infections. When methods for performing these studies become available (Sect. 3.2), it will be of great interest to determine (1) whether the IgG-bound AVs found in the plasma of chronically infected people are capable of starting an infection in vitro and in vivo and, should they be unable to do so, (2) the stability of the immune complexes formed.

It should be noted, in any case, that the high prevalence in all the populations studied of chronic infections in which different genogroups and/or genotypes of AVs coexist in the circulation (Maggi et al. 2005a; Ninomiya et al. 2008) is unlikely to result only from simultaneous exposures to multiple viruses. Thus, as also implied from the superinfection described by our group (Maggi et al. 2006) and discussed above, it appears that the fact of possessing antibodies directed against certain AVs does not protect the host from other AVs. Some degree of cross-protection is, nevertheless, likely to exist at least among the AVs that are less genetically distant. Again, however, the methods currently available do not shed light on these aspects. On the other hand, a role for cell-mediated immunity in the control of AV infections can be postulated but has never been investigated in any detail.

## *Immunomodulating Activity of the AVs*

Many viruses, including some structurally akin to the AVs such as the circoviruses (Todd et al. 2001), especially when they produce persistent systemic infections as many if not all the AVs do, have the ability to perturb the immune reactivity of their hosts. This results in altered humoral and/or cell-mediated responses upon contact with superimposed immunogens and/or by changes in the functions and relative proportions of the cells that participate in such responses (Specter et al. 1989; Todd et al. 2001; Darwich et al. 2004; Segales et al. 2004). In Sect. 2, we have seen that the AVs, or at least selected viruses within them, have the ability to modulate the function of crucial mediators involved in innate immunity. Do the AVs also have the potential to modulate the host's adaptive immune responses? The question is momentous since the AVs are so ubiquitous in the human population that even a marginal effect of this kind might have significant implications (at the level of general population and/or of selected individuals). However, exactly for the same reason, the question is also exceedingly hard to answer due to the difficulty in finding appropriate subgroups of totally AV-free individuals, who might serve as uninfected controls against whom the immune reactivity of those AV-infected might be compared (Maggi et al. 2005a; Ninomiya et al. 2008).

Indeed, the only comparison currently feasible is between groups of individuals who carry substantially different burdens of AVs in their blood, as an indicator of different replicative activities of the infecting viruses, and/or carry specific anellovirus species, genogroups or genotypes. By using the first approach, in an investigation of children with acute respiratory diseases, our group not only observed that the ones with more severe diseases also tended to have the highest loads of TTV in nasal fluid and plasma but also found that the sizes of the TTV loads carried were inversely correlated with the percentages of circulating total and helper T lymphocytes and directly correlated with the percentages of circulating B cells (Maggi et al. 2003b), thus opening the possibility that florid TTV replication leads to lymphocyte imbalances and, possibly, produces local or generalised immunomodulating effects. However, similar to what was discussed in the previous section with regard to the studies showing a correlation between AV loads and severity of HIV disease, these findings are open to the opposite interpretation, namely that the lymphocyte imbalances produced by the underlying diseases might have permitted an increased replication of the AVs. Less susceptible to this alternative explanation are findings, in the same study, pointing to the possibility that specific TTV genogroups may impact on lymphocyte subpopulations more than others. By assigning the TTVs detected to the different genogroups in which these viruses are classified by means of a panel of genogroup-specific PCR assays, it was observed that the children who carried more than one TTV genogroup displayed lower percentages of total and CD4 T cells than the ones with one genogroup alone, and those who carried genogroup-3 TTV were the most affected. Furthermore, Sagir et al. (2005a) in a retrospective study of 217 German HIV patients with PCR assays specific for the SEN-D and -H viruses, both belonging to genogroup-3 TTV, found that the prevalence of these viruses was higher than in 122 healthy blood donors, and the authors also observed that those with high titres of SEN-H virus had significantly shorter average survivals than those with low titres of this virus.

Also of interest are recent findings obtained by studying TTV viraemia in patients given high-dose chemotherapy and autologous haematopoietic stem cell transplantation as treatment for multiple myeloma. In the 19 consecutive patients that were enrolled in the study, the viral loads underwent a substantial increase starting several weeks post-transplant, and this increase was found to be associated with an increase in the percentages of $CD8^{+}57^{+}$ T lymphocytes circulating in blood (Maggi et al. 2008). An increase in the levels of $CD8^{+}57^{+}$ T lymphocytes is a frequent finding in a variety of diverse clinical conditions having immune cell activation and immunological dysfunctions as common denominators, including cytomegalovirus infections, AIDS, rheumatoid arthritis, and myelodysplastic syndromes, and has, therefore, been proposed as a marker for evaluating the host's immune status (Gutierrez et al. 2003; Alvaro et al. 2006; Focosi and Petrini 2006).

The mechanisms whereby viruses can modulate the host's immune reactivity have proved difficult to disentangle, even for those viruses that have been most intensively investigated in this respect, including HIV-1, but they are believed to be numerous and to act in concert with varying factors. One factor most often invoked is, however, the frequent ability of viruses to invade one or more cell types that compose the immune system and to replicate within these cells more or less productively, with consequent impairment of their anatomical and/or functional integrity (Specter et al. 1989). The breadth of cell types supporting the replication of AVs in infected hosts has yet to be precisely defined due to the problems encountered in propagating these viruses in culture and to the limits many molecular methods currently have in distinguishing the virions passively adsorbed to cells from the surrounding fluids and the ones truly replicating inside the cells. Given the heterogeneity, in the AVs, of the capsid protein believed to mediate their interaction with the cell surface (Bendinelli et al. 2001; Biagini 2004), it is also plausible that the spectrum of permissive cells varies depending on the AV considered.

In the case of TTV there is, however, substantial evidence that immunocytes may be one major site of viral replication. Freshly harvested white blood cells (WBCs) or peripheral mononuclear blood cells (PBMCs) have repeatedly been reported to harbour TTV DNA (Okamura et al. 1999; Okamoto et al. 1999a, 2000b; Maggi et al. 2001b; Mariscal et al. 2002; Takahashi et al. 2002; Yu et al. 2002;

Zhong et al. 2002; Desai et al. 2005), sometimes at titres and as genotypes difficult to explain on the basis of the viral loads and genotypes found in the corresponding plasma samples (Okamoto et al. 2000b). This demonstrates that these cells may act at least as reservoirs for this virus. Nevertheless, studies trying to establish which cell type(s) within the WBCs is preferentially infected have provided conflicting results. Yu et al. (2002) identified such cells as B lymphocytes; Takahashi et al. (2002) found decreasing viral contents in the order in neutrophils, monocytes, natural killer (NK) cells, and lymphocytes; and our group detected no substantial differences between monocyte-enriched, B lymphocyte-enriched, and T lymphocyte-enriched PBMCs (Maggi et al. 2001b).

That TTV may infect and replicate in proliferating haematopoietic cells is also suggested by observations showing that in bone marrow transplant patients the pre-existing loads of circulating TTV decreased to undetectable levels during the myelosuppressive period (Kanda et al., 1999). In addition, Kikuchi et al (2000) found evidence of TTV replication in bone marrow cells but not in the hepatocytes of a patient with subacute hepatitis and aplastic anaemia. Moreover, two terminally ill AIDS patients were found to harbour an exceptionally high load of TTV in their bone marrow [more than 1,000 copies of viral DNA per cell; Simmonds (2002)]

The reported detection of replicative forms of TTV DNA in PBMCs and bone marrow cells (Okamoto et al. 2000a, c; Zhong et al. 2002; Bendinelli and Maggi 2005) also would appear to support the notion that TTV replicates in cells of the immune system, although the significance of these findings is somewhat weakened by the fact that the methods used for detecting such intermediates are known to be prone to serious artefacts. They are corroborated, however, by studies in which in vitro cultured human T lymphocytes were found to release into the supernatant fluid increased copy numbers of TTV DNA when the cells were activated with a polyclonal stimulus such as phytohaemagglutinin relative to what observed with unstimulated cultures (Maggi et al. 2001b; Mariscal et al. 2002; Desai et al. 2005), showing that not only human T lymphocytes can support TTV replication, but that this occurs optimally only when these cells become activated. Accordingly, by monitoring TTV in people undergoing vaccinations against influenza and hepatitis B, our group found evidence that TTV DNA levels in blood increased as a result of immune stimulation. It may be worth recalling in this context that other single-stranded DNA viruses, including parvoviruses and circoviruses, have a marked preference for or replicate exclusively in DNA-synthesising cells (Maggi et al. 2005b).

As a matter of fact, on the basis of what is discussed above and other evidence, it is not unreasonable to propose that TTV is a lymphotropic virus rather than an hepatotropic virus, as was precipitately suggested soon after it was first identified (Nishizawa et al. 1997; Okamoto and Mayumi 2001). Given that viruses lacking an external lipid envelope are usually cytolytic for the cells in which they replicate, and should this contention be supported by future studies, the overall impact of AVs on the immune system of their hosts might be considerable.

## Concluding Remark

The AVs are a relatively recent addition to the growing list of viruses that can infect humans. Furthermore, their investigation is hindered by multiple difficulties, only some of which have been mentioned in this chapter. It is therefore inevitable that what we have come to know so far is only a small fraction of the information that would be needed for a satisfactory understanding of these widespread agents of infection. This is true for their immunobiology as well as for many other aspects of their biology and natural history that are discussed in other chapters of this volume. What we have grasped to date suffices, however, at stimulating scientific curiosity about these, in many respects, amazing viruses and about the relationships they establish with their hosts. Future research is not only warranted but also looked forward to with considerable expectation.

## References

Alvaro T, Lejeune M, Salvado MT, Lopez C, Jaen J, Bosh R, Pons LE (2006) Immunohistochemical patterns of reactive microenvironment are associated with clinicobiologic behavior in follicular lymphoma patients. J Clin Oncol 24:5350–5357

Bendinelli M, Maggi F (2005) TT virus and other anelloviruses. In: Mahy B, ter Meulen V (eds) Topley and Wilson—microbiology and microbial infections. Arnold Press, London

Bendinelli M, Pistello M, Maggi F, Fornai C, Freer G, Vatteroni ML (2001) Molecular properties, biology and clinical implications of TT virus, a recently identified widespread infectious agent of humans. Clin Microbiol Rev 14:98–113

Biagini P (2004) Human circoviruses. Vet Microbiol 98:95–101

Biagini P, Charrel RN, de Micco P, de Lamballerie X (2003) Association of TT virus primary infection with rhinitis in a newborn. Clin Infect Dis 36:128–129

Biagini P, Todd D, Bendinelli M, Hino S, Mankertz A, Mishiro S, Niel C, Okamoto H, Raidal S, Ritchie BW, Teo GC (2004) Anellovirus. In: Fauquet CM, Mayo MA, Maniloff J, Desselberger U, Ball LA (eds) Virus taxonomy, 8th report of the International Committee for the Taxonomy of Viruses. Elsevier/Academic Press, New York

Christensen JK, Eugen-Olsen J, Sorensen M, Ullum H, Gjedde SB, Pedersen BK, Nielsen JO, Krogsgaard K (2000) Prevalence and prognostic significance of infection with TT virus in patients infected with human immunodeficiency virus. J Infect Dis 181:1796–1799

Darwich L, Segales J, Mateu E (2004) Pathogenesis of postweaning multisystemic wasting syndrome caused by porcine circovirus 2: an immune riddle. Arch Virol 149:857–874

Desai M, Pal R, Deshmukh R, Banker D (2005) Replication of TT virus in hepatocyte and leucocyte cell lines. J Med Virol 77:136–143

Devalle S, Niel C (2004) Distribution of TT virus genomic groups 1–5 in Brazilian blood donors, HBV carriers, and HIV-1 infected patients. J Med Virol 72:166–173

Focosi D, Petrini M (2007) CD57 expression on lymphoma microenvironment as a new prognostic marker related to immune dysfunction. J Clin Oncol 25:1289–1291

Gergely PJr, Pullmann R, Stancato C, Otvos L Jr, Koncz A, Blazsek A, Poor G, Brown KE, Phillips PE, Perl A (2005) Increased prevalence of transfusion-transmitted virus and cross-reactivity with immunodominant epitopes of the HRES-1/p28 endogenous retroviral autoantigen in patients with systemic lupus erythematosus. Clin Immunol 116:124–134

Gutierrez A, Munoz I, Solano C, Benet I, Gimeno C, Marugan I, Gea MD, Garcia-Conde J, Navarro D (2003) Reconstitution of lymphocyte populations and cytomegalovirus viremia or disease after allogeneic peripheral blood stem cell transplantation. J Med Virol 70:399–403

Handa A, Dickstein B, Young NS, Brown KE (2000) Prevalence of the newly described human circovirus, TTV, in United States blood donors. Transfusion 40:245–251

Hijikata M, Takahashi K, Mishiro S (1999) Complete circular DNA genome of a TT virus variant (isolate name SANBAN) and 44 partial ORF2 sequences implicating a great degree of diversity beyond genotypes. Virology 260:17–22

Hino S, Miyata H (2007) Torque teno virus (TTV): current status. Rev Med Virol 17:45–57

Hiscott J (2007) Convergence of the NF-κB and IRF pathways in the regulation of the innate antiviral response. Cytokine Growth Factor Rev 18:483–490

Inami T, Obara T, Moriyama M, Arakawa Y, Abe K (2000) Full-length nucleotide sequence of a simian TT virus isolate obtained from a chimpanzee: evidence for a new TT virus-like species. Virology 277:330–335

Itoh Y, Takahashi M, Fukuda M, Shibayama T, Ishikawa T, Tsuda F, Tanaka T, Nishizawa T, Okamoto H (2000) Visualization of TT virus particles recovered from the sera and feces of infected humans. Biochem Biophys Res Commun 279:718–724

Jelcic I, Hotz-Wagenblatt A, Hunziker A, Zur Hausen H, de Villiers EM (2004) Isolation of multiple TT virus genotypes from spleen biopsy tissue from a Hodgkin's disease patient: genome reorganization and diversity in the hypervariable region. J Virol 78:7498–7507

Jones MS, Kapoor A, Lukashov VV, Simmonds P, Hecht F, Delwart E (2005) New DNA viruses identified in patients with acute viral infection syndrome. J Virol 79:8230–8236

Kakkola L, Hedman K, Vanrobaeys H, Hedman L, Soderlund-Venermo M (2002) Cloning and sequencing of TT virus genotype 6 and expression of antigenic open reading frame 2 proteins. J Gen Virol 83:979–990

Kanda Y, Tanaka Y, Kami M, Saito T, Asai T, Izutsu K, Yuji K, Ogawa S, Honda H, Mitani K, Chiba S, Yazaki Y, Hirai H (1999) TT virus in bone marrow transplant recipients. Blood 93:2485–2490

Kikuchi K, Miyakawa H, Abe K, Kako M, Katayama K, Fukushi S, Mishiro S (2000) Indirect evidence of TTV replication in bone marrow cells, but not in hepatocytes, of a subacute hepatitis/aplastic anemia patient. J Med Virol 61:165–170

Koch G, van Roozelaar DJ, Verschueren CA, van der Eb AJ, Noteborn MH (1995) Immunogenic and protective properties of chicken anaemia virus proteins expressed by baculovirus. Vaccine 13:763–770

Lefrere JJ, Roudot-Thoraval F, Lefrere F, Kanfer A, Mariotti M, Lerable J, Thauvin M, Lefevre G, Rouger P, Girot R (1999) Natural history of the TT virus infection through follow-up of TTV DNA-positive multiple-transfused patients. Blood 95:347–351

Lundberg P, Welander P, Han X, Cantin E (2003) Herpes simplex type 1 DNA is immunostimulatory in vitro and in vivo. J Virol 77:11158–11169

Luo K, He H, Liu Z, Liu D, Xiao H, Jiang X, Liang W, Zhang L (2002) Novel variants related to TT virus distributed widely in China. J Med Virol 67:118–126

Madsen CD, Eugen-Olsen J, Kirk O, Parner J, Christensen JK, Brasholt MS, Nielsen JO, Krogsgaard K (2002) TTV viral load as a marker for immune reconstitution after initiation of HAART in HIV-infected patients. HIV Clin Trials 3:287–295

Maggi F, Fornai C, Vatteroni ML, Siciliano G, Menichetti F, Tascini C, Specter S, Pistello M, Bendinelli M (2001a) Low prevalence of TT virus in the cerebrospinal fluid of viremic patients with central nervous system disorders. J Med Virol 65:418–422

Maggi F, Fornai C, Zaccaro G, Morrica A, Vatteroni ML, Isola P, Marchi S, Ricchiuti A, Pistello M, Bendinelli M (2001b) TT virus (TTV) loads associated with different peripheral blood cell types and evidence for TTV replication in activated mononuclear cells. J Med Virol 64:190–194

Maggi F, Pistello M, Vatteroni ML, Presciuttini S, Marchi S, Isola P, Fornai C, Fagnani S, Andreoli E, Antonelli G, Bendinelli M (2001c) Dynamics of persistent TT virus infection, as estimated in patients treated with interferon-alpha for concomitant hepatitis C. J Virol 75:11999–12004

Maggi F, Pifferi M, Fornai C, Andreoli E, Tempestini E, Vatteroni ML, Presciuttini S, Pietrobelli A, Boner A, Pistello M, Bendinelli M (2003a) TT virus in the nasal secretions of children with acute respiratory diseases: relations to viremia and disease severity. J Virol 77:2418–2425
Maggi F, Pifferi M, Tempestini E, Fornai C, Lanini L, Andreoli E, Vatteroni ML, Presciuttini S, Pietrobelli A, Boner A, Pistello M, Bendinelli M (2003b) TT virus loads and lymphocyte subpopulations in children. J Virol 77:9081–9083
Maggi F, Pifferi M, Tempestini E, Lanini L, De Marco E, Fornai C, Andreoli E, Presciuttini S, Vatteroni ML, Pistello M, Ragazzo V, Macchia PA, Pietrobelli A, Boner A, Bendinelli M (2004) Correlation between torque tenovirus infection and serum levels of eosinophil cationic protein in children hospitalized for acute respiratory diseases. J Infect Dis 190:971–974
Maggi F, Andreoli E, Lanini L, Fornai C, Vatteroni ML, Pistello M, Bendinelli M (2005a) Relationships between total plasma load of torquetenovirus (TTV) and TTV genogroups carried. J Clin Microbiol 43:4807–4810
Maggi F, Tempestini E, Lanini L, Andreoli E, Fornai C, Giannecchini S, Vatteroni ML, Pistello M, Marchi S, Ciccorossi P, Specter S, Bendinelli M (2005b) Blood levels of TT virus following immune stimulation with influenza or hepatitis B vaccine. J Med Virol 75:358–365
Maggi F, Andreoli E, Lanini L, Meschi S, Rocchi J, Fornai C, Vatteroni ML, Pistello M, Bendinelli M (2006) Rapid increase in total torquetenovirus (TTV) plasma viremia load reveals an apparently transient superinfection by a TTV of a novel group 2 genotype. J Clin Microbiol 44:2571–2574
Maggi F, Andreoli E, Lanini L, Rocchi J, Ricci V, Vatteroni ML, Albani M, Ceccherini Nelli L, Pistello M, Bendinelli M (2007a) TTV and the host's immune system: light and shade of a new virus-host interaction. 5th International Geminivirus Symposium and 3rd International ssDNA Comparative Virology Workshop. http://www.ufv.br/dfp/virologia/op2007/index_files/Abstract%20book,%205th%20IGS%20and%203rd%20IssDNACVW.pdf. Cited 21 June 2008
Maggi F, Andreoli E, Riente L, Meschi S, Rocchi J, Delle Sedie A, Vatteroni ML, Ceccherini Nelli L, Specter S, Bendinelli M (2007b) Torquetenovirus in patients with arthritis. Rheumatology 46:885–886
Maggi F, Focosi D, Ricci V, Paumgardhen E, Ghimenti M, Papineschi F, Bendinelli M, Petrini M, Ceccherini Nelli L (2008) Changes in CD8+57+ T lymphocyte expansions after autologous hematopoietic stem cell transplantation correlate with changes in torquetenovirus viremia. Transplantation 85:1867–1868
Mariscal LF, Lopez-Alcorocho JM, Rodriguez-Inigo E, Ortiz-Movilla N, de Lucas S, Bartolome J, Carreno V (2002) TT virus replicates in stimulated but not in nonstimulated peripheral blood mononuclear cells. Virology 301:121–129
Miyata H, Tsunoda H, Kazi A, Yamada A, Khan MA, Murakami J, Kamahora T, Shiraki K, Hino S (1999) Identification of a novel GC-rich 113-nucleotide region to complete the circular, single-stranded DNA genome of TT virus, the first human circovirus. J Virol 73:3582–3586
Moen EM, Sleboda J, Grinde B (2002) Serum concentrations of TT virus and TT virus-like mini virus in patients developing AIDS. AIDS 16:1679–1682
Moen EM, Sagedal S, Bioro K, Degrè M, Opstad PK, Grinde B (2003) Effect of immune modulation on TT virus (TTV) and TTV-like-mini-virus (TLMV) viremia. J Med Virol 70:177–182
Mushahwar IK, Erker JC, Muerhoff AS, Leary TP, Simons JN, Birkenmeyer LG, Chalmers ML, Pilot-Matias TJ, Desai SM (1999) Molecular and biophysical characterization of TT virus. Evidence for a new family infecting humans. Proc Natl Acad Sci USA 96:3177–3182
Ninomiya M, Takahashi M, Shimosegawa T, Okamoto H (2007) Analysis of the entire genomes of fifteen torque teno midi virus variants classifiable into a third group of genus Anellovirus. Arch Virol 152:1961–1975
Ninomiya M, Takahashi M, Nishizawa T, Shimosegawa T, Okamoto H (2008) Development of PCR assays with nested primers specific for differential detection of three human anelloviruses and early acquisition of dual or triple infection during infancy. J Clin Microbiol 46:507–514
Nishizawa T, Okamoto H, Konishi K, Yoshizawa H, Miyakawa Y, Mayumi M (1997) A novel DNA virus (TTV) associated with elevated transaminase levels in posttransfusion hepatitis of unknown etiology. Biochem Biophys Res Commun 241:92–97

Nishizawa T, Okamoto H, Tsuda F, Aikawa T, Sugai Y, Konishi K, Akahane Y, Ukita M, Tanaka T, Miyakawa Y, Mayumi M (1999) Quasispecies of TT (TTV) with sequence divergence in hypervariable regions of the capsid protein in chronic TTV infection. J Virol 73:9604–9608

Noteborn MH, Verschueren CA, Koch G, van der Eb AJ (1998) Simultaneous expression of recombinant baculovirus-encoded chicken anaemia virus (CAV) proteins VP1 and VP2 is required for formation of the CAV-specific neutralizing epitope. J Gen Virol 79:3073–3077

Okamoto H, Mayumi M (2001) TT virus: virological and genomic characteristics and disease associations. J Gastroenterol 36:519–529

Okamoto H, Kato N, Iizuka H, Tsuda F, Miyakawa Y, Mayumi M (1999a) Distinct genotypes of a nonenveloped DNA virus associated with posttransfusion non-A to G hepatitis (TT virus) in plasma and peripheral blood mononuclear cells. J Med Virol 57:252–258

Okamoto H, Nishizawa T, Ukita M, Takahashi M, Fukuda M, Iizuka H, Miyakawa Y, Mayumi M (1999b) The entire nucleotide sequence of a TT virus isolate from the United States (TUS01): comparison with reported isolates and phylogenetic analysis. Virology 259:437–448

Okamoto H, Nishizawa T, Tawara A, Takahashi M, Kishimoto J, Sai T, Sugai Y (2000a) TT virus mRNAs detected in the bone marrow cells from an infected individual. Biochem Biophys Res Commun 279:700–707

Okamoto H, Takahashi M, Kato N, Fukuda M, Tawara A, Fukuda S, Tanaka T, Miyakawa Y, Mayumi M (2000b) Sequestration of TT virus of restricted genotypes in peripheral blood mononuclear cells. J Virol 74:10236–10239

Okamoto H, Takahashi M, Nishizawa T, Tawara A, Sugai Y, Sai T, Tanaka T, Tsuda F (2000c) Replicative forms of TT virus DNA in bone marrow cells. Biochem Biophys Res Commun 270:657–662

Okamoto H, Nishizawa T, Takahashi M, Asabe S, Tsuda F, Yoshikawa A (2001) Heterogeneous distribution of TT virus of distinct genotypes in multiple tissues from infected humans. Virology 288:358–368

Okamura A, Yoshioka M, Kubota M, Kikuta H, Ishiko H, Kobayashi K (1999) Detection of a novel DNA virus (TTV) sequence in peripheral blood mononuclear cells. J Med Virol 58:174–177

Ott C, Duret L, Chemin I, Trepò C, Mandrand B, Komurian-Pradel F (2000) Use of a TT virus ORF1 recombinant protein to detect anti-TT virus antibodies in human sera. J Gen Virol 81:2949–2958

Peters MA, Jackson DC, Crabb BS, Browning GF (2002) Chicken anemia virus VP2 is a novel dual specificità protein phosphatase. J Biol Chem 277:39566–39573

Pifferi M, Maggi F, Andreoli E, Lanini L, De Marco E, Fornai C, Vatteroni ML, Pistello M, Ragazzo V, Macchia P, Boner AL, Bendinelli M (2005) Relationships between nasal TT virus burdens and spirometric indices in asthmatic children. J Infect Dis 192:1141–1148

Pifferi M, Maggi F, Caramella D, De Marco E, Andreoli E, Meschi S, Macchia P, Bendinelli M, Boner AL (2006) High torquetenovirus loads are correlated with bronchiectasis and peripheral airflow limitation in children. Pediatr Infect Dis J 25:804–808

Pifferi M, Maggi F, Di Cristofano C, Cangiotti AM, Ragazzo V, Ceccherini Nelli L, Bevilacqua G, Macchia P, Bendinelli M, Boner AL (2008) Torquetenovirus infection and ciliary dysmotility in children with recurrent pneumonia. Pediatr Infect Dis J 27:413–418

Pistello M, Morrica A, Maggi F, Vatteroni ML, Freer G, Fornai C, Casula F, Marchi S, Ciccorossi P, Rovero P, Bendinelli M (2001) TT virus levels in the plasma of infected individuals with different hepatic and extrahepatic pathologies. J Med Virol 63:189–195

Puig-Basagoiti F, Cabana M, Guilera M, Gimenez-Barcons M, Sirera G, Tural C, Clotet B, Sanchez-Tapias JM, Rodes J, Saiz JC, Martinez MA (2000) Prevalence and route of transmission of infection with a novel DNA virus (TTV), hepatitis C virus, and hepatitis G virus in patients infected with HIV. J Acquir Immune Defic Syndr 23:89–94

Qiu J, Kakkola L, Cheng F, Ye C, Soderlund-Venermo M, Hedman K, Pintel DJ (2005) Human circovirus TT virus genotype 6 expresses six proteins following transfection of a full-length clone. J Virol 79:6505–6510

Sagir A, Kirschberg O, Heintges T, Erhardt A, Haussinger D (2004) SEN virus infection. Rev Med Virol 14:141–148

Sagir A, Adams O, Antakyali M, Oette M, Erhardt A, Heintges T, Haussinger D (2005a) SEN virus has an adverse effect on the survival of HIV-positive patients. AIDS 19:1091–1096

Sagir A, Adams O, Oette M, Erhardt A, Heintges T, Haussinger D (2005b) SEN virus seroprevalence in HIV positive patients: association with immunosuppression and HIV-replication. J Clin Virol 33:183–187

Segales J, Domingo M, Chianini F, Majo N, Dominguez J, Darwich L, Mateu E (2004) Immunosuppression in postweaning multisystemic wasting syndrome affected pigs. Vet Microbiol 98:151–158

Shang D, Lin YH, Rigopoulou I, Chen B, Alexander GJM, Allain JP (2000) Detection of TT virus DNA in patients with liver disease and recipients of liver transplant. J Med Virol 61:455–461

Sherman KE, Rousters SD, Feinberg J (2001) Prevalence and genotypic variability of TTV in HIV-infected patients. Dig Dis Sci 46:2401–2407

Shibayama T, Masuda G, Ajisawa A, Takahashi M, Nishizawa T, Tsuda F, Okamoto H (2001) Inverse relationship between the titre of TT virus DNA and CD4 cell count in patients infected with HIV. AIDS 15:563–570

Simmonds P (2002) TT virus infection: a novel virus-host relationship. J Med Microbiol 51:455–458

Sospedra M, Zhao Y, zur Hausen H, Muraro PA, Hamashin C, de Villiers EM, Pinilla C, Martin R (2005) Recognition of conserved amino acid motifs of common viruses and its role in autoimmunity. PLoS Pathog 1:335–348

Specter S, Bendinelli M, Friedman H (1989) Virus-induced immunosuppression. Plenum Press, New York

Takahashi K, Ohta Y, Mishiro S (1998) Partial ~ 2.4-kb sequences of TT virus (TTV) genome from eight Japanese isolates: diagnostic and phylogenetic implications. Hepatol Res 12:111–120

Takahashi K, Iwasa Y, Hijikata M, Mishiro S (2000) Identification of a new human DNA virus (TTV-like mini virus, TLMV) intermediately related to TT virus and chicken anemia virus. Arch Virol 145:979–993

Takahashi M, Asabe S, Gotanda Y, Kishimoto J, Tsuda F, Okamoto H (2002) TT virus is distributed in various leukocyte subpopulations at distinct levels, with the highest viral load in granulocytes. Biochem Biophys Res Commun 290:242–248

Tawara A, Akahane Y, Takahashi M, Nishizawa T, Ishikawa T, Okamoto H (2000) Transmission of human TT virus of genotype 1a to chimpanzees with fecal supernatant or serum from patients with acute TTV infection. Biochem Biophys Res Commun 278:470–476

Thom K, Petrik J (2007) Progression towards AIDS leads to increased torque teno virus and torque teno minivirus titers in tissues of HIV infected individuals. J Med Virol 79:1–7

Todd D, McNulty MS, Adair BM, Allan GM (2001) Animal circoviruses. Adv Virus Res 57:1–70

Tokita H, Murai S, Kamitsukasa H, Yagura M, Harada H, Takahashi M, Okamoto H (2002) High TT virus load as an independent factor associated with the occurrence of hepatocellular carcinoma among with hepatitis C virus-related chronic liver disease. J Med Virol 67:501–509

Touinssi M, Gallian P, Biagini P, Attoui H, Vialettes B, Berland Y, Tamalet C, Dhiver C, Ravaux I, De Micco P, De Lamballerie X (2001) TT virus infection: prevalence of elevated viraemia and arguments for the immune control of viral load. J Clin Virol 21:135–141

Tsuda F, Okamoto H, Ukita M, Tanaka T, Akahane Y, Konishi K, Yoshizawa H, Miyakawa Y, Mayumi M (1999) Determination of antibodies to TT virus (TTV) and application to blood donors and patients with post-transfusion non-A to G hepatitis in Japan. J Virol Methods 77:199–206

Tsuda F, Takahashi M, Nishizawa T, Akahane Y, Konishi K, Yoshizawa H, Okamoto H (2001) IgM-class antibodies to TT virus (TTV) in patients with acute TTV infection. Hepatol Res 19:1–11

Ukita M, Okamoto H, Nishizawa T, Tawara A, Takahashi M, Iizuka H, Miyakawa Y, Mayumi M (2000) The entire nucleotide sequences of two distinct TT virus (TTV) isolates (TJN01 and TJN02) remotely related to the original TTV isolates. Arch Virol 145:1543–1559

Welch J, Bienek C, Gomperts E, Simmonds P (2006) Resistance of porcine circovirus and chicken anemia virus to virus inactivation procedures used for blood products. Transfusion 46:1951–1958

Yu Q, Shiramizu B, Nerurkar VR, Hu N, Shikuma CM, Melish ME, Cascio K, Imrie A, Lu Y, Yanagihara R (2002) TT virus: preferential distribution in CD19+ peripheral blood mononuclear cells and lack of viral integration. J Med Virol 66:276–284

Yusufu Y, Mochida S, Matsui A, Okamoto H, Fujiwara K (2001) TT virus infection in cases of fulminant hepatic failure-evaluation by clonality based on amino acid sequence of hypervariable regions. Hepatol Res 21:85–96

Zheng H, Ye L, Fang X, Li B, Wang Y, Xiang X, Kong L, Wang W, Zeng Y, Ye L, Wu Z, She Y, Zhou X (2007) Torque teno virus (SANBAN isolate) ORF2 protein suppresses NF-kappaB pathways via interaction with IkappaB kinases. J Virol 81:11917–11924

Zhong S, Yeo W, Tang MW, Lin XR, Mo F, Ho WM, Hui P, Johnson PJ (2001) Gross elevation of TT virus genome load in the peripheral blood mononuclear cells of cancer patients. Ann N Y Acad Sci 945:84–92

Zhong S, Yeo W, Tang M, Liu C, Lin X, Ho WM, Hui P, Johnson PJ (2002) Frequent detection of the replicative form of TT virus DNA in peripheral blood mononuclear cells and bone marrow cells in cancer patients. J Med Virol 66:428–434

# Intragenomic Rearrangement in TT Viruses: A Possible Role in the Pathogenesis of Disease

**E.-M. de Villiers(✉), R. Kimmel, L. Leppik, and K. Gunst**

**Contents**



**Abstract** A role for the ubiquitous Torque teno (TT) viruses in the pathogenesis of disease has not been resolved. In vivo and in vitro intragenomic rearrangement of TT virus genomes has been demonstrated. Replication in cell culture of a subviral molecule (411 bp) occurs through oligomerisation of RNA transcripts. Although the functions of the respective TT viral genes, as well as the newly formed genes in the rearranged subviral molecules, are largely unknown, certain similarities to genes of plant viruses of the family *Geminiviridae* will be described. A degree of similarity to certain cellular genes poses the question as to a role of molecular mimicry in the pathogenesis of autoimmune disease and diabetes.

E.-M. de Villiers
Division for the Characterisation of Tumour Viruses, Deutsches Krebsforschungszentrum, Im Neuenheimer Feld 242, 69120 Heidelberg, Germany
e.devilliers@dkfz.de

E.-M. de Villiers, H. zur Hausen (eds.), *TT Viruses: The Still Elusive Human Pathogens,* 

**Abbreviations** NTR: Non-translated region; ORF: Open reading frame; PCR: Polymerase chain reaction; TTV: Torque teno virus.

## Introduction

Peripheral blood mononuclear cells act as a reservoir for TTVs (Okamoto et al. 2000a; Yu et al. 2002), whereas viral transcription and replication of these single-stranded DNA viruses occur in bone-marrow cells (Kanda et al. 1999; Okamoto et al. 2000b, c; Zhong et al. 2002). Circular double-stranded forms of the viral DNA have also been demonstrated in the liver (Okamoto et al. 2000d). Despite several attempts to clarify the primary route of infection, it remains unclear. TTV infections occur already within the first days after birth (Davidson et al. 1999; Saback et al. 1999; Sugiyama et al. 1999; Kazi et al. 2000; Peng et al. 2002) and close to 100% of infants are persistently infected at 1 year of age (Ninomiya et al. 2008). Vertical transmission from mother to child is still debatable, as TTV DNA could be demonstrated in cord blood samples in a number of reports (Gerner et al. 2000; Goto et al. 2000; Morrica et al. 2000; Matsubara et al. 2001; Xiu et al. 2004), whereas others failed to do so (Ohto et al. 2002; Ninomiya et al. 2008).

The plurality of Anelloviruses (Khudyakov et al. 2000; Okamoto et al. 2001) became evident very soon after their discovery (Nishizawa et al. 1997). The genetic variability existing within and between viral genomes increases the difficulties in defining a complete viral genome or a strain/type of this virus group (Jelcic et al. 2004; Leppik et al. 2007; Ninomiya et al. 2007a, b). It also complicates efforts to associate a specific virus type/strain with a specific disease or disease complex.

The recent isolation of intragenomic rearranged subviral DNA molecules from serum samples taken from mothers of leukaemic children initiated the notion that these viruses may share similarities to the plant viruses of the family *Geminiviridae* (Leppik et al. 2007). Plants exhibiting disease are infected with a complex mixture of viral, subviral and recombinant DNA components (Briddon and Stanley 2006; H. Jeske, this volume). Deletions and recombination occurring at repetitive sequences within or between the DNA-A and DNA-B components of begomoviruses lead to defective DNA components which in turn cause a modification of disease symptoms (Patil et al. 2007). Begomoviruses associate with single-stranded DNA satellites to form a disease-inducing complex (Saunders et al. 2000; Briddon and Stanley 2006).

Several other virus infections result in a heterogeneous viral progeny. The high degree of heterogeneity in the human immunodeficiency virus RNA genome results from slippage-mediated mutations involving single nucleotides or small regions of the genome leading to progression of disease (Hamburgh et al. 2006). Hepatitis C virus also represents a quasi-species in that it continuously adapts to its host by cell defence evasion mechanisms (Simmonds 2004).

TTVs also represent a quasi-species (Nishizawa et al. 1999; Jelcic et al. 2004). The hypervariable region [ca. 90 amino acids (aa)] of the TTV ORF1 varies as much as 31% in amino acid composition between TTV isolates displaying only 2%

sequence divergence in the rest of the genome. Recombinational events occurring among single-stranded DNA viruses infecting humans have not been described. Intragenomic rearrangement within a viral genome resulting in smaller subviral molecules was recently reported to occur in TTVs (Leppik et al. 2007). The following deals with analyses of full-length and rearranged smaller TT viral isolates. Replication and transcription in tissue culture (in vitro) of full-length TTV (tth25; Jelcic et al. 2004) genome and of subviral molecules provide evidence for the origin and selection of the latter. The subviral molecules described here as examples were isolated from serum samples (Leppik et al. 2007). All are related most closely to TTV strain tth25. In addition, an in silico analysis of the putative proteins points to similarities that genes of TTVs and the subviral molecules share with geminiviruses, as well as with human cellular genes. This may hint at a function and possible association with disease.

## Full-Length TT Viruses

### *Rearranged Genomes*

Analyses of full-length TTV sequences available in the databank have revealed a number of viral sequences apparently representing rearranged genomes, as the putative protein sizes expected for the various ORFs do not correlate with other TTV genomes within the same phylogenetic group (Table 1). At present the main obstacle concerning TTV research is the failure to produce large amounts of viral particles in cell culture or to isolate adequate quantities from human samples. It is conceivable that many of the observations made on anellovirus genomes relate back to artefacts originating from PCR amplification of viral DNA from tissue samples. Unfortunately, this unavailability of intact virus particles and the exceptionally large number of currently known anellovirus isolates will make it an almost impossible task to verify present data in future.

TTV prototype ta278 is an example in which ORF2 codes for a putative protein of 202aa in length, whereas in other strains of the same phylogenetic group it is divided between an ORF2 varying between 148 and 156aa and a truncated ORF2a (49aa). Similarly, ORF3 is 105aa compared to 53aa of other isolates. This variation is not ORF-dependent (Table 1; all accession numbers available in Jelcic et al. 2004). The largest variation in putative ORF sizes occurs in group 4, where ORF1 is also generally smaller (coding capacity 610–657aa) than in others (720–770aa). It was previously noted that one nucleotide change in ORF1 could lead to an “interrupted” ORF1 (Jelcic et al. 2004; Erker et al. 1999; Khudyakov et al. 2000; Luo et al. 2002). Given the high number (up to 15) of start codons present in the majority of TTV ORF1 s, two or more smaller putative ORF1 s then result. The general notion is that ORF1 encodes the capsid protein (see S. Hino and A.A. Prasetyo, this volume), but conclusive evidence is still lacking. Difficulties have been encountered in expressing

**Table 1** Open reading frames of a number of full-length TTV genomes

| Phylo group | Strain | Length (bp) | ORF1 (aa) | ORF2 | ORF2a | ORF3 | ORF4 | ORF5 | ORF in NTR | Antisense (aa) |
|---|---|---|---|---|---|---|---|---|---|---|
| 1 | hel32 | 3,748 | 736 | 156 | 49 | 53 | 108 | 59 | 60 | 84,111,128 |
| | p1c1 | 3,756 | 733 | 149 | 49 | - | 107 | 59 | 56 | 74,128 |
| | tth25 | 3,757 | 736 | 148 | 49 | 53 | 107 | 59 | 56,57 | 82,105 |
| | ta278 | 3,873 | *770* | *202* | | *105* | - | 57 | 100 | 88,105 |
| 2 | kt–08f | 3,790 | 745 | 152 | 49 | 56 | 96 | *131* | 123 | |
| | vipi04 | 3,774 | 732 | 153 | 49 | 53 | *119* | *129* | 125,111 | 81,88 |
| | tth6 | 3,740 | 720 | 154 | 96 | 53 | 83 | 88 | 133,129 | 95,114 |
| | kav | 3,705 | 719 | *161* | 49 | 53 | 79 | 86 | | 80,137 |
| 3 | tus01 | 3,818 | 761 | 156 | 49 | *76* | 84 | 58 | | 81,89 |
| | tjn01 | 3,787 | 754 | 157 | 49 | 59 | 88 | 58 | 100 | 87,113,97 |
| | tth4 | 3,772 | 746 | *150* | 49 | 57 | 81 | | 170,66 | 135 |
| | sanban | 3,808 | 745 | *163* | 49 | 55 | 83 | 56 | 66 | 93,128,143 |
| | saf–09 | 3,155 | 742 | *141* | | 50 | 98 | 59 | | 123,187 |
| | saa–01 | 3,246 | 766 | 156 | | 55 | *104* | *186* | | 153 |
| | tchna | 3,312 | *61,689* | 154 | 49 | 56 | 90 | - | | 152 |
| 4 | jt41f | 3,727 | 648 | 128 | 50 | 75 | *164* | - | 69,64 | 183,123,96 |
| | ct23f | 3,729 | 60,610 | 128 | | 76 | *92* | 60 | 59 | 88,124,158 |
| | jt19f | 3,676 | 652 | *144* | 50 | *120* | *145* | - | | 77,113 |
| | ct43f | 3,629 | 657 | *101* | | *114* | *115* | - | 64 | 122,136 |
| 5 | ct39f | 3,759 | 743 | 120 | 69 | 50 | - | - | 73,98 | 141,72 |
| | tth8 | 3,753 | 743 | 154 | | 50 | 131 | 65 | 58,93 | 93,107 |
| | jt34f | 3,770 | 745 | *166* | | 50 | - | 65 | 74,166 | 72,98,104 |
| | tth5 | 3,700 | 726 | 154 | | 50 | - | - | 83 | 94 |

full-length ORF1 in vitro (L. Kakkola et al., this volume; Qiu et al. 2005; L. Leppik and E.-M. de Villiers, unpublished data). It is therefore an open question as to which start codon is actually used for the translation of the viral capsid. Only 5 of 10 closely related TTV full-length genomes (homology 98%) which we isolated from serum samples, harboured the intact full-length ORF1 (736aa) (Table 2 and Fig. 1). A number of smaller ORFs resulting from parts of the ORF1 were present in the others. Interestingly, the N-terminal arginine-rich 79aa part of the putative ORF1 protein represented a separate ORF in two of these—this stretch of amino acids is of particular interest as it hampers in vitro expression of the full-length ORF1 and has been reported to be recognised by T cell clones from patients with multiple sclerosis (MS) (Sospedra et al. 2005).

**Table 2** TT virus tth25-related full-length isolates and their open reading frames

| Phylo group | Strain | Length (bp) | ORF1 | ORF2 | ORF2a | ORF3 | ORF4 | ORF5 | ORF in NTR |
|---|---|---|---|---|---|---|---|---|---|
| 1 | sle1931 | 3,760 | 488,286 | 148 | | 53 | 107 | 59 | 112 |
| | sle1932 | 3,759 | 736 | 148 | | 53 | 107 | 59 | 112 |
| | sle1957 | 3,773 | 142,138,355 | 149 | 49 | 53 | 107 | *138* | |
| | sle2045[a] | 3,759 | 736 | 148 | 49 | 53 | 107 | 59 | 112 |
| | sle2057[a] | 3,757 | 736 | 148 | 49 | 53 | 107 | 59 | |
| | sle2058[a] | 3,758 | 736 | 148 | 49 | 53 | 107 | 59 | |
| | sle2061[a] | 3,759 | 736 | 148 | 49 | 53 | 107 | 59 | |
| | sle2065 | 3,759 | 249,455 | 148 | 49 | 53 | 107 | 59 | |
| | sle2072 | 3,759 | 79,659 | 148 | 49 | 53 | 107 | 59 | |
| | *sle2070*[b] | *3,231* | *79,355,255* | - | - | - | *107* | *59* | |

[a]These isolates differed from each other in single nucleotides scattered throughout the genome, but clustered in the hypervariable region

[b]This isolate represents a 3,231-bp genome, but with a large part of the complete genome deleted

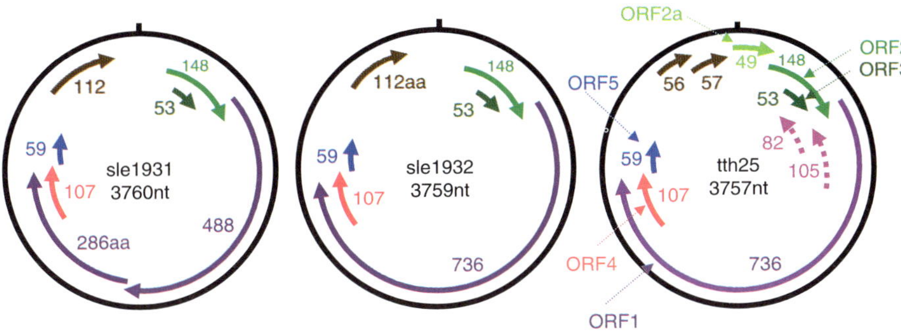


**Fig. 1** Examples of full-length TT virus genomes (nucleotides). The putative protein sizes of each open reading frame are given (amino acids). Open reading frames present in the non-translated region are presented in *brown*

### *Replication and Transcription*

In vitro replication of full-length TTV genomes has been reported for two different TTV strains, tth8 (group 5) (Leppik et al. 2007) and HEL32–6a (group 1) (Kakkola et al. 2007). Transcription of these viruses has also been described (Leppik et al. 2007; Qiu et al. 2005; L. Kakkola et al., this volume). Several off-sized transcripts were identified in the case of tth8, in which the splice-donor and/or -acceptor sites were replaced by repetitive sequences, indicating intragenomic rearrangement events. These data were mirrored in off-sized DNA molecules appearing shortly after replication initiation of the re-circularised TTV DNA molecule (Leppik et al. 2007). Off-sized rearranged transcripts were similarly present during tissue culture replication of TTV tth25 (group 1) (L. Leppik and E.-M. de Villiers, unpublished data). This observation seems to be cell type-independent, as it occurred in both a Hodgkin's lymphoma cell line L428 and in the kidney cell line 293. Interferon-γ treatment of tth25-transfected cells inhibited TTV DNA replication in vitro. An inhibitory effect was measured independent of time point and period of interferon-γ treatment of the cells.

## Subviral TT Genomes

### *Complete or Incomplete Genomes?*

Human Anellovirus isolates have been reported to range between 2.8 kb (TT mini viruses, TTMV) and 3.9 kb (TTV) in size. Recently a third group (TT midi virus, TTMDV) was isolated with genomes of 3.2 kb in length (Ninomiya et al. 2007a, b). The reported small anelloviruses SAV1 (2,249 bp) and SAV2 (2,635 bp) (Jones et al. 2005) clearly represent rearranged molecules of the related TTMDV isolates MD1–073 and MD2–013, respectively (Ninomiya et al. 2007a; E.-M. de Villiers, unpublished data). The so-called non-translated region (NTR) between ORF5 and ORF2 is largely deleted. We have isolated a number of rearranged subviral genomes from serum samples which harbour sufficient information to be regarded as "full-length" genomes. Examples (Table 3 and Fig. 2, all accession numbers available in Leppik et al. 2007 and Jelcic et al. 2004) range between 2,100 and 2,700 bp. They can be divided into two groups: 2,100–2,300 bp versus 2,400–2,700 bp in size. Isolate sle2054 is an example of an intact genome at first glance, i.e. to be grouped with the small anellovirus isolates. It differs, however, from TTV tth25 (3,758 bp) by less than 2% in nucleotide sequence homology of the overlapping regions, and its NTR is deleted. A similar situation was noted for the small anellovirus isolates (Ninomiya et al. 2007a). In general, ORF2 and ORF3, as well as ORF2a, remained intact in our isolates, whereas the NTR was either completely deleted as in the majority of clones, or only partially present, with the exception of clone sle1803 having a large NTR. The region spanning ORF4 and ORF5 was also frequently

**Table 3** Intermediate length TTV tth25-related isolates and their open reading frames

| Phylogenetic group | Strain | Length (bp) | ORF1 | ORF2 | ORF2a | ORF3 | ORF4 | ORF5 |
|---|---|---|---|---|---|---|---|---|
| 1 | sle1785 | 2,105 | 107,*291rev* | 148 | 49 | 53 | - | |
| | sle1793 | 2,275 | 143,113,346 | 148 | - | 53 | - | - |
| | sle1804 | 2,170 | 143,62,*324rev* | 149 | 49 | 53 | *137rev* | - |
| | sle1789 | 2,130 | 125,*380rev* | 148 | 49 | 53 | *104rev* | - |
| | sle1797 | 2,405 | 73,*355rev* | 67 | 49 | 53 | *107rev* | 188 |
| | sle1803 | 2,404 | 73,*355rev* | 148 | 49 | 53 | *107rev* | 59rev |
| | sle2050 | 2,688 | 429,282 | 148 | 49 | 53 | 93 | - |
| | sle2051 | 2,687 | 330,110,242 | 148 | 49 | 53 | 53 | - |
| | sle2052 | 2,686 | 256,168,282 | 148 | 49 | 53 | 95 | - |
| | sle2054 | 2,688 | 732 | 148 | 49 | 53 | 95 | - |

rev, reverse orientation of all other ORFs located on the positive strand

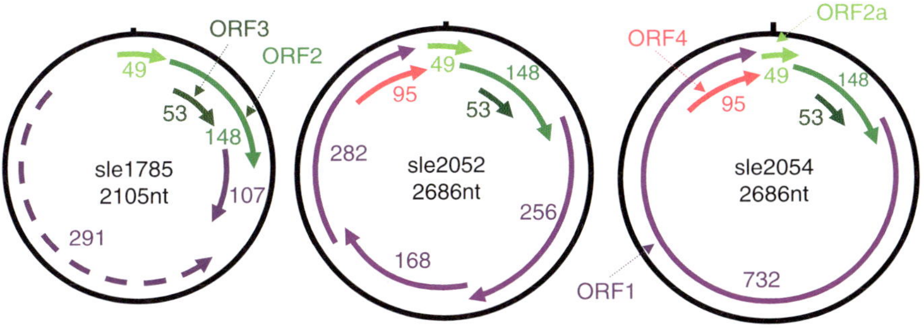


**Fig. 2** Examples of intermediate length TT virus genomes (nucleotides). The putative protein sizes of each open reading frame are given (amino acids)

affected. Another feature of these isolates was the division of ORF1 into several smaller ORFs. Interestingly, the smaller ORFs resulting from the last part of the ORF1 as well as ORF4 and ORF5, were often present in antisense orientation. Further functional analyses of the putative proteins will determine whether this region of the subviral genome is translated.

## *Subviral Molecules Smaller Than 2,000 bp*

A large number of additional rearranged subviral molecules were isolated from serum samples. These were identified (by cloning and sequencing) as subviral molecules originating from different full-length known TTVs or TTMVs (Leppik et al. 2007). The pattern of rearrangement was independent of related TTV or TTMV strain. A number of isolates ranging between 367 bp and 1,845 bp and sharing closest homology to TTV tth25 are selected here and presented as examples in Table 4 and Fig. 3. Isolates consisting of merely 737 bp comprised ORF2, ORF2a, ORF3 and parts of ORF1 (e.g. isolate sle2319). The smallest isolate presented here, sle2018, consisting of 367 bp, harboured only ORF2a. Increase in overall nucleotide length did not necessarily imply the presence of a larger ORF1 (e.g. sle1998 of 1,098 bp vs sle2008 of 1,478 bp or sle2009 of 1,624 bp).

## *Tissue Culture Replication of Subviral Molecules*

We transfected a small number of rearranged subviral DNA molecules into the L428 cell line and monitored for replication and transcription through DNA and RNA blot hybridisation. Indications are that certain subviral molecules do replicate at least for up to 9 days post-transfection. Difficulties were encountered in distinguishing newly replicated from input DNA in the case of certain smaller clones, as the transfected DNA persists up to 9 days post-infection. Replication of subviral isolate sle2160 (411 bp, acc. No. AM712017 and with closest homology to TTV tchn-a, acc. No. AF345526; Fig. 4) replicated up to 23 days in both L428 and 293 cell lines. Replicated subviral DNA was separated from high molecular weight cellular DNA by selective extraction (Hirt 1967) prior to blot hybridisation. Newly formed sle2160 DNA molecules were resistant to DNase 1 digestion, and Northern blot analysis demonstrated multimeric forms of sle2160 RNA (Fig. 4). These results point to a circularisation and/or a tight secondary structure of the newly formed subviral DNA molecule.

Geminiviruses replicate via a rolling circle mechanism, as well as recombinant-dependent replication. Multimeric forms are also present (H. Jeske, this volume; Jeske et al. 2001). Our sle2160 rearranged DNA molecule harbours ORF2a and three ORFs each partially homologous to ORF1. The putative proteins coded for by the latter partial ORF1s share about 30aa with the protein coded for by tth25 ORF1 (Fig. 4). A rearranged putative protein (108aa) in sle2019, resulting from an ORF located in the same region of the sle2019 genome as ORF2a, displayed a degree of similarity to amino acid sequences characteristic for the replicator enhancer C3 of geminiviruses (Settlage et al. 2005). Further investigations are needed to determine whether the putative proteins of each of these ORFs are functional, as well as to analyse the mechanism through which replication takes place. Selection for certain rearranged subviral molecules seems to occur in vivo (Leppik et al. 2007). The fact that only certain subviral molecules replicate in vitro may indicate the existence of both defective subviral molecules, as well as—in analogy to geminiviruses—subviral molecules acting as satellites.

**Table 4** Intragenomic rearranged subviral DNA molecules all related to TTV tth25

| Isolate | Length (bp) | ORF1 | ORF2 | ORF2a | ORF3 | ORF4 | ORF5 |
|---|---|---|---|---|---|---|---|
| sle2018 | 367 | - | - | 49 | - | - | - |
| sle1988 | 402 | 48 | - | - | - | - | - |
| sle2016 | 539 | 90 | *101rev* | *49rev* | - | - | - |
| sle2644 | 561 | - | 81 | 131 | - | - | - |
| sle1748 | 631 | - | 148 | 49 | 53 | - | - |
| sle1751 | 679 | - | 169 | 49 | - | - | - |
| sle1755 | 691 | 150 | 148 | 67 | 53 | - | - |
| sle2019 | 729 | 79 | 148 | 49 | 53 | - | - |
| sle1753 | 764 | - | 148 | 49 | 53 | - | - |
| sle2319 | 765 | 155 | 148 | - | 53 | - | - |
| sle1752 | 782 | 153 | 148 | - | 53 | - | - |
| sle1756 | 872 | 125 | 148 | 49 | 53 | - | - |
| sle1999 | 922 | 122 | 148 | 49 | 53 | - | - |
| sle1996 | 965 | 245,174 | - | - | - | - | - |
| sle2022 | 874 | 106 | 175 | 49 | 53 | - | - |
| sle1963 | 1,080 | 196 | 148 | 49 | 53 | - | - |
| sle1998 | 1,098 | 143 | 148 | 139 | 53 | - | - |
| sle1995 | 1,068 | 48,82 | 148 | 56 | 53 | - | - |
| sle1983 | 1,079 | 41,88 | 331 | 49 | 53 | - | - |
| sle2015 | 1,095 | 200 | 148 | - | 53 | - | - |
| sle2017 | 1,140 | 224 | 148 | 49 | 53 | - | - |
| sle1916 | 1,113 | 183 | 148 | - | 53 | - | - |
| sle1850 | 1,113 | 183 | 148 | 49 | 53 | - | - |
| sle1857 | 1,109 | 68,97 | 148 | - | - | - | - |
| sle1981 | 1,244 | 306 | 148 | - | 53 | - | - |
| sle2002 | 1,287 | 52 | 161,*109rev* | 49 | 53 | - | - |
| sle2001 | 1,293 | 73 | 148 | - | 53 | - | - |
| sle2004 | 1,316 | 68,57 | 148 | 49 | 53 | - | *59rev* |
| sle1762 | 1,361 | 288 | 148 | 49 | 53 | - | - |
| sle2003 | 1,361 | 55,180 | 148 | 49 | 53 | - | - |
| sle2025 | 1,378 | 142,238 | 148 | - | 53 | 64 | - |
| sle2026 | 1,405 | 119,237 | 148 | 45 | 53 | - | - |
| sle1760 | 1,408 | 229 | 148 | - | 53 | - | - |
| sle1770 | 1,432 | 242 | 148 | 49 | 53 | - | - |
| sle1761 | 1,432 | 163,*115rev* | 148 | - | 53 | - | *87rev* |
| sle1768 | 1,473 | 335 | 148 | 49 | 53 | - | - |
| sle2008 | 1,478 | 143,61,*129rev* | 148 | - | 53 | - | 57 |
| sle1763 | 1,530 | 271,110 | 148 | 49 | 53 | - | - |
| sle1769 | 1,554 | 79,235 | 148 | 49 | 53 | - | - |
| sle2009 | 1,624 | 142,59,147 | 148 | - | 53 | - | - |
| sle1779 | 1,726 | 410 | 65 | - | 53 | - | - |
| sle1780 | 1,734 | 447 | 148 | 83 | 53 | - | - |
| sle2046 | 1,763 | 69,315 | 148 | 49 | 53 | - | - |
| sle1975 | 1,764 | 69,315 | 148 | - | 53 | - | - |
| sle1782 | 1,775 | 113,*305rev* | 148 | 49 | 53 | - | - |
| sle1772 | 1,845 | 112,*305rev* | 148 | - | 53 | - | - |

rev, reverse orientation of all other ORFs located on the positive strand

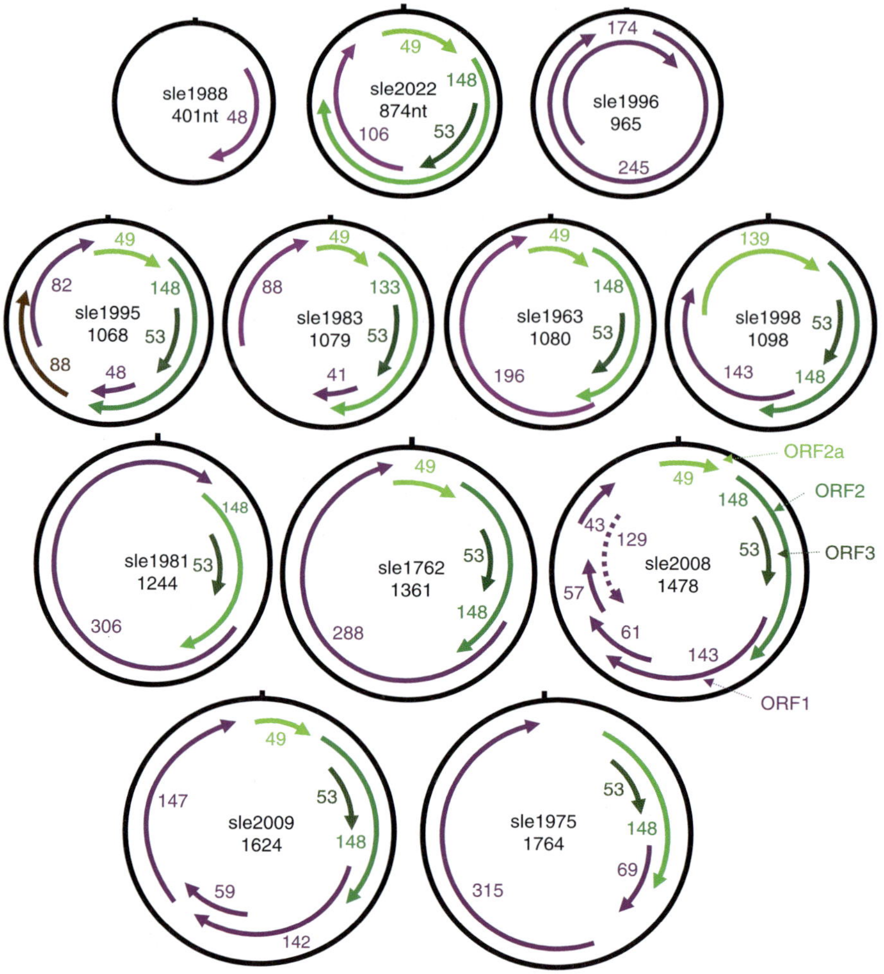


**Fig. 3** Examples of intragenomic rearranged subviral TT virus genomes (nucleotides) related to TTV tth25. The putative protein sizes of each open reading frame are given (amino acids)

# Putative Gene Functions

## *Viral Gene Functions*

Putative functions for the individual genes of TTVs have been discussed based mainly on transcription analyses. Previous studies have used a variety of cell lines and either free full-length genomes or vector-based genomes to study their transcription in vitro (Kamahora et al. 2000; Qiu et al. 2005; Leppik et al. 2007). It is

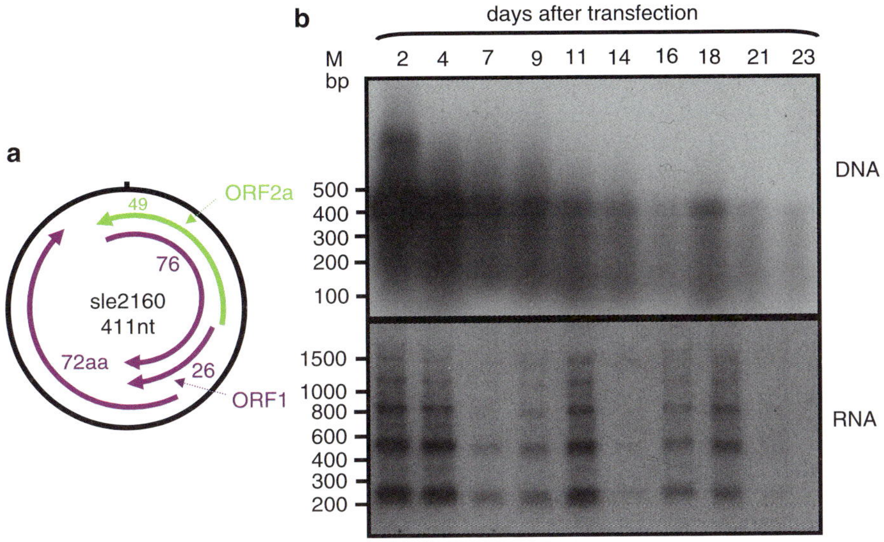


**Fig. 4** **A** Schematic presentation of intragenomic rearranged subviral sle2160. Putative proteins sizes for each open reading frame are given in amino acids. **B** Replication of sle2160 DNA after transfection into L428 cells. Aliquots of cells were harvested on days indicated. Small and supercoiled DNA was selectively extracted (Hirt 1967) and separated by gel electrophoresis. Hybridization was performed on Southern (DNA) and Northern (RNA) blots using radio-labelled sle2160 DNA

evident that transcription patterns vary depending on cell type, indicating a differential regulation for virus replication in different tissues. Relatively low sequence homology between TTV strains may complicate this even more. In silico analyses of available genome sequences revealed—for the TTV strains analysed (Table 1)—relatively large ORFs present in the NTR. This region has an exceptionally high GC content which hampers investigations. Long PCR amplification, performed to isolate complete genomes, is often initiated within the NTR and additional other experimental restrictions may hamper the isolation of messenger RNA (mRNA) from this region. Taken together, the ORFs present in the NTR may be functional under in vivo conditions of viral infection.

Anelloviruses are known to be single-stranded DNA viruses having a double-stranded replicative intermediate (Okamoto et al. 2000d). Almost all TTV strains analysed harbour a number of large ORFs in antisense orientation (Table 1). We investigated whether antisense mRNA is transcribed during tissue culture replication of TTV tth25. Preliminary data revealed the presence of these antisense transcripts. Parts of ORF1 are often located on the antisense strand in several of the rearranged subviral molecules. This often entails the C-terminal 300–400aa coded for by ORF1. At present it is not known whether these proteins are being expressed in patients.

A number of functions have previously been associated with ORFs 1, 2 and 3 of TTV (reviewed in Hino and Miyata 2007; S. Hino and A.A. Prasetyo, this volume).

ORF1 is regarded as coding for the viral capsid protein. However, attempts to express full-length ORF1 in tissue culture have been unsuccessful. The hypervariable region, located roughly between nucleotides 1400 and 1700 of this ORF, seems to undergo constant variation as mirrored in the analyses of eight full-length TTV isolates from the spleen of a patient with Hodgkin's disease (Jelcic et al. 2004). Isolates varied between 3% and 31% in encoded amino acids within this region, whereas the rest of the genomes differed by less than 2% in their nucleotide sequences. A single amino acid change in the viral capsid VP1 of chicken anaemia virus (CAV) has been reported to modify the pathogenicity of this virus (K.A. Schat, this volume; Yamaguchi et al. 2001). Given the variation identified between TTV isolates, not only in the hypervariable region but throughout the entire genome, it is quite likely that minor changes may also be responsible for pathogenetic changes. No data are available on the usage of start codons in ORF1. The large number of start codons present, as well as the smaller ORF1-derived putative proteins in the intragenomic rearranged molecules, opens many possibilities. Pathological changes of renal epithelial cells in mice transgenic for TTV ORF1 were reported. A transcript isolated from these animals encoded the N-terminal arginine-rich region of ORF1 spliced to ORF4 (Yokoyama et al. 2002). Interestingly, this region of ORF1 often appears as an independent ORF in rearranged subviral molecules varying with a coding capacity between 50 and 79aa (Table 4 and Fig. 3). The positions of the start codons used in the majority of the ORF1 in the described subviral genomes related to amino acid 1, 108, 118, 282, 382, 451 and 490 of the putative tth25 ORF1 protein. Computer analyses also pointed to a degree of similarity of this ORF to DNA mismatch endonuclease, which may indicate an important function in the replication of the viral DNA, as well as to ubiquitin-conjugating enzyme E2 involved in location or trafficking of proteins.

TTV ORF2, similar to VP2 of CAV, possesses dual-specificity protein phosphatase (DSP) activity (Peters et al. 2002). DSP activity may contribute to the virus-induced immunosuppression of CAV. This was confirmed when single mutations in the signature motif in CAV VP2 protein led to marked reduction in proliferation and pathogenicity of the virus (Peters et al. 2005). DSPs inactivate mitogen-activated protein kinases (MAPK) (Farooq and Zhou 2004) which play a crucial role in cell proliferation. Inappropriate functions of MAPK have been identified in cancer, inflammatory disease and diabetes (Lawrence et al. 2008).

ORF2a is present as an individual ORF in the majority of TTV strains, in contrast to the prototype TTV ta278 in which it has been identified as truncated ORF2a (Tanaka et al. 2000; Okamoto et al. 2000d). In silico analysis has indicated a degree of similarity to the tubby protein.

ORF3 shares similarities to apoptin (Kooistra et al. 2004; M.H. de Smit and M. H.M. Noteborn, this volume). The apoptin function of TTVs is relatively low in contrast to CAV apoptin, which may indicate that apoptosis is not a general function of the virus protein. Interestingly, computer analysis points to a similarity between this putative protein and the movement protein (MP) of geminiviruses. Although it does have a nuclear signal, it appears to be expressed in the cytoplasm (M.H. de Smit and M.H.M. Noteborn, this volume). The MP of geminiviruses interacts with the nuclear shuttle protein (NSP) which is bound to double-stranded DNA and anchors the complex to plant plasma membranes and microsomes for cell-to-cell movement (Hehnle et al. 2004).

## *Possible Molecular Mimicry*

Cross-recognition of foreign agents and self-proteins, i.e. molecular mimicry, is a mechanism by which infectious agents can cause autoimmune diseases (Fujinami and Oldstone 1985). Amino acid homologies between cellular and virus proteins may induce this autoimmune reaction. An example where TTVs may be involved was recently demonstrated in a case of MS. In vivo clonally expanded $CD4^{+}$ T cells isolated from the cerebrospinal fluid of an MS patient responded to the poly-arginine motif in the first 76aa of TTV ORF1 (Sospedra et al. 2005). Computer analyses of the respective ORFs of TTVs and the subviral molecules indicated a degree of similarity to a series of cellular proteins.

Different regions of ORF1 protein shared signature motifs with a number of proteins known to be involved in autoimmune diseases. These include galanin, a neurotransmitter in the peripheral and central nervous system. Its action in the central nervous system stimulates feeding behaviour and release of growth hormone, whereas one of its functions in the peripheral system is inhibiting glucose-induced insulin release (Lundström et al. 2005). Neuropeptide Y receptor is another G-coupled receptor which is abundant in the mammalian brain and stimulates food intake (Gruninger et al. 2007). The autoimmune regulator (AIRE) is the protein responsible for a rare inherited disease, autoimmune polyglandular syndrome with phenotypes such as insulin-dependent diabetes mellitus, autoimmune thyroid disease and chronic gastritis (Notarangelo et al. 2006). The formation of antibodies against gliadin residues of wheat is the cause of coeliac disease or food intolerance syndrome (McGough and Cummings 2005; Amantea et al. 2006). ORF1 also shares signature motifs with the leukotriene B4 type 1 receptor. Leukotriene B4 is a chemoattractant which directs T cell migration to the airways and co-operates with chemokines in asthma (Luster and Tager 2004). It also plays a major role in chronic joint inflammation, as in rheumatoid arthritis (Mathis et al. 2007). TTV ORF3 shares signature motifs with plant lipoxygenase, which plays a role in metabolism of leukotrienes. TTVs have been implicated as players in chronic respiratory disease (Maggi et al. 2003) and arthritis (Maggi et al. 2007).

Computer analyses of the putative ORF2 protein revealed a degree of similarity to signature motifs of tubulin-tyrosine ligase, collagen and the G-coupled glucose-dependent insulinotropic polypeptide (GIP) receptor. The putative protein contains collagen triple helix repeats. Interestingly, a collagen-like protein of *Herpesvirus saimiri*, saimiri transformation-associated protein (stpC) stimulates NF-kappa-B and interleukin-2 gene expression and thereby facilitates viral induced cell transformation (Merlo and Tsygankov 2001). Tubulin-tyrosine ligase induces nitrotyrosination of alpha-tubulin which in turn induces alterations in cell morphology by changes in the microtubule organisation (Eiserich et al. 1999). The GIP-receptor is a glucose-dependent insulinotropic polypeptide important in insulin secretion and pro-insulin gene expression of pancreatic beta-cells (Volz et al. 1995). This ORF2 protein also shares similarity to orexin/hypocretin signature motifs. This cellular protein has been demonstrated as an important regulator of feeding behaviour, sleep and wakefulness and is expressed in the hypothalamus, as well as in the gastrointestinal tract where it affects insulin secretion (Heinonen et al. 2008; Sakurai 2007).

TTV ORF2a is a very interesting protein in that it shares some degree of similarity to the tubby protein. Mutations in tubby lead to obesity, retinal and cochlear degeneration (reviewed in Carroll et al. 2004). It is released from the membrane on G protein-coupled receptor activation and translocated to the nucleus where it is probably involved as transcriptional activator. G protein-coupled receptors which may act upstream of Tub are e.g. melanocortin and dopamine receptors (TTV ORF3 shares signature motifs of dopamine receptors). Tub is expressed in the brain and is tyrosine-phosphorylated in response to insulin. Dysregulation of Tub leads to altered concentrations of galanin and neuropeptide Y. A genetic interplay between Tub and microtubule-associated protein 1A has been reported.

## Conclusions

The associations with cellular proteins discussed above are vague, but it is notable that the majority of these cellular proteins play a major role in diseases involving insulin regulation, neurological functions and autoimmune disorders. The ubiquitous nature of TTVs, their remarkable susceptibility not only to intragenomic rearrangement, but also to single nucleotide modifications, and an apparent selection for these modified molecules, triggers the question whether such infections in an individual may, over time, lead to expression of proteins similar to host cellular proteins. Further investigation is needed to prove the molecular mimicry role of these viral proteins in the pathogenesis of disease.

**Acknowledgements** This study was financed in part by the Ministry of Health, Berlin and by the European Union (grant LSHC-CT-2004–503465).

## References

Amantea G, Cammarano M, Zefferino L, Martin A, Romito G, Piccirillo M, Gentile V (2006) Molecular mechanisms responsible for the involvement of tissue transglutaminase in human diseases: celiac disease. Front Biosci 11:249–255

Briddon RW, Stanley J (2006) Subviral agents associated with plant single-stranded DNA viruses. Virology 344:198–210

Carroll K, Gomez C, Shapiro L (2004) Tubby proteins: the plot thickens. Nat Rev Mol Cell Biol 5:55–63

Davidson F, MacDonald D, Mokili JL, et al (1999) Early acquisition of Tt virus (TT virus) in an area endemic for TT infection. J infect Dis 179:1070–1076

Eiserich JP, Estévez AG, Bamberg TV, et al (1999) Microtubule dysfunction by posttranslational nitrotyrosination of alpha-tubulin: a nitric oxide-dependent mechanism of cellular injury. Proc Natl Acad Sci USA 96:6365–6370

Erker JC, Leary TP, Desai SM, et al (1999) Analyses of TT virus full-length genome sequences. J Gen Virol 80:1743–1750

Farooq A, Zhou MM (2004) Structure and regulation of MAPK phosphatases. Cell Signal 16:769–779

Fujinami RS, Oldstone MB (1985) Amino acid homology between the encephalitogenic site of myelin basic protein and virus: mechanism for autoimmunity. Science 230:1043–1045

Gerner P, Oettinger R, Gerner W, et al (2000) Mother-to-infant transmission of TT virus: prevalence, extent and mechanism of vertical transmission. Pediatr Infect Dis J 19:1074–1077

Goto K, Sugiyama K, Ando T, et al (2000) Detection rates of TT virus DNA in serum of umbilical cord blood, breast milk and saliva. Tohoku J Exp Med 191:203–207

Gruninger TR, LeBoeuf B, Liu Y, Garcia LR (2007) Molecular signaling involved in regulating feeding and other motivated behaviors. Mol Neurobiol 35:1–20

Hamburgh ME, Curr KA, Monaghan M, et al (2006) Structural determinants of slippage-mediated mutations by human immunodeficiency virus type 1 reverse transcriptase. J Biol Chem 281:7421–7428

Hehnle S, Wege C, Jeske H (2004) Interaction of DNA with the movement proteins of geminiviruses revisited. J Virol 78:7698–7706

Heinonen MV, Purhonen AK, Mäkelä KA, Herzig KH (2008) Functions of orexins in peripheral tissues. Acta Physiol Oxf 192:471–485

Hino S, Miyata H (2007) Torque teno virus (TTV): current status. Rev Med Virol 17:45–57

Hirt B (1967) Selective extraction of polyoma DNA from infected mouse cell cultures. J Mol Biol 26:365–369

Jelcic I, Hotz-Wagenblatt A, Hunziker A, et al (2004) Isolation of multiple TT virus genotypes from spleen biopsy tissue from a Hodgkin's disease patient: genome reorganization and diversity in the hypervariable region. J Virol 78:7498–7507

Jeske H, Lütgemeier M, Preiß W (2001) DNA forms indicate rolling circle and recombination-dependent replication of Abutilon mosaic virus. EMBO J 20:6158–6167

Jones MS, Kapoor A, Lukashov VV, et al (2005) New DNA viruses identified in patients with acute viral infection syndrome. J Virol 79:8230–8236

Kakkola L, Tommiska J, Boele LC, et al (2007) Construction and biological activity of a full-length molecular clone of human torque teno virus (TTV) genotype 6. FEBS J 274: 4719–4730

Kamahora T, Hino S, Miyata H (2000) Three spliced mRNAs in TT virus transcribed from a plasmid containing the entire genome in COS1 cells. J Virol 74:9980–9986

Kanda Y, Tanaka Y, Kami M, et al (1999) TT virus in bone marrow transplant recipients. Blood 93:2485–2490

Kazi A, Miyata H, Kurokawa K, et al (2000) High frequency of postnatal transmission of TT virus in infancy. Arch Virol 145:535–540

Khudyakov YE, Chong M, Nichols B, et al (2000) Sequence heterogeneity of TT virus and closely related viruses. J Virol 74:2990–3000

Kooistra K, Zhang YH, Henriquez NV, Weiss B, Mumberg D, Noteborn MH (2004) TT virus-induced apoptosis-inducing protein induces apoptosis preferentially in hepatocellular carcinoma-derived cells. J Gen Virol 85:1445–1450

Lawrence MC, Jivan A, Shao C, et al (2008) The soles of MAPKs in disease. Cell Res 18:436–442

Leppik L, Gunst K, Lehtinen M, et al (2007) In vivo and in vitro intragenomic rearrangement of TT viruses. J Virol 81:9346–9356

Lundström L, Elmquist A, Bartfai T, Langel U (2005) Galanin and its receptors in neurological disorders. Neuromolecular Med 7:157–180

Luo K, He H, Liu Z, et al (2002) Novel variants related to TT virus distributed widely in China. J Med Virol 67:118–126

Luster AD, Tager AM (2004) T-cell trafficking in asthma: lipid mediators grease the way. Nat Rev Immunol 4:711–724

Maggi F, Pifferi M, Fornai C, et al (2003) TT virus in the nasal secretions of children with acute respiratory diseases: relations to viremia and disease severity. J Virol 77:2418–2425

Maggi F, Andreoli E, Riente L, et al (2007) Torquetenovirus in patients with arthritis. Rheumatology 46:885–886

Mathis S, Jala VR, Haribabu B (2007) Role of leukotriene B4 receptors in rheumatoid arthritis. Autoimmun Rev 7:12–17

Matsubara H, Michitaka K, Horiike N, et al (2001) Existence of TT virus DNA and TTV-like mini virus DNA in infant cord blood: mother-to-neonatal transmission. Hepatol Res 21:280–287

McGough N, Cummings JH (2005) Coeliac disease: a diverse clinical syndrome caused by intolerance of wheat, barley and rye. Proc Nutr Soc 64:434–450

Merlo JJ, Tsygankov AY (2001) Herpesvirus saimiri oncoproteins Tip and StpC synergistically stimulate NF-kappaB activity and interleukin-2 gene expression. Virology 279:325–338

Morrica A, Maggi F, Vatteroni ML, et al (2000) TT virus: evidence for transplacental transmission. J Infect Dis 181:803–804

Ninomiya M, Nishizawa T, Takahashi M, et al (2007a) Identification and genomic characterization of a novel human Torque teno virus of 3.2 kb. J Gen Virol 88:1939–1944

Ninomiya M, Takahashi M, Shimosegawa T, et al (2007b) Analysis of the entire genomes of fifteen Torque teno midi virus variants classifiable into a third group of Anellovirus. Arch Virol 152:1961–1975

Ninomiya M, Takahashi M, Nishizawa T, et al (2008) Development of PCR assays with nested primers specific for differential detection of three human anelloviruses and early acquisition of dual and triple infection during infancy. J Clin Microbiol 46:507–514

Nishizawa T, Okamoto H, Konishi K, et al (1997) A novel DNA virus (TTV) associated with elevated transaminase levels in posttransfusion hepatitis of unknown etiology. Biochem Biophys Res Commun 241:92–97

Nishizawa T, Okamoto H, Tsuda F, et al (1999) Quasi-species of TT virus (TTV) with sequence divergence in hypervariable regions of the capsid protein in chronic TTV infection. J Virol 73:9604–9608

Notarangelo LD, Gambineri E, Badolato R (2006) Immunodeficiencies with autoimmune consequences. Adv Immunol 89:321–370

Ohto H, Ujiie N, Takeuchi A, et al (2002) TT virus infection during childhood. Transfusion 42:892–898

Okamoto H, Takahashi M, Kato N, et al (2000a) Sequestration of TT virus of restricted genotypes in peripheral blood mononuclear cells. J Virol 74:10236–10239

Okamoto H, Takahashi M, Nishizawa T, et al (2000b) Replicative forms of TT virus DNA in bone marrow cells. Biochem Biophys Res Commun 270:657–662

Okamoto H, Tawara A, Takahashi M, et al (2000c) TT virus mRNAs detected in bone marrow cells from an infected individual. Biochem Biophys Res Commun 279:700–707

Okamoto H, Ukita M, Nishizawa T, et al (2000d) Circular double-stranded forms of TT virus DNA in liver. J Virol 74:5161–5167

Okamoto H, Nishizawa T, Takahashi M, et al (2001) Heterogeneous distribution of TT virus of distinct genotypes in multiple tissues from infected humans. Virology 288:358–368

Patil BL, Dutt N, Briddon RW, et al (2007) Deletion and recombination events between the DNA-A and DNA-B components of Indian cassava-infecting geminiviruses generate defective molecules in Nicotiana benthamiana. Virus Res 124:59–67

Peng YH, Nishizawa T, Takahashi T, et al (2002) Analysis of the entire genomes of thirteen TT virus variants classifiable into the fourth and fifth genetic groups, isolated from viremic infants. Arch Virol 147:21–41

Peters MA, Jackson DC, Crabb BS, et al (2002) Chicken anemia virus VP2 is a novel dual specificity protein phosphatase. J Biol Chem 277:39566–39573

Peters MA, Jackson DC, Crabb BS, et al (2005) Mutation of chicken anemia virus VP2 differentially affects serine/threonine and tyrosine protein phosphatase activities. J Gen Virol 86:623–630

Qiu J, Kakkola L, Cheng F, et al (2005) Human circovirus TT virus genotype 6 expresses six proteins following transfection of a full-length clone. J Virol 79:6505–6510

Saback FL, Gomes SA, de Paula VS, et al (1999) Age-specific prevalence and transmission of TT virus. J Med Virol 59:318–322

Sakurai T (2007) The neural circuit of orexin (hypocretin): maintaining sleep and wakefulness. Nat Rev Neurosci 8:171–181

Saunders K, Bedford ID, Briddon RW, et al (2000) A unique virus complex causes Ageratum yellow vein disease. Proc Natl Acad Sci USA 97:6890–6895

Settlage SB, See RG, Hanley-Bowdoin L (2005) Geminivirus C3 protein: replication enhancement and protein interactions. J Virol 79:9885–9895

Simmonds P (2004) Genetic diversity and evolution of hepatitis C virus: 15 years on. J Gen Virol 85:3173–3188

Sospedra M, Zhao Y, zur Hausen H, et al (2005) Recognition of conserved amino acid motifs of common viruses and its role in autoimmunity. PLoS Pathog 1:e41

Sugiyama K, Goto K, Ando T, et al (1999) Route of TT virus infection in children. J Med Virol 59:204–207

Tanaka Y, Orito E, Ohno T, et al (2000) Identification of a novel 23-kDa protein encoded by putative open reading frame 2 of TT virus (TTV) genotype 1 different from the other genotypes. Arch Virol 145:1385–1398

Volz A, Goke R, Lankat-Buttgereit B, et al (1995) Molecular cloning, functional expression, and signal transduction of GIP-receptor cloned from a human insulinoma. FEBS Lett 373:23–29

Xiu X, Xiaoguang Z, Ninghu Z, et al (2004) Mother-to-infant vertical transmission of transfusion transmitted virus in South China. J Perinat Med 32:404–406

Yamaguchi S, Imada T, Kaji N, et al (2001) Identification of genetic determinant of pathogenicity in chicken anaemia virus. J Gen Virol 82:1233–1238

Yokoyama H, Yasuda J, Okamoto H, et al (2002) Pathological changes of renal epithelial cells in mice transgenic for the TT virus ORF1 gene. J Gen Virol 83:141–150

Zhong S, Yeo M, Tang M, et al (2002) Frequent detection of the replicative form of TT virus DNA in peripheral blood mononuclear cells and bone marrow cells in cancer patients. J Med Virol 66:428–434

# TT Viruses: Oncogenic or Tumor-Suppressive Properties?

**H. zur Hausen(✉) and E.-M. de Villiers**

**Contents**



**Abstract** Torque teno (TT) viruses have been more frequently reported in malignant biopsies when compared to normal control tissue. The possible contribution of TT virus infection to human carcinogenesis or the potential oncolytic functions of these virus infections are being discussed based on available experimental evidence. The data could suggest an involvement of TT virus infections as an indirect carcinogen by modulating T cell immune responses. Significant oncolytic functions, potentially mediated by the inhibition of nuclear factor (NF)-κB transcription factor or by apoptin-like gene activities, are emerging to be less likely.

## Introduction

As outlined in the previous chapters, Torque teno (TT) viruses are ubiquitous, occurring in all human populations, but a number of different genotypes have also been observed in nonhuman primates, cats, dogs, and pigs (see H. Okamoto, this volume). Previous publications from our own and other groups noted a higher rate of TT virus positivity in different types of cancer cells in comparison to noncancerous tissues (de Villiers et al. 2002, 2007; Camci et al. 2002; Zhong et al. 2002; Jelcic et al. 2004; Girard et al. 2007). This raised the suspicion that this infection may

H. zur Hausen
Deutsches Krebsforschungszentrum, Im Neuenheimer Feld 280,69120 Heidelberg, Germany
zurhausen@dkfz-heidelberg.de

E.-M. de Villiers, H. zur Hausen (eds.), *TT Viruses: The Still Elusive Human Pathogens,* 

somehow contribute to human cancer development. Pre- and perinatal infections with TT-like viruses were discussed in their possible relationship to childhood leukemias and lymphomas (zur Hausen and de Villiers 2005). A model had been developed postulating a role of the viral load in pre- or perinatal infections as risk factor for these hematopoietic malignancies. This was based on the hypothesis that these infections result in adaptive immunotolerance while still responding to innate immune functions. The frequently observed protective effect of multiple infections within the first year of life for leukemia development was interpreted to be due to the reduction of viral load by intermittent infection-induced interferon synthesis. Although the basic features of this model would also fit to other pre- or perinatal infections (e.g., reactivation of endogenous retroviruses or herpesvirus infections), the occasionally recorded absence of an adaptive immune response to TT virus infections (Tsuda et al. 1999; Handa et al. 2000) and the reported responsiveness to interferon treatment (Chayama et al. 1999; Toyoda et al. 1999; Nishizawa et al. 2000; Maggi et al. 2001) resulted in a preferential consideration of TT virus-like agents as candidates for these malignancies. The hypervariable region in the large open reading frame of TT viruses could provide one mechanism by which these viruses may escape from adaptive immune surveillance (Nishizawa et al. 1999; Jelcic et al. 2004).

This review attempts to summarize the potential role of TT virus infections as indirect carcinogens and, at the same time, discusses possible oncosuppressive functions of these infections.

## TT Virus Infections as Indirect Carcinogens?

Tumor viruses contribute to human malignancies in different ways. They have been characterized as making direct or indirect contributions to tumorigenesis (zur Hausen 2006). Direct modes of interaction imply the persistence and expression of viral oncogenes within the respective tumor cells where they act as essential components for the maintenance of the malignant phenotype. Alternatively, integration of viral DNA into host cell DNA may lead to activation of previously silent oncogenes or to the functional destruction of effective tumor-suppressor genes. In both situations long term persistence of viral DNA in each individual tumor cell seems to be the requirement for a direct oncogenic effect.

As far as TT viruses are concerned, current evidence for a direct carcinogenic effect is missing. No oncogenic functions of specific TT virus genes have yet been defined, nor does definite proof exist of viral DNA integration into the host cell genome. Studies analyzing such potential events are limited and further investigations are desirable along such lines.

There exist, however, some suggestions that TT virus infection may contribute to human carcinogenesis by an indirect mode. Indirect contributions to carcinogenesis by infectious agents could be due to a number of different interactions: in human carcinogenesis the best-documented mode is the induction of immunosuppression by human immunodeficiency virus (HIV) infections. Many of the subsequently arising tumors are caused by other tumor virus infections, such as Epstein-Barr virus, human herpesvirus type 8, high-risk papillomaviruses (zur Hausen 2006) and probably also by the recently identified human Merkel cell polyomavirus, MCPyV (Feng et al. 2008). Failure of immunosurveillance mechanisms is in such cases responsible for the nonelimination and outgrowth of virus-transformed cell clones.

A second, probably also very important mode of indirect contribution of infections to carcinogenesis is the induction of chronic inflammatory events, concomitant induction of oxygen radicals and nitric oxide synthase (NOS), resulting in mutational modifications of the host cell genome and activation of the nuclear factor (NF)-κB signaling pathway. This seems to represent the most common contribution of hepatitis B and C viruses, *Helicobacter pylori* infections of the stomach, and parasitic infections (e.g., *Schistosoma haematobium* to bladder cancer and *Clonorchis viverinni* to cholangiocarcinomas; reviewed in zur Hausen 2006).

A potentially additional indirect contribution to carcinogenesis is the prevention of apoptosis after DNA-damaging events. This has been postulated to play a role in the possible involvement of cutaneous papillomavirus infections in squamous cell carcinomas of the skin at sun-exposed skin sites (Jackson and Storey 2000).

Other indirect contributions to cancer development include the induction of chromosomal aberrations and DNA mutations in the host cell genome by infections and the amplification of silently persisting tumor virus genomes (papilloma- or polyomavirus DNA) by herpes-, adeno-, or poxvirus infection, although no evidence has been found thus far for the involvement of these factors in human cancers (reviewed in zur Hausen 2006).

Do TT virus infections contribute by any of these modes to human carcinogenesis? Replicative forms of TT virus DNA have been observed in human bone marrow cells (Okamoto et al. 2000a, b; Zhong et al. 2002; Mariscal et al. 2002; Desai et al. 2005; Bendinelli and Maggi 2005). During ablative myelosuppression, pre-existing TT virus titers decreased to undetectable (Kanda et al. 1999), pointing in addition to an important role of bone marrow cells in TT virus replication. A high rate of TT virus replication within these cells may influence the function of their differentiated progeny. There exist some hints for a TT virus contribution to immunosuppression. It has been reported that the TT virus load is inversely correlated with the percentages of circulating total and helper T lymphocytes (Maggi et al. 2003). Several studies in HIV-infected patients also published a correlation between the degree of immunosuppression and the TT virus load (Christensen et al. 2000; Touinssi et al. 2001; Shibayama et al. 2001; Sherman et al. 2001; Thom and Petrik 2007). TT viruses seem to infect CD4 as well as CD8 cells (Maggi et al. 2003). A reduction of T lymphocytes has been noted in patients with a high TT virus load (Sagir et al. 2005). HIV patients with high TT virus loads reveal an average reduced survival of CD4 lymphocytes when compared to similar patients with a low TT virus load (Maggi et al. 2008). In terminally ill AIDS patients, exceptionally high TT virus loads have been noted in the bone marrow (Simmonds 2002). In addition, high-dose chemotherapy and hematopoietic stem cell transplantation result in increase in

CD8$^{+}$57$^{+}$ T lymphocytes and in a substantial increases of TT virus viremia (Maggi et al. 2008). This is also observed in other conditions with immune cell activation and immunological dysfunction (Gutierrez et al. 2003; Alvaro et al. 2006; Focosi and Petrini 2007). Kidney transplant patients receiving cyclosporine and corticosteroids have revealed a remarkable increase in TT virus load (Moen et al. 2003).

Taken together these data may point to a role of TT virus infections as immunomodulators and contributing to immunosuppression. They could also provide an explanation why TT virus detection in malignant tumors exceeds that of normal control tissue (de Villiers et al. 2002, 2007a; Camci et al. 2002; Zhong et al. 2002; Jelcic et al. 2004; Girard et al. 2007). In particular, tumors associated with inflammatory infiltrates seem to provide an excellent target for TT virus replication and may simultaneously lead to a dysfunction of the immune reactivity, as observed in Hodgkin's disease (reviewed in Roullet and Bagg 2007). Indeed, up to 25 different TT virus genotypes have been demonstrated in one heavily infiltrated spleen biopsy from a patient with Hodgkin's disease (Jelcic et al. 2004). Thus, a high TT virus load in malignant tissue may permit a more efficient escape of tumor cells from immune surveillance mechanisms. A role of specific TT virus intragenomic recombinants may in addition deserve attention as a contributing factor (Leppik et al. 2007). Clearly this hypothesis requires further investigation. If confirmed it would point to an important role of TT virus infections in the disturbance of immune regulatory mechanisms controlling tumor growth and classify these infections as indirect carcinogens. Another important question that needs to be resolved concerns the influence of TT virus infection on differentiation of lymphatic cell pathways. Based on available data one could speculate that this infection may have a negative effect on CD4 cell differentiation. This may underline the requirement of studies investigating TT virus replication in thymus cells and lymphoid organs.

## Tumor-Suppressive Properties of TT Viruses

A number of small single-stranded DNA viruses have been identified possessing tumor-suppressive properties. In particular, several types of helper-independent and helper-dependent parvoviruses reveal oncolytic or tumor-suppressive properties. This accounts in particular for the murine viruses H-1 and the minute virus of mice (MVM) and human adeno-associated viruses (see review Raykov and Rommelaere 2008). Both murine viruses preferentially replicate in malignant cells, resulting in a lytic response and tumor destruction, whereas normal cells remain unaffected. Adeno-associated viruses, in contrast, do not lytically infect cells. They require helper viruses for their own effective replication. Apparently, however, a protein product of their Rep gene interferes with the proliferation of malignant cells.

Two recent discoveries could point to a tumor-suppressive or oncolytic function of at least several types of TT viruses. The description of an apoptin-like protein coded for by specific TT virus types (Kooistra et al. 2004), and functionally similar to apoptin of chicken anemia virus (see M.H. de Smit and M.H.M. Noteborn, this volume), may point to a negative interference of TT virus infections with tumor growth. It has, however, been noted that the apoptotic property of TT virus apoptin is lower when compared to chicken anemia virus and seems to be most pronounced in hepatocellular carcinoma cells only. Based on these data it is somewhat unlikely that these TT virus infections significantly interfere with tumor growth.

The second hint of a potential antitumor activity with TT virus infections originates from observations of an inhibitory effect of these infections on NF-κB expression (Zheng et al. 2007). NF-κB signaling is elevated in the vast majority of malignant proliferations (see review Naugler and Karin 2008). Its inhibition results regularly in a negative effect on tumor proliferation. Thus, in this respect TT virus infection may contribute to decreased tumor growth.

It should be noted, however, that all inflammatory reactions are accompanied by high NF-κB activity. In view of the preferential presence of TT viruses in lymphatic tissues (Maggi et al. 2003), this seems to rather support an opposite function of TT virus infection: the interference with lymphocyte functions, specifically with the T cell compartment (F. Maggi and M. Bendinelli, this volume; Maggi et al. 2008), should render inflammatory reactions less effective in controlling tumor growth.

Thus, the available data, even though they are scarce, rather support an indirect tumor-growth stimulating role of TT virus infection than a significant oncosuppressive function. Future studies, however, will have to clarify the accurate role of these viruses in malignant proliferations.

## The Target Cell Conditioning Concept Revisited

In 2005 we published the virus target cell conditioning model to explain some epidemiologic characteristics of childhood leukemias and lymphomas (zur Hausen and de Villiers 2005; zur Hausen 2006). A high load with a persisting pre-or perinatally infecting ubiquitous virus was hypothesized to represent the major risk factor for childhood leukemias and lymphomas. The respective infection should lead to tolerance against adaptive immune reactions by retaining susceptibility to innate immune functions. High levels of the persisting initial infection would be reduced by intermittent other infections resulting in interferon induction and, as a consequence, in the reduction of the load of the persisting virus and concomitantly to a reduction of the leukemogenic or lymphomagenic risk. TT virus-like infections had been proposed as possible candidates, since early infections occur relatively frequently in the perinatal period (Bagaglio et al. 2002; Xin et al. 2004; Indolfi et al. 2007) and TT virus titers have been shown to be reduced upon interferon treatment (Chayama et al. 1999; Toyoda et al. 1999; Nishizawa et al. 2000; Maggi et al. 2001; E.-M. de Villiers et al., this volume). Although fitting by these criteria, TT viruses do not represent the sole infections that could be accommodated by this model. Prenatal herpesvirus infections, e.g., human herpesvirus type 6 or cytomegalovirus, human endogenous retroviruses, or polyomaviruses could be considered as well.

In the light of the preceding discussion, as far as TT virus-like infections are concerned, the model seems to require a modification: a direct carcinogenic function of genes of TT virus-like agents becomes increasingly unlikely. An indirect role of such infections, however, remains an interesting possibility. The negative interference with immune functions of the infected host could substantially increase the risk for the clonal outgrowth of modified cells. Whether such modification rests on a different persistent infection or relates to specific chromosomal aberrations, or depends on two such events, remains to be elucidated. The observed higher rate of TT virus positivity in human leukemias (de Villiers et al. 2002), may support the view of an indirect contribution of these infections to leukemogenesis.

## Conclusions

Available data seem to support the concept that TT virus infections may contribute indirectly to carcinogenesis by negative interference with cell-mediated immune functions. Results pointing to potential inhibitory functions of these virus infections on tumor growth, relying as they do on the reported inhibition of NF-κB activity or the production of apoptin-like TT virus proteins, increasingly appear unlikely to contribute to significant oncolytic effects via members of this virus family.

## References

Alvaro T, Lejeune M, Salvado MT, et al (2006) Immunohistochemical patterns of reactive microenvironment are associated with clinicobiologic behaviour in follicular lymphoma patients. J Clin Oncol 25:5350–5357

Bagaglio S, Sitia G, Prati D, Cella D, Hasson H, Novati R, Lazzarin A, Morsica G (2002) Mother-to-child transmission of TT virus: sequence analysis of non-coding region of TT virus in infected mother-infant pairs. Arch Virol 147:803–812

Bendinelli M, Maggi F (2005) TT virus and other anelloviruses. In: Mahy B, ter Meulen V (eds) Topley and Wilson: microbiology and microbial infections. Arnold Press, London

Camci C, Guney C, Balkan A, et al (2002) The prevalence of TT virus in cancer patients. New Microbiol 25:463–468

Chayama K, Kobayashi M, Tsubota A, et al (1999) Susceptibility of TT virus to interferon therapy. J Gen Virol 80:631–634

Christensen JK, Eugen-Olsen J, Sorensen M, et al (2000) Prevalence and prognostic significance of infection with TT virus in patients infected with human immunodeficiency virus. J Infect Dis 181:1796–1799

de Villiers EM, Schmidt R, Delius H, et al (2002) Heterogeneity of TT virus related sequences isolated from human tumour biopsy specimens. J Mol Med 80:44–50

de Villiers EM, Bulajic M, Nitsch C, et al (2007) TTV infection in colorectal cancer tissues and normal mucosa. Int J Cancer 121:2109–2112

Desai M, Pal R, Deshmukh R, et al (2005) Replication of TT virus in hepatocyte and leucocyte cell lines. J Med Virol 77:136–143

Feng H, Shuda M, Chang Y, et al (2008) Clonal integration of a polyomavirus in human Merkel cell carcinoma. Science 319:1096–1100

Focosi D, Petrini M (2007) CD57 expression on lymphoma microenvironment as a new prognostic marker to immune dysfunction. J Clin Oncol 25:1289–1291

Girard C, Ottomani L, Ducos J, et al (2007) High prevalence of Torque teno (TT) virus in classical Kaposi's sarcoma. Acta Derm Venereol 87:14–17

Gutierrez A, Munoz I, Solano C, et al (2003) Reconstitution of lymphocyte populations and cytomegalovirus viremia or disease after allogeneic peripheral blood stem cell transplantation. J Med Virol 70:399–403

Handa A, Dickstein B, Young NS, et al (2000) Prevalence of the newly described human circovirus, TTV, in United States blood donors. Transfusion 40:245–251

Indolfi G, Moriondo M, Galli L, et al (2007) Mother-to-infant transmission of multiple blood-borne viral infections from multi-infected mothers. J Med Virol 79:743–747

Jackson S, Storey A (2000) E6 proteins from diverse cutaneous HPV types inhibit apoptosis in response to UV damage. Oncogene 19:592–598

Jelcic I, Hotz-Wagenblatt A, Hunzicker A, et al (2004) Isolation of multiple TT virus genotypes from spleen biopsy tissue from a Hodgkin's disease patient: genome reorganization and diversity of the hypervariable region. J Virol 78:7498–7507

Kanda Y, Tanaka Y, Kami M, et al (1999) TT virus in bone marrow transplant recipients. Blood 93:2485–2490

Kooistra K, Zhang YH, Henriquez NV, et al (2004) TT virus-derived apoptosis inducing protein induces apoptosis preferentially in hepatocellular carcinoma cell lines. J Gen Virol 85:1445–1450

Leppik L, Gunst K, Lehtinen M, Dillner J, Streker K, de Villiers EM (2007) In vivo and in vitro intragenomic rearrangement of TT viruses. J Virol 81:9346–9356

Maggi F, Pistello M, Vatteroni M, et al (2001) Dynamics of persistent TT virus infection, as determined in patients treated with alpha interferon for concomitant hepatitis C virus infection. J Virol 75:11999–12004

Maggi F, Pifferi M, Tempestini E, et al (2003) TT virus loads and lymphocyte subpopulations in children with acute respiratory diseases. J Virol 77:9081–9083

Maggi F, Focosi D, Ricci V, et al (2008) Changes in $CD8^{+}57^{+}$ T lymphocyte expansions after autologous hematopoietic stem cell transplantation correlate with changes in torquetenovirus viremia. Transplantation (in press)

Mariscal LF, Lopez-Alcorocho JM, Rodriguez-Inigo E, et al (2002) TT virus replicates in stimulated but not in nonstimulated peripheral blood mononuclear cells. Virology 301:121–129

Moen EM, Sagedal S, Bjoro K (2003) Effect of immune modulation on TT virus (TTV) and TT-like-mini-virus (TMLV) viremia. J Med Virol 70:177–182

Naugler WE, Karin M (2008) NF-kappaB and cancer-identifying targets and mechanisms. Curr Opin Genet Dev 18:19–26

Nishizawa T, Okamoto H, Tsuda F, et al (1999) Quasispecies of TT virus (TTV) with sequence divergence in hypervariable regions of the capsid protein in chronic TTV infection. J Virol 73:9604–9608

Nishizawa Y, Tanaka E, Orii K, et al (2000) Clinical impact of genotype 1 TT virus infection in patients with chronic hepatitis C and response of TT virus to alpha-interferon. J Gastroenterol Hepatol 15:1292–1297

Okamoto H, Nishizawa C, Tawara A, et al (2000a) TT virus mRNAs detected in the bone marrow cells from an infected individual. Biochem Biophys Res Commun 279:700–707

Okamoto H, Takahashi M, Nishizawa C, et al (2000b) Replicative forms of TT virus DNA in bone marrow cells. Biochem Biophys Res Commun 270:657–662

Raykov Z, Rommelaere J (2008) Potential of tumour cells for delivering oncolytic viruses. Gene Ther 15:704–710

Roullet MR, Bagg A (2007) Recent insights into the biology of Hodgkin lymphoma: unraveling the mysteries of the Reed-Sternberg cell. Expert Rev Mol Diagn 7:805–820

Sagir A, Adams O, Oette M, et al (2005) SEN virus seroprevalence in HIV positive patients: association with immunosuppression and HIV replication. J Clin Virol 33:183–187

Sherman KE, Rousters SD, Feinberg J (2001) Prevalence and genotypic variability of TTV in HIV-infected patients. Dig Dis Sci 46:2401–2407

Shibayama T, Masuda G, Ajisawa A, et al (2001) Inverse relationship between the titre of TT virus DNA and CD4 cell count in patients infected with HIV. AIDS 15:563–570

Simmonds P (2002) TT virus infection: a novel virus-host relationship. J Med Microbiol 51:455–458

Thom K, Petrik J (2007) Progression towards AIDS leads to increased torque teno virus and torque teno minivirus titers in HIV infected individuals. J Med Virol 79:1–7

Touinssi M, Gallian P, Biagini P, et al (2001) TT virus infection: prevalence of elevated viraemia and arguments for the immune control of viral load. J Clin Virol 21:135–141

Toyoda H, Fukuda Y, Yokozaki S, et al (1999) Interferon treatment of two patients with quadruple infection with hepatitis C virus (HCV), human immunodeficiency virus (HIV), hepatitis G virus (HGV), and TT virus (TTV). Liver 19:438–444

Tsuda S, Okamoto H, Ukita M, et al (1999) Determination of antibodies to TT virus (TTV) and application to blood donors and patients with post-transfusion non-A to G hepatitis in Japan. J Virol Methods 77:199–206

Xin X, Xiaoguang Z, Ninghu Z, et al (2004) Mother-to-infant vertical transmission of transfusion transmitted virus in South China. J Perinat Med 32:404–406

Zheng H, Ye L, Fang X, et al (2007) Torque teno virus (SANBAN isolate) ORF2 protein suppresses NF-kappaB pathways via interaction with IkappaB kinases. J Virol 81:11917–11924

Zhong S, Yeo W, Tang M, et al (2002) Frequent detection of the replicative form of TT virus DNA in peripheral blood mononuclear cells and bone marrow cells in cancer patients. J Med Virol 66:428–434

zur Hausen H (2006) Infections causing human cancer. Wiley-VCH, Weinheim, New York

zur Hausen H, de Villiers EM (2005) Virus target cell conditioning model to explain some epidemiologic characteristics of childhood leukemias and lymphomas. Int J Cancer 115:1–5

# Relationship of Torque Teno Virus to Chicken Anemia Virus

**S. Hino(✉) and A.A. Prasetyo**

**Contents**



**Abstract** This chapter examines the correlation between Torque teno virus (TTV) and chicken anemia virus (CAV). Each has a circular single-stranded (ss)DNA genome with every one of its known open reading frames (ORF) on its antigenomic strand. This structure is distinct from those of circoviruses. The genomic sizes of TTV and CAV are different, 3.8 kb and 2.3 kb, respectively. While the spectrum of the TTV genome is enormously diverse, that of the CAV genome is quite narrow. Although a 36-nt stretch near the replication origin of TA278 TTV possesses more than 80% similarity to that of CAV, the sequence of the other genomic regions does not exhibit a significant similarity. Nevertheless, the relative allocation of ORFs on each frame in these viruses mimics each other. Three or more messenger RNA

S. Hino, A.A. Prasetyo
Division of Virology, Faculty of Medicine, Tottori University, Yonago, 683-8503, Japan
Current address of SH: Medical Scanning, 4-3-1F, Kanda-Surugadai, Chiyoda, Tokyo 101-0062, Japan
shg.hino@gmail.com

E.-M. de Villiers, H. zur Hausen (eds.), *TT Viruses: The Still Elusive Human Pathogens*, (mRNAs) are generated by transcription in both of them. The structural protein with the replicase domain is coded for by frame 1 in each virus, and a nonstructural protein with a phosphatase domain is coded for by frame 2. A protein on frame 3 in each virus induces apoptosis in transformed cells. Recently, we confirmed that apoptin is necessary for the replication of CAV. TTV has been proposed to constitute a new family, Anelloviridae. Considering these similarities and dissimilarities between CAV and TTV, it seems more reasonable to place CAV, the only member of genus *Gyrovirus*, into *Anelloviridae* together with TTV, or into a new independent family.

## Torque Teno Virus

### *History of TTV*

Torque teno virus (TTV) was found in 1997 by differential display technology in serum from a hepatitis patient with the initials T.T. who did not possess any markers of known hepatitis viruses (Nishizawa et al. 1997). Some researchers refer to TTV as "transfusion-transmitted virus," even though it has been clearly stated in the literature that the acronym TTV is based on the initials of the index patient (Takahashi et al. 2000). The virus is nonenveloped and its genome is composed of a circular, 3.8-kb ssDNA (Miyata et al. 1999; Mushahwar et al. 1999; Fig. 1). It was the first human virus with a circular ssDNA genome, and was arbitrary placed in the Circoviridae family (Miyata et al. 1999).

Although the virus was originally introduced as a novel hepatitis virus, the evidence has not been available for its causative role in any of the significant pathological entities (Hino and Miyata 2007). The vast majority of virological efforts have been focused on the extraordinary diverse spectra of the TTV genome and its ubiquitous molecular epidemiology (Peng et al. 2002). Over 10 years have passed since the discovery of TTV, but the nature of this virus still remains unclear. The major obstacles for investigating the biology of TTV include its indistinct pathogenicity, enormous diversity in the spectrum of TTVs, extreme ubiquity when all of the TTVs are combined, and lack of a suitable system to replicate TTV.

### *Classification of TTV*

Although the virus possessing a circular ssDNA genome was a rookie in the world of human viruses, those members have been well known for some time in the society of bacterial, plant, and animal viruses. Due to the recent accumulation of knowledge of these viruses, their taxonomic groups have been reformed recently, e.g., the bacterial viruses Inoviridae and Microviridae, the plant viruses Geminiviridae and Nanoviridae, and the animal viruses Circoviridae (International Committee on Taxonomy of Viruses 2007; Table 1). Members of Geminiviridae and Nanoviridae

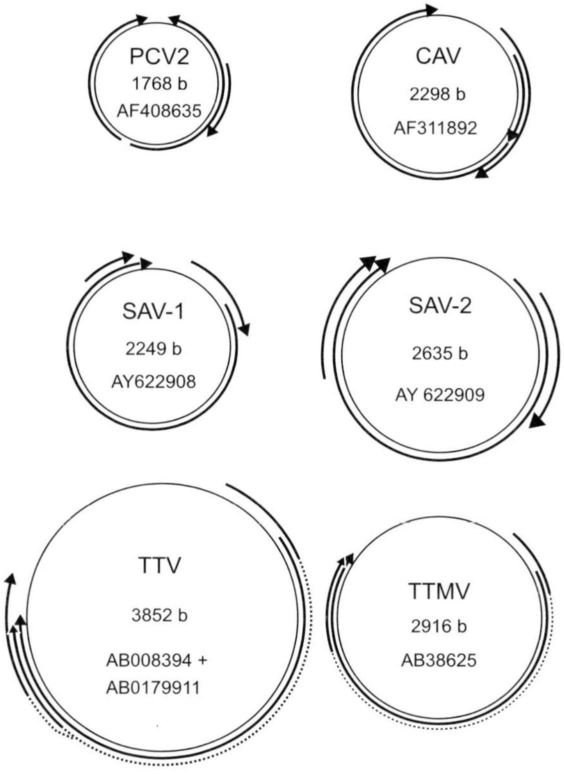


**Fig. 1** Strands of TTV, Torque teno mini virus (TTMV), and small anelloviruses (SAVs) in comparison to CAV (genus *Gyrovirus*) and porcine circovirus type 2 (PCV-2, genus *Circovirus*). The *sizes* of the circles correspond to the genome size of each virus. *Clockwise arrows*, translations on the antigenomic strand; *counter clockwise arrows*, those on the genomic strand

have more than one piece of circular DNA. However, a single circular component of geminivirus was shown to be necessary and sufficient for infection (Dry et al. 1993; Lazarowitz 1988).

Within the Circoviridae, the genus *Circovirus* includes animal viruses, such as the porcine circovirus (PCV), psittacine beak and feather disease virus (PBFDV). The genome of *Circovirus* is ambisense and has a common 9-nt stem-loop structure at its replication origin. The largest Rep protein of *Circovirus* is coded for by the genomic strand, while the other two proteins are coded for by the antigenomic strand (Niagro et al. 1998; Fig. 1). The genus *Gyrovirus* contains only one member, chicken anemia virus (CAV), which has a negative strand circular DNA distinct from that of the other members in the Circoviridae. Plant viruses, such as banana bunchy top virus (BBTV) formally classified in the genus *Circovirus*, have been removed from the Circoviridae and placed into the Nanoviridae, a new family with a genome consisting of six segments of sense-strand circular DNA (International Committee on Taxonomy of Viruses 2007; Table 1).

**Table 1** Viruses with a single-stranded circular DNA genome

| Family | Genus | Representative viruses | Strandedness | No. of circles |
|---|---|---|---|---|
| Unassigned[a] | *Anellovirus* | TT virus (human, nonhuman primates, and other animals), TTV-like mini-virus (TTMV), small anellovirus (SAV) | Antisense | 1 |
| Circoviridae | *Circovirus* | Porcine circovirus (PCV); psittacine beak and feather disease virus (BFDV); pigeon circovirus (PiCV); goose circovirus (GoCV); canary circovirus (CaCV) | Ambisense | 1 |
| | *Gyrovirus* | Chicken anemia virus (CAV) | Antisense | 1 |
| Geminiviridae | *Begomovirus*; *Curtovirus*; *Mastrevirus*; *Topocuvirus*; unclassified Geminiviridae | Maize streak virus (plant) | Ambisense | 1–2 |
| Inoviridae | *Inovirus*; *Plectrovirus* | (Bacteria) | Sense | 1 |
| Microviridae | *Chlamydiamicrovirus*; *Microvirus*; *Spiromicrovirus* | jX174 (Bacteria) | Sense | 1 |
| Nanoviridae | *Babuvirus*; *Nanovirus* | Banana bunchy top virus (BBTV); coconut foliar decay virus (plant) | Sense | 6 |

[a] A new family, Anelloviridae, is proposed, and the National Center for Biotechnology Information has incorporated the family name into the taxonomic table

## *TTV and TTMV*

One-third of the TTV genome is a untranslated region (UTR), and has a high degree of similarity within extremely divergent TTVs (Takahashi et al. 1998). The 36-nt region (nt 3816–3851 of TA278) within the 113-nt GC-rich stretch and its immediate vicinity constitutes a stem-loop structure, serving as the origin of DNA replication (Mushahwar et al. 1999; Okamoto et al. 1999). Moreover, multiple transcription modifier motifs, such as ATF/CREB, AP-2, SP-1, and NF-κB binding sites, are found in this region (Miyata et al. 1999).

Although the genomic sizes of TTV and Torque teno mini virus (TTMV) are different, they share several common features (Fig. 2). The UTRs occupying approximately one-third of the genome inclusive of the GC-rich stretch are similar to each other. The structures of the coding regions in these two virus groups are also similar to each other (Fig. 2). The largest, ORF1, which accounts for approximately

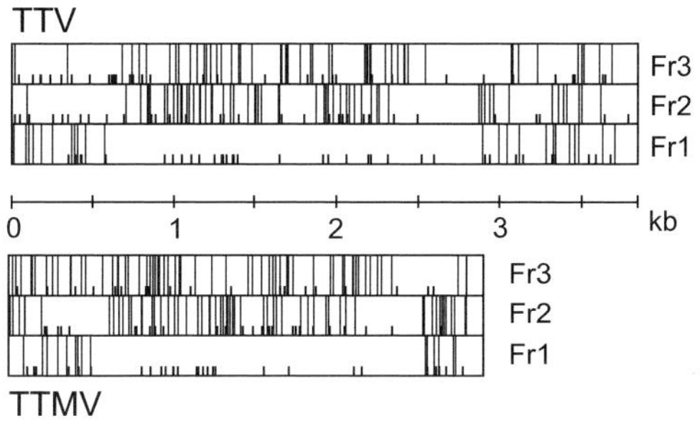


**Fig. 2** Frames and ORF structures of TTV (AB008394+AB017911) and TTMV (AB38625). *Short vertical lines* represent ATG while *long vertical lines* correspond to stop codons

two-thirds of the viral genome, is also common. A single host can be co-infected with TTV and TTMV. However, the replication of these virus genomes should be independent, because double infection is not a prerequisite. They transcribe a long mRNA coding for the ORF1, using a single TATA box and a polyA signal. Two additional spliced mRNAs are transcribed using the same TATA box and polyA signal as the long mRNA (Kamahora et al. 2000; Okamoto et al. 2001).

## *Classification of TTV*

The current spelled-out names for TTV and TTV-like mini virus (TLMV) (Takahashi et al. 2000) were approved as Torque (necklace) teno (thin) virus and Torque teno mini virus (TTMV) (International Committee on Taxonomy of Viruses 2007). TTV is currently classified into an unassigned genus *Anellovirus* (ring) within ssDNA viruses. However, moving it up to a family Anelloviridae has been proposed [ICTV Study Group 2007 (for details, see the chapter by P. Biagini, this volume); Table 1]. Recently isolated 3.2-kb TTVs (TTMDV) exhibiting a 76%–99% identity with SAV (2.2-kb) genomes (Jones et al. 2005) were phylogenetically distinguishable from reported TTVs and TTMVs. SAVs were also proposed as deletion mutants of TTMDVs (Ninomiya et al. 2007).

## *Transcriptional Control of TTV*

Kamada et al. (2004) reported that the region −154/−76 contains the critical regulatory element for the functioning of the TA278 TTV (group 1) promoter, as deduced by dual luciferase assays (Fig. 3). Suzuki et al. (2004) observed similar results with the SANBAN TTV (group 3). Interestingly, the group 1 promoter was more active in HepG2 cells than Huh7, while preference of the group 3 promoter was the reverse, possibly suggesting distinct cell tropisms for these two TTVs.

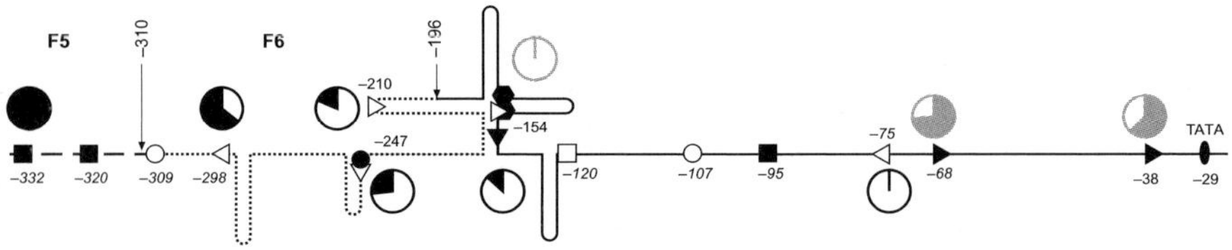


**Fig. 3** Enhancer and promoter region of TTV in relation to the tertiary structure in the NCR. Binding sites for transcription factors conserved in various TTVs (*open circle*, CREB; *closed circle*, E2F; *open square*, c-Ets; *closed square*, USF; *hexagon*, SP-1), and their positions (in *italics*) are shown. *Open triangles* indicate the boundary of 5′-deletion mutants, and *closed triangles* indicate the boundary of 3′-deletion mutants. Their luciferase signals relative to that of the pGL3B (1.2-kb NCR) construct are shown on the graphs as *circles* (*black* for 5′-deletions, and *gray* for 3′-deletions). The initiation site (nt 114) is denoted as +1

## *Proteins of TTV*

On the antigenomic strand of TTV, an ORF resides on each frame starting in the upstream region (Fig. 4). A TATA box is located at nt 86–90 (TA278) just upstream of these ORFs. A polyA signal (nt 3073–3079) follows the downstream end of the largest ORF on the frame 1. Upstream of this polyA signal, two relatively large ORFs are located on the 1.2-kb and 1.0-kb mRNAs, respectively (Fig. 4). However, no legitimate signals for transcription could be found in the near upstream of these ORFs.

Kamahora et al. (2000) found three species of mRNA in COS1 cells transfected with a plasmid containing the promoter region through the polyA signal (nt 2762–3852 connected tandem with nt 1–3770; Fig. 4). Each mRNA, at 3.0 kb, 1.2 kb, and 1.0 kb, respectively, had the 5′-terminus at nt 114 adjacent to the TATA box and the 3′-terminus at nt 3087, 6–9 nt downstream of the polyA signal. All the three species of mRNA exhibited a common splicing at nt 186–276 (Fig. 4).

The largest ORF on the 3.0-kb mRNA starts with an initiation codon at $A^{589}$TG and codes for the Rep protein with 770 amino acids (aa) (Fig. 4). The same 3.0-kb mRNA can also code for two additional small proteins, VP2 and VP3 starting from nt 353 and 372 on frame 2 and 3, respectively. The second splicing of the 1.2-kb mRNA connects $ORF1^{N}$ with $ORF1^{1.2}$ and $ORF2^{N}$ with $ORF2^{1.2}$ without frame-shift. The second splicing of the 1.0-kb mRNA causes frame-shift connecting $ORF1^{N}$ with $ORF2^{1.0}$ and $ORF2^{N}$ with $ORF3^{1.0}$. The ORF for VP3 is also open in both 1.2-kb and 1.0-kb mRNAs. Though these three mRNAs potentially code for seven different proteins, the real role of these proteins in TTV replication has to be elucidated (Kooistra et al. 2004; Qiu et al. 2005). The nomenclature for ORFs in this chapter is modified from the system used in Miyata et al. (1999) and Kamahora et al. (2000), trying to accommodate connecting ORFs developed by splicing. The system would explain the similarity of TTV and CAV more easily than the system used by others.

The ORF1 protein probably represents the major structural protein of TTV, since it contains a highly basic stretch in its N-terminus and several conserved Rep protein motifs (FTL and YXXK) consistent with other circoviral capsid proteins (Niagro et al. 1998; Takahashi et al. 1998). The posttranslational modification should be inves-

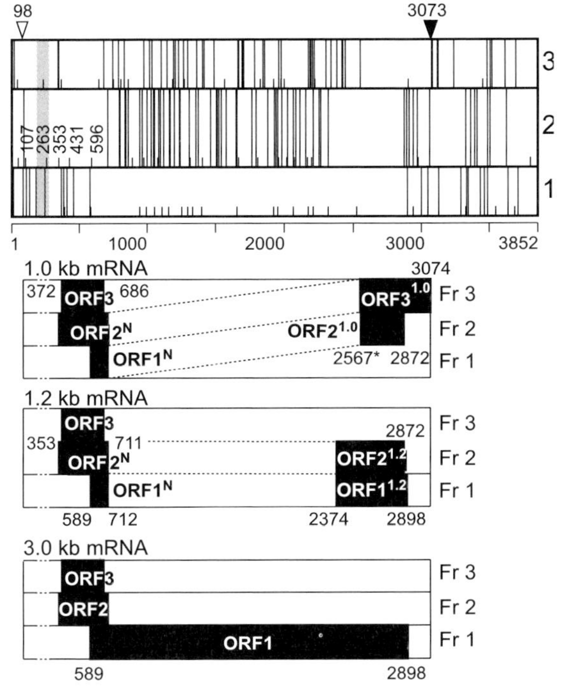


**Fig. 4** Schematic diagram of the TTV genome (AB008394+AB017911) and its mRNAs. Three reading frames of the genome are shown in the *upper panel*. The *open triangle* indicates the position of the cap site while the *closed triangle* indicates that of the polyA signal. The *short* and *long vertical lines* indicate ATGs and stop codons, respectively. The *shaded area* represents the first splicing common to all mRNAs of three different sizes. The *lower panel* indicates the frames used and configurations of 3.0-kb, 1.2-kb, and 1.0-kb mRNAs. The *solid lines* indicate exons, the *dotted lines* represent introns, and the *boxes* indicate coding regions. Because of the alternative splicing for the 1.0-kb mRNA, the 5′-terminal nt 2567 is labeled with an *asterisk*

tigated further. Phosphorylation of the $ORF2^{N}$–$ORF2^{1.2}$ product on frame 2, derived from the 1.2-kb spliced mRNA using an expression plasmid, has a similarity to NS5A of hepatitis C virus (HCV) (Asabe et al. 2001). Apoptotic ORF3 protein will be discussed later with apoptin of CAV.

## Chicken Anemia Virus

Chicken anemia virus was first isolated in Japan (Yuasa et al. 1979). CAV causes anemia by apoptosis of hemocytoblasts in the bone marrow (Taniguchi et al. 1983). CAV induces apoptosis of cortical thymocytes and lymphoblastoid cell lines after in vitro infection and in vivo infection, respectively (Jeurissen et al. 1992). CAV has a circular ssDNA genome of approximately 2.3 kb in size (Noteborn et al. 1991).

## *Similarity of CAV to TTV and Their Distinction*

CAV has an antisense genome as in the case of TTV, distinct from the ambisense genome of *Circovirus* (Hino and Miyata 2007; Miyata et al. 1999; Fig. 1). Configurations of ORFs in CAV and TTV were similar to each other (Hino and Miyata 2007; Fig. 5). While TTVs are extraordinarily diverse, CAVs are strikingly homogeneous. A 36-nt stretch in the origin of replication of CAV has approximately 80% identity to TTV (Miyata et al. 1999). However, no other parts of the two genomes show any significant similarities in their nucleotide sequences. TTV lacks a unique region of CAV possessing four or five near-perfect direct repeats of 21 bp (Noteborn et al. 1994b).

At least three proteins—a capsid and polymerase polyprotein, phosphatase, and apoptin—have been identified as the translational products of CAV (Noteborn et al. 1994a). Nevertheless, CAV has long been believed to produce a single mRNA 2.0 kb in length, consistent with a single set comprising the promoter, TATA-box, and polyA signal (Noteborn et al. 1991; Noteborn et al. 1992; Phenix et al. 1994). However, CAV also transcribes multiple mRNAs, just like TTV (Kamada et al. 2006; Kamahora et al. 2000; Qiu et al. 2005; see Fig. 2 in the chapter by K.A. Schat, this volume). The actual biological importance of the putative proteins computed from these spliced mRNAs has not been studied.

## *Classification of CAV*

CAV is currently the only virus in the genus *Gyrovirus* of the Circoviridae. While circoviruses have ambisense genomes, CAV has an antisense genome. CAV lacks the 9-nt stem-loop structure common in *Circovirus*.

TTV is currently classified in a floating genus, *Anellovirus*, in ssDNA viruses (Bendinelli et al. 2001; Hino and Miyata 2007; International Committee on Taxonomy of Viruses 2007). However, because of the enormous diversity of TTVs, a new family, Anelloviridae, will probably be established in the near future (ICTV Study Group 2007). In contrast, the redesignation of CAV together with TTV has not been discussed extensively. Considering the features described here, CAV might be transferred into a new family—Gyroviridae or Anelloviridae—in the near future.

## *Proteins of CAV*

CAV replicates via a circular double-stranded replicative form (RF) (Noteborn et al. 1991). The major transcript of CAV is an unspliced polycistronic mRNA that possesses three overlapping ORFs encoding for VP1 (52 kDa), VP2 (24 kDa), and

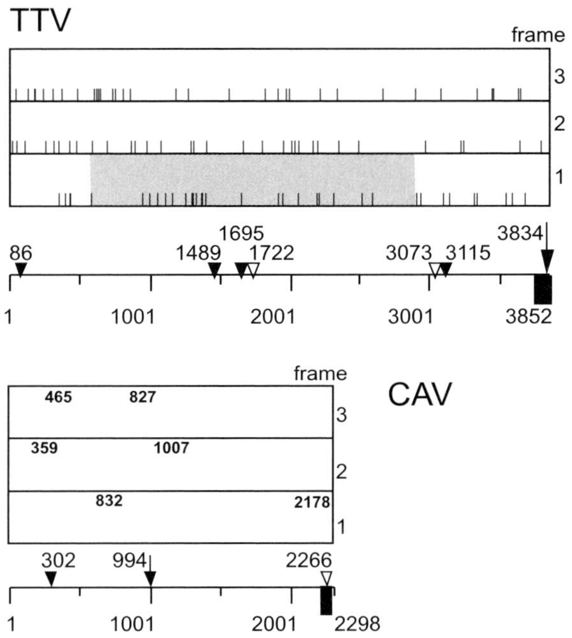


**Fig. 5** Comparative arrangement of ORFs expected from the genomes of TTV (AB008394+AB017911) and CAV (B031296). SP-1 site (*arrow*), TATA box (*closed triangle*), PolyA signal (*open triangle*), and GC-rich region (*closed box*) are indicated

VP3 (14 kDa) (Claessens et al. 1991; Noteborn et al. 1991). The capsid of CAV contains only the VP1 protein (Noteborn et al. 1998). A single amino acid change in VP1, glutamine to histidine at position 394, abrogates the pathogenicity of CAV (Yamaguchi et al. 2001). Cells expressing either VP1 or VP2 alone do not generate neutralizing antibodies, suggesting that the nonstructural protein VP2 may act as a scaffold protein in the virion assembly (Noteborn et al. 1998). VP2 and VP3 proteins are probably early proteins because they appear at 12 h after infection, while VP1 appears only after 24 h (Douglas et al. 1995). VP3, also named apoptin, is a nonstructural protein with a 121aa length (Fig. 1). Although additional proteins can be coded for by spliced mRNAs, their biological meanings have not been elucidated (Kamada et al. 2006).

## *VP3 protein of CAV, Apoptin*

It has been well established that apoptin induces apoptosis by itself in various transformed and/or tumorigenic cell lines, but not in normal diploid cells (Noteborn 2004). However, the function of apoptin in viral replication has not been elucidated.

Apoptin-induced apoptosis is independent of p53 and Bcl-2. A number of proteins were shown to interact with apoptin in transformed cells. Threonine$^{108}$ of apoptin is phosphorylated in transformed cells, and the apoptotic character is significantly reduced by loss of the hydroxyl group of $T^{108}$ for phosphorylation (Lee et al. 2007; Maddika et al. 2007; Poon et al. 2005; Rohn et al. 2002). Nuclear localization is required for apoptotic capacity, but not sufficient (Zhang et al. 2004). VP3 of TTV, named TTV-derived apoptosis-inducing protein (TAIP), also has an apoptotic character for hepatocellular carcinoma (HCC) cells, but induced only a low-level of apoptosis in several non-HCC human cancer cell lines (Kooistra et al. 2004).

## *Apoptin Is Indispensable for CAV Replication*

Recently, Prasetyo et al. (A. A. Prasetyo, T. Kamahora, A. Kuroishi, K. Murakami, and S. Hino, submitted for publication) found that apoptin is indispensable for the replication of CAV DNA and CAV virion formation. Not only virion formation but also DNA replication was abolished in apoptin knocked out virus, CAV/Ap(−) (Table 2). The defect of CAV/Ap(−) was fully recovered by the reverse mutation, CAV/ApRM. The DNA replication of CAV/Ap(−) was restored by supplying wild-type apoptin by co-transfection of the wild-type apoptin-expressing plasmid, but production of infectious virions did not follow. While the intracellular CAV DNA replication of CAV/ApT$^{108}$I was as high as that of wild-type, the production of infectious virus, CAV/ApT$^{108}$I, was 1/40 that of wild-type virus. When the wild-type apoptin was supplemented in cells by co-transfection with pAp/WT, both the DNA replication and the infectious virus titer of CAV/ApT$^{108}$I were restored to the level of the wild-type CAV.

Localization of apoptin was compared between the wild-type apoptin and ApT$^{108}$I by transfection of expression plasmids into COS1 cells. Cells expressing more than 50% of apoptin signals within nuclei were approximately 60% and 40% for the wild-type apoptin and ApT$^{108}$I, respectively, which was consistent with the idea that nuclear localization of apoptin is important for DNA replication of CAV. Surprisingly, the DNA replication of CAV/Ap(−) was fully complemented by TTV-VP3 (Table 3). The level of cells with dominant nuclear staining with TTV-VP3 was approximately 60%, nearly equal to that of wild-type apoptin.

## *Why TTV-VP3 Can Complement CAV/Ap(−) Replication?*

Both CAV-apoptin and TTV-VP3 were found to complement the DNA replication of CAV/Ap(−). However, even though Kooistra et al. (2004) pointed out the similarity of these proteins according to their apoptotic activity, proline-rich nature, and hydrophobic amino acid stretch, significant similarity of their amino sequences could not be confirmed. Although *N*-myristoylation sites were located at aa 39–44

**Table 2** DNA replication and infectious virus of CAV within cells transfected with RF DNA of CAV and its mutants at 72h after transfection. MDCC-MSB1 cells of an established chicken lymphoblastoid cell line from Marek's disease were transfected with a circularized, double-stranded, replicative form DNA of wild-type (WT) CAV, apoptin knock-out [Ap(–)] CAV, reverse mutant (ApRM) CAV or point mutated (ApT$^{108}$I) CAV, and co-transfected with apoptin expression plasmid, pAp/WT

| Transfection with | Co-transfection with | CAV DNA copies/culture (relative to WT) | Virus titer $TCID_{50}$/culture (relative to WT) |
|---|---|---|---|
| RF DNA of | | | |
| CAV/WT | None | 1.00 | 1.00 |
| CAV/Ap(–) | None | <0.01 | <0.001 |
| | pAp/WT | 1.00 | <0.001 |
| CAV/ApRM | None | 1.00 | 1.00 |
| CAV/ApT$^{108}$I | None | 1.00 | 0.025 |
| | pAp/WT | 1.00 | 1.00 |

$TCID_{50}$, median tissue culture infective dose

**Table 3** Complementation of CAV/Ap(–) DNA replication by apoptin and its mutant and by TTV-VP3

| Transfection with | | Relative CAV DNA (copies/culture) | Virus titer |
|---|---|---|---|
| RF DNA | Expression plasmid | | |
| CAV/Ap(–) | None | <0.01 | Nd |
| | ApWT | 1.00 | Nd |
| | ApT$^{108}$I | 0.10 | Nd |
| | TTV-VP3 | 1.00 | Nd |

Nd, not detectable

(GITITL) and aa 48–53 (GCANAR) on apoptin of CAV, and aa 57–62 (GCRLSR) on VP3 of TTV, the *N*-myristoylation site probably is not a key factor for this complementation because an apoptin mutant completely lacking this domain still showed the significant capacity of complementation. The reason for the comparable capacity of complementing CAV/Ap(–) DNA replication remains unanswered.

## *Complementation with Apoptin Will Lead to Virion Formation*

To test if complementation of DNA replication may lead to virion formation, CAV virions in transfected cells were partially purified by sucrose density gradient centrifugation. CAV DNA that was de novo synthesized within cells transfected with CAV/WT was mixed with infectious CAV and co-centrifuged. The DNA

**Table 4** Virion formation of CAV after complementation of apoptin knocked-out CAV by apoptin

| | Fraction No. | | 3 | 4 | 5 |
|---|---|---|---|---|---|
| | Density (g/cm$^3$) | | 1.24 | 1.22 | 1.19 |
| Virus | Complemented with | Relative amount of CAV[a] | | | |
| Wild type | | Infectivity | 0.06 | 1.00 | 0.06 |
| | | DNA | 0.08 | 1.00 | 0.08 |
| CAV/Ap(–) | None | DNA | <0.02 | <0.02 | <0.02 |
| | ApWT | DNA | 0.08 | 1.00 | 0.08 |
| | TTV-Vp3 | DNA | 0.08 | 1.00 | 0.08 |

[a]Relative amount of CAV in infectivity (TCID$_{50}$/culture) and in DNA (copies/culture) are shown to make the wild-type virus as 1.00

appeared in the same fraction as the infectious virions (Table 4). In cells co-transfected with RF DNA of CAV/Ap(–) and a plasmid expressing either ApWT or TTV-VP3, the CAV-specific DNA did co-migrate with the virion (Table 4). Neither cells transfected with RF DNA of CAV/Ap(–) alone, nor those sham-transfected with RF DNA of CAV/WT in the absence of the transfection reagent showed the viral DNA activity in the virion fraction. These data suggest that complementation by apoptin or TTV-VP3 is effective in DNA replication and virion formation. Although the actual mechanism of complementation of CAV/Ap(–) by TTV-VP3 is still unanswered, these functional similarities between TTV-VP3 and apoptin may provide further evidence of the common ancestry of these two viruses.

**Acknowledgements** We greatly appreciate the critical reading and editing of this manuscript by Dr. K.A. Schat.

## References

Asabe S, Nishizawa T, Iwanari H, Okamoto H (2001) Phosphorylation of serine-rich protein encoded by open reading frame 3 of the TT virus genome. Biochem Biophys Res Commun 286:298–304

Bendinelli M, Pistello M, Maggi F, Fornai C, Freer G, Vatteroni ML (2001) Molecular properties, biology, and clinical implications of TT virus, a recently identified widespread infectious agent of humans. Clin Microbiol Rev 14:98–113

Claessens JAJ, Schrier CC, Mockett APA, Jagt EHJM, Sondermeijer PJA (1991) Molecular cloning and sequence analysis of the genome of chicken anaemia agent. J Gen Virol 72:2003–2006

Douglas AJ, Phenix K, Mawhinney KA, Todd D, Mackie DP, Curran WL (1995) Identification of a 24 kDa protein expressed by chicken anaemia virus. J Gen Virol 76:1557–1562

Dry IB, Rigden JE, Krake LR, Mullineaux PM, Rezaian MA (1993) Nucleotide sequence and genome organization of tomato leaf curl geminivirus. J Gen Virol 74:147–151

ICTV Study Group (2007) Third International ssDNA Comparative Virology Workshop. Ouro Preto, Brazil

Hino S, Miyata H (2007) Torque teno virus (TTV): current status. Rev Med Virol 17:45–57

International Committee on Taxonomy of Viruses (2007) Index of viruses. Taxonomic table. http://www.ncbi.nlm.nih.gov/ICTVdb/Ictv/index.htm. Cited 22 Jun 2008

Jeurissen SH, Wagenaar F, Pol JM, van der Eb AJ, Noteborn MH (1992) Chicken anemia virus causes apoptosis of thymocytes after in vivo infection and of cell lines after in vitro infection. J Virol 66:7383–7388

Jones MS, Kapoor A, Lukashov VV, Simmonds P, Hecht F, Delwart E (2005) New DNA viruses identified in patients with acute viral infection syndrome. J Virol 79:8230–8236

Kamada K, Kamahora T, Kabat P, Hino S (2004) Transcriptional regulation of TT virus: promoter and enhancer regions in the 1.2-kb noncoding region. Virology 321:341–348

Kamada K, Kuroishi A, Kamahora T, Kabat P, Yamaguchi S, Hino S (2006) Spliced mRNAs detected during the life cycle of chicken anemia virus. J Gen Virol 87:2227–2233

Kamahora T, Hino S, Miyata H (2000) Three spliced mRNAs of TT virus transcribed from a plasmid containing the entire genome in COS1 cells. J Virol 74:9980–9986

Kooistra K, Zhang YH, Henriquez NV, Weiss B, Mumberg D, Noteborn MH (2004) TT virus-derived apoptosis-inducing protein induces apoptosis preferentially in hepatocellular carcinoma-derived cells. J Gen Virol 85:1445–1450

Lazarowitz SG (1988) Infectivity and complete nucleotide sequence of the genome of a South African isolate of maize streak virus. Nucleic Acids Res 16:229–249

Lee YH, Cheng CM, Chang YF, Wang TY, Yuo CY (2007) Apoptin T108 phosphorylation is not required for its tumor-specific nuclear localization but partially affects its apoptotic activity. Biochem Biophys Res Commun 354:391–395

Maddika S, Wiechec E, Ande SR, Poon IK, Fischer U, Wesselborg S, Jans DA, Schulze-Osthoff K, Los M (2007) Interaction with PI3-kinase contributes to the cytotoxic activity of Apoptin. Oncogene 27:3060–3065

Miyata H, Tsunoda H, Kazi A, Yamada A, Khan MA, Murakami J, Kamahora T, Shiraki K, Hino S (1999) Identification of a novel GC-rich 113-nucleotide region to complete the circular, single-stranded DNA genome of TT virus, the first human circovirus. J Virol 73:3582–3586

Mushahwar IK, Erker JC, Muerhoff AS, Leary TP, Simons JN, Birkenmeyer LG, Chalmers ML, Pilot Matias TJ, Dexai SM (1999) Molecular and biophysical characterization of TT virus: evidence for a new virus family infecting humans. Proc Natl Acad Sci USA 96:3177–3182

Niagro FD, Forsthoefel AN, Lawther RP, Kamalanathan L, Ritchie BW, Latimer KS, Lukert PD (1998) Beak and feather disease virus and porcine circovirus genomes: intermediates between the geminiviruses and plant circoviruses. Arch Virol 143:1723–1744

Ninomiya M, Nishizawa T, Takahashi M, Lorenzo FR, Shimosegawa T, Okamoto H (2007) Identification and genomic characterization of a novel human Torque teno virus of 3.2 kb. J Gen Virol 88:1939–1944

Nishizawa T, Okamoto H, Konishi K, Yoshizawa H, Miyakawa Y, Mayumi M (1997) A novel DNA virus (TTV) associated with elevated transaminase levels in posttransfusion hepatitis of unknown etiology. Biochem Biophys Res Commun 241:92–97

Noteborn MH (2004) Chicken anemia virus induced apoptosis: underlying molecular mechanisms. Vet Microbiol 98:89–94

Noteborn MH, de Boer GF, van Roozelaar DJ, Karreman C, Kranenburg O, Vos JG, Jeurissen SH, Hoeben RC, Zantema A, Koch G (1991) Characterization of cloned chicken anemia virus DNA that contains all elements for the infectious replication cycle. J Virol 65:3131–3139

Noteborn MH, Kranenburg O, Zantema A, Koch G, de Boer GF, van der Eb AJ (1992) Transcription of the chicken anemia virus (CAV) genome and synthesis of its 52-kDa protein. Gene 118:267–271

Noteborn MH, Todd D, Verschueren CA, de Gauw HW, Curran WL, Veldkamp S, Douglas AJ, McNulty MS, van der EA, Koch G (1994a) A single chicken anemia virus protein induces apoptosis. J Virol 68:346–3451

Noteborn MH, Verschueren CA, Zantema A, Koch G, van der Eb AJ (1994b) Identification of the promoter region of chicken anemia virus (CAV) containing a novel enhancer-like element. Gene 150:313–318

Noteborn MH, Verschueren CA, Koch G, Van der Eb AJ (1998) Simultaneous expression of recombinant baculovirus-encoded chicken anaemia virus (CAV) proteins VP1 and VP2 is required for formation of the CAV-specific neutralizing epitope. J Gen Virol 79:3073–3077
Okamoto H, Nishizawa T, Ukita M (1999) A novel unenveloped DNA virus (TT virus) associated with acute and chronic non-A to G hepatitis. Intervirology 42:196–204
Okamoto H, Nishizawa T, Takahashi M, Tawara A, Peng Y, Kishimoto J, Wang Y (2001) Genomic and evolutionary characterization of TT virus (TTV) in tupaias and comparison with species-specific TTVs in humans and nonhuman primates. J Gen Virol 82:2041–2050
Peng YH, Nishizawa T, Takahashi M, Ishikawa T, Yoshikawa A, Okamoto H (2002) Analysis of the entire genomes of thirteen TT virus variants classifiable into the fourth and fifth genetic groups, isolated from viremic infants. Arch Virol 147:21–41
Phenix KV, Meehan BM, Todd D, McNulty MS (1994) Transcriptional analysis and genome expression of chicken anaemia virus. J Gen Virol 75:905–909
Poon IK, Oro C, Dias MM, Zhang J, Jans DA (2005) Apoptin nuclear accumulation is modulated by a CRM1-recognized nuclear export signal that is active in normal but not in tumor cells. Cancer Res 65:7059–7064
Qiu J, Kakkola L, Cheng F, Ye C, Soderlund-Venermo M, Hedman K, Pintel DJ (2005) Human circovirus TT virus genotype 6 expresses six proteins following transfection of a full-length clone. J Virol 79:6505–6510
Rohn JL, Zhang YH, Aalbers RI, Otto N, Den Hertog J, Henriquez NV, Van De Velde CJ, Kuppen PJ, Mumberg D, Donner P, Noteborn MH (2002) A tumor-specific kinase activity regulates the viral death protein Apoptin. J Biol Chem 277:50820–50827
Suzuki T, Suzuki R, Li J, Hijikata M, Matsuda M, Li TC, Matsuura Y, Mishiro S, Miyamura T (2004) Identification of basal promoter and enhancer elements in an untranslated region of the TT virus genome. J Virol 78:10820–10824
Takahashi K, Hoshino H, Ohta Y, Yoshida N, Mishiro S (1998) Very high prevalence of TT virus (TTV) infection in general population of Japan revealed by a new set of PCR primers. Hepatol Res 12:233–239
Takahashi K, Iwasa Y, Hijikata M, Mishiro S (2000) Identification of a new human DNA virus (TTV-like mini virus, TLMV) intermediately related to TT virus and chicken anemia virus. Arch Virol 145:979–993
Taniguchi T, Yuasa N, Maeda M, Horiuchi T (1983) Chronological observations on hemato-pathological changes in chicks inoculated with chicken anemia agent. Natl Inst Anim Health Q (Tokyo) 23:1–12
Yamaguchi S, Imada T, Kaji N, Mase M, Tsukamoto K, Tanimura N, Yuasa N (2001) Identification of a genetic determinant of pathogenicity in chicken anaemia virus. J Gen Virol 82:1233–1238
Yuasa N, Taniguchi T, Yoshida I (1979) Isolation and some characteristics of an agent inducing anemia in chicks. Avian Dis 23:366–385
Zhang YH, Kooistra K, Pietersen A, Rohn JL, Noteborn MH (2004) Activation of the tumor-specific death effector apoptin and its kinase by an N-terminal determinant of simian virus 40 large T antigen. J Virol 78:9965–9976

# Apoptosis-Inducing Proteins in Chicken Anemia Virus and TT Virus

M.H. de Smit and M.H.M. Noteborn(✉)

**Contents**



**Abstract** Torque teno viruses (TTVs) share several genomic similarities with the chicken anemia virus (CAV). CAV encodes the protein apoptin that specifically induces apoptosis in (human) tumor cells. Functional studies reveal that apoptin induces apoptosis in a very broad range of (human) tumor cells. A putative TTV open reading frame (ORF) in TTV genotype 1, named TTV apoptosis inducing protein (TAIP), it induces, like apoptin, p53-independent apoptosis in various human hepatocarcinoma cell lines to a similar level as apoptin. In comparison to apoptin, TAIP action is less pronounced in several analyzed human non-hepatocarcinoma-derived

M.H.M. Noteborn
Department of Molecular Genetics, Leiden Institute of Chemistry, Leiden University, Einsteinweg 55, 2333 CC, Leiden, The Netherlands
m.noteborn@chem.leidenuniv.nl

E.-M. de Villiers, H. zur Hausen (eds.), *TT Viruses: The Still Elusive Human Pathogens,* 

cell lines. Detailed sequence analysis has revealed that the TAIP ORF is conserved within a limited group of the heterogeneous TTV population. However, its N-terminal half, N-TAIP, is rather well conserved in a much broader set of TTV isolates. The similarities between apoptin and TAIP, and their relevance for the development and treatment of diseases is discussed.

**Abbreviations** AIP: Apoptin-interacting protein; Apoptin: CAV-derived apoptosis-inducing protein; APC1: Subunit 1 of the anaphase promoting complex/cyclosome; CAV: Chicken anemia virus; DAPI: 2,4 Diamidino-2-phenylindole; DEDAF: Death effector domain-associated factor; Hippi: Huntingtin interacting protein 1 protein interactor; LT: Large T antigen; MDV: Marek's disease virus; NMI: N-myc interacting protein; ORF: Open reading frame; PML: Promyelocytic leukemia protein; PP2A: Protein phosphatase 2A; RA: Rheumatoid arthritis; st Small t antigen; SV40: Simian virus 40; TTMV: Torque teno mini virus; TAIP Torque teno virus apoptosis-inducing protein; TTV: Torque teno virus.

## Introduction

Certain virus proteins are capable of inducing transformation processes in human normal cells resulting in tumor formation (Psyrri and DiMaio 2008; Bellon and Nicot 2008), whereas others can induce apoptosis in normal cells (Cossarizza 2008). A third group of proteins induces apoptosis specifically in tumor cells. Among these proteins are e.g., Parvovirus-H1 protein NS1 (Nüesch and Rommelaere 2007), Adenovirus protein E4orf4 (Landry et al. 2006), and the chicken anemia virus (CAV) protein apoptin (Backendorf et al. 2008). The latter group of proteins is especially intriguing for they can specifically kill tumor cells, which harbor tumor-related blockages within their apoptosis machinery. Paradoxically, tumor-related processes are relevant for the activation of these tumor-specific apoptosis-inducing proteins (Kooistra et al. 2007).

Several recent reports have noted that the presence of Torque teno virus (TTV) is more frequent in human tumors than in normal tissue (De Villiers et al. 2007; Bando et al. 2008). In this respect, an interesting observation was reported by Miyata et al. (1999) that some similarities in genomic organization exist between a specific TTV genotype (TA278, GenBank accession No. AB017610) and CAV. In particular, TTV ORF3/TAIP seems to be related to CAV VP3/apoptin.

Here, we describe the apoptosis characteristics of VP3/apoptin in comparison with ORF3/TAIP, and its occurrence within the heterogeneous TTV population. The relevance of VP3/apoptin and ORF3/TAIP for the development of diseases and their treatment will be discussed.

## Apoptin Specifically Induces Apoptosis in a Broad Range of Human Tumor Cells

Early after infection, the nonspliced CAV messenger RNA (mRNA) transcript encodes three proteins (Noteborn et al. 1992), of which VP2 and VP3 both harbor apoptosis-inducing activity (Zhuang et al. 1995; Noteborn and Van der Eb 1999). Recently, Kamada et al. (2006) have reported that late in infection, due to splicing events, three other mRNAs are synthesized encoding, besides VP3, several putative proteins among them being a VP3 consisting of the first 58 amino acids only. The different ORFs for the known and putative CAV proteins are shown in Fig. 1a. The main apoptosis activity can be generated by expression of the 121-amino acid protein VP3 (Noteborn et al. 1994). Therefore, VP3 is named in this review further as apoptin, which is an abbreviation of apoptosis inducing.

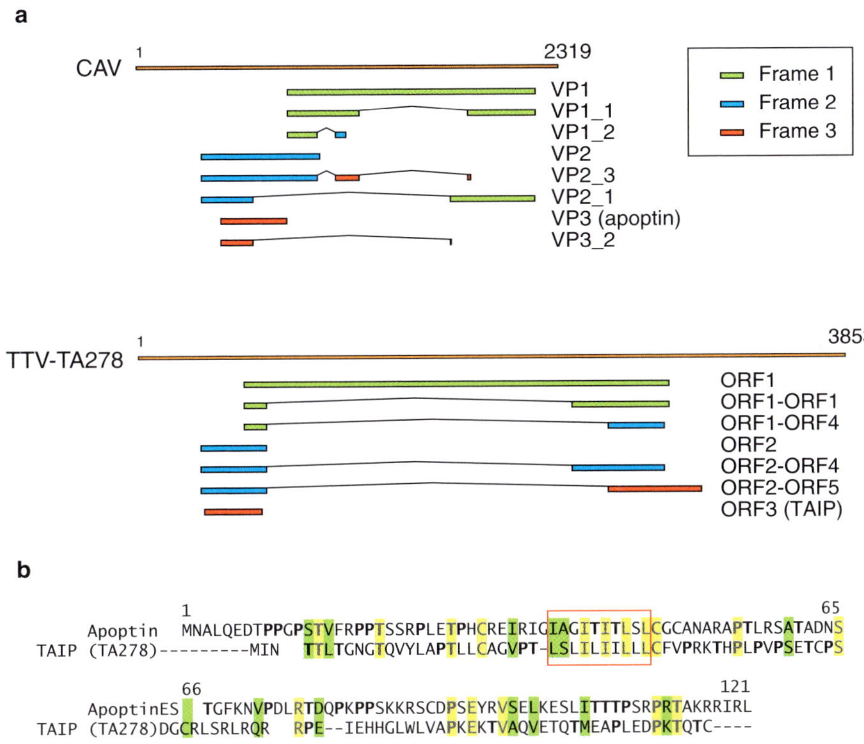


**Fig. 1** **a, b** Comparison of CAV-apoptin and TTV-TAIP. **a** Genomes of CAV and TTV (TA278) and putative peptides encoded by the various nonspliced and spliced mRNAs (after Kamada et al. 2006; Hino and Miyata 2007). **b** Tentative alignment of the apoptin and TAIP amino-acid sequences. The conserved hydrophobic region is *boxed in red*, prolines and threonines are shown in *bold*

Apoptin's primary protein structure is characterized by its high percentage of prolines, serines, and threonines, a hydrophobic region within its N-terminal half and a positively charged C-terminal region (Noteborn et al. 1991). Biophysical analysis shows that (bacterially produced) apoptin forms distinct functionally active protein multimers consisting of 30–40 monomers (Leliveld et al. 2003a). In vitro studies have provided evidence that apoptin multimers can bind to DNA and cooperatively form distinct superstructures upon binding to DNA (Leliveld et al. 2003b). Apoptin adopts little if any regular secondary structure within the aggregates. This surprising result classifies apoptin as a protein for which, rather than the formation of a well-defined tertiary and quaternary structure, semi-random aggregation is sufficient for activity (Leliveld et al. 2003a).

Remarkably, expression of apoptin results in induction of apoptosis in human tumor cells but not in their normal counterparts (Danen-Van Oorschot et al. 1997; Russo et al. 2006). Various studies have shown that apoptin induces apoptosis in a broad range of different human cancer cell lines derived from various tumor types. Among these are breast tumors, cervix carcinoma, colon carcinomas, gliomas, hepatocarcinomas, leukemias, lung tumors, lymphomas, melanomas, osteosarcomas, ovarian cancers, prostate cancers, and squamous cell carcinomas (Noteborn 2005). Apoptin does not induce apoptosis in normal, nontransformed human diploid cells, such as primary fibroblasts, keratinocytes, smooth muscle cells, T cells, or endothelial cells (Noteborn 2005). Long-term expression of apoptin in normal human fibroblasts revealed that it has no toxic or transforming activity in these cells and that apoptin does not interfere with cell proliferation (Danen-Van Oorschot et al. 1997).

## Molecular Aspects of Apoptin-Induced Apoptosis

### *Apoptin Induces Apoptosis Independent of Tumor Suppressor p53*

Most of the known tumors occur due to inactivation of tumor suppressor p53, which is a known apoptosis mediator (Halazonetis et al. 2008). Apoptin induces apoptosis in tumor cells whether or not they contain functional tumor suppressor p53, which implies that apoptin induces apoptosis in a p53-independent manner (Rohn and Noteborn 2004). Recently, Klanrit et al. (2008) have reported that in p53-deficient cells specific p73 isoforms play a crucial role in apoptin-induced apoptosis. Liu et al. (2006) have reported that antiapoptotic factors such as Bcl-2, survivin, FLIP(S), and XIAP cannot prevent apoptin-induced apoptosis in human tumor cells. One of the apoptin-activated apoptosis pathways is based on apoptin-induced elevation of the tumor suppressor lipid ceramide (Liu et al. 2006), causing cytochrome c release from mitochondria (Liu et al. 2006; Danen-Van Oorschot et al. 2000), resulting finally in the activation of caspase 3 and execution of apoptosis. A more detailed description of apoptin-induced apoptosis can be found in a recent review from our research group (Backendorf et al. 2008).

## *Cellular Localization of Apoptin*

One of the main tumor-specific characteristics of apoptin is its nuclear localization. In all (human) tumor cells analyzed thus far, apoptin is translocated into the nucleus and subsequently the tumor cells will undergo apoptosis (Danen-Van Oorschot et al. 2003; Guelen et al. 2004). In contrast, apoptin is localized in the cytoplasm of normal cells, rapidly neutralized, and eventually degraded (Zhang et al. 2003).

Apoptin also induces apoptosis in transformed fibroblast-like synoviocytes derived from rheumatoid arthritis (RA) patients. Surprisingly, apoptin is mainly localized in the cytoplasm of these RA-derived cells (Tolboom et al. 2006) but these observations can be explained by studies of Danen-Van Oorschot et al. (2003). Both the N-terminal and the C-terminal regions of apoptin contain distinct apoptosis-inducing activities in human tumor cells. An artificially truncated apoptin (consisting of amino acids 1–69) was found to be located in the cytoplasm of human tumor cells, but harbors a distinct apoptosis activity, which is enhanced upon nuclear translocation (Danen-Van Oorschot et al. 2003). Further truncation of the amino acids 61–69 results in a nuclear and cytoplasmic located apoptin with a significant level of apoptosis induction in human tumor cells, but not in human normal cells (A.A. Danen-Van Oorschot and M.H. Noteborn, unpublished results). These observations suggest that within the apoptin amino acids 61–69 a cytoplasmic retardation signal might be located. In this respect, the recent publication from Kamada et al. (2006) is interesting for they report that late in CAV infection a spliced mRNA is produced encoding a putative apoptin protein consisting of the amino acids 1–58.

In conclusion, apoptin harbors at least two domains able to induce the apoptotic machinery in transformed cells. These studies show the intriguing complexity of apoptin's apoptosis induction in transformed cells.

## *Transforming Proteins Activate Apoptin in Human Normal Cells*

The fact that apoptin induces apoptosis in a very broad range of tumor cells and related transformed cells suggests that apoptin uses an early transformation event to become activated (Backendorf et al. 2008). DNA tumor virus SV40 and especially its transforming proteins LT/st have been used as a model for studying the (early) developmental steps of tumor formation (Ahuja et al. 2005). Transient expression of transforming nuclear SV40 LT/st proteins in human fibroblasts results in nuclear localization of apoptin and apoptosis induction comparable to established tumor cell lines (Deppert 2000; Zhang et al. 2004).

In human fibroblasts, SV40 LT transfers very quickly into the nucleus resulting in initiation of transforming processes. In human mesenchymal stem cells, SV40 LT transfers very slowly into the nucleus and its transforming activity is also delayed. Co-expression of SV40 LT and apoptin in these mesenchymal stem cells resulted in a strongly delayed nuclear localization and apoptosis induction by apoptin (Y.H. Zhang, personal communication).

These results strengthen the hypothesis that apoptin becomes activated by very early (nuclear) transforming events that seem to be present in a very broad panel of transformed cells.

## *Tumor-Specific Phosphorylation of Apoptin*

Lysates derived from various human tumor cell lines, and from tumor tissue obtained from cancer patients, are able to phosphorylate recombinant apoptin protein, whereas lysates from normal cells and healthy tissue are not. Ectopic expression of apoptin in human tumor cells, but not in normal cells, results in phosphorylation of apoptin at position T108 (Rohn et al. 2002; Lee et al. 2007). In tumor cells, mutation of this tumor-specific phosphorylation site only partially affects apoptin's apoptosis activity. In transformed RA cells, apoptin induces apoptosis but does not become phosphorylated (Tolboom et al. 2006). These results are in agreement with Danen-Van Oorschot and colleagues' (2003) observation that apoptin contains two independent apoptosis domains, as described above. A C-terminal apoptin fragment with a mutated phosphorylation site has completely lost its apoptotic activity in human tumor cells (Rohn et al. 2005). Apoptin-induced apoptosis activation seems to occur via at least two different signaling routes, one of which seems to involve tumor-related kinase activity.

Zhang et al. (2004) have shown that transient expression of SV40 LT/st in normal cells also induces tumor-related kinase activity resulting in phosphorylation of apoptin at position T108. SV40 LT/st proteins contain several domains with transforming activity. The p53-binding domain of SV40 LT, which inactivates p53, and the SV40 st binding domain for protein phosphatase 2A (PP2A), which inactivates PP2A, are prominent transforming domains (Ahuja et al. 2005). Mutagenesis of the p53-binding domain within SV40 LT shows that inactivation of p53 is not per se needed for phosphorylation of apoptin, which is in agreement with the observation that apoptin induces apoptosis in a p53-independent manner (Zhuang et al. 1995). Mutagenesis studies confirmed that both the J domain and the PP2A binding site within SV40 st are essential for phosphorylation of apoptin. As recent crystallographic data have shown that both domains are involved in the inactivation of PP2A by SV40 st (Cho et al. 2007; Chen et al. 2007), it appears that tumor-related suppression of PP2A phosphatase activity might be sufficient for activation of apoptin. This view was recently confirmed by a small-interfering RNA (siRNA) analysis (D.J. Peng, personal communication). We postulate that biochemical identification of this tumor-related kinase phosphorylating apoptin will lead to the unraveling of one of the early steps within cell transformation (Backendorf et al. 2008).

In this respect, it is of interest that CAV infection is often associated with co-infection by transforming viruses such as MDV (Noteborn et al. 1998), which may upregulate the kinase of apoptin during cellular transformation, thereby offering an opportunistic advantage to CAV. Interestingly, Peters et al. (2002, 2005) have shown that the CAV-derived protein VP2 functions as a dual-specificity phosphatase.

They have recently provided evidence that a mutation within this atypical dual-specificity protein phosphatase signature motif affects CAV replication and cytopathogenicity (Peters et al. 2006). These observations are consistent with the idea that phosphorylation pathways can be of importance for apoptin's activity and for CAV's replication cycle. The various apoptin characteristics in human normal cells versus transformed cells are summarized in Table 1.

## *Apoptin-Interacting Proteins*

As protein-protein interactions are crucial for the functionality of a specific protein within a living cell (Collura and Boissy 2007), it is important to identify apoptin's interacting partners that play a role in its tumor-related apoptosis activity. A panel of such apoptin-interacting proteins has been identified by us and other research groups (Backendorf et al. 2008). On one hand, interacting proteins such as N-myc interacting protein (NMI), promyelocytic leukemia protein (PML), anaphase promoting complex/cyclosome subunit 1 (APC1), and death effector domain-associated factor (DEDAF) can be related to apoptin's tumor-specific apoptosis activities (Sun et al. 2002; Janssen et al. 2007; Heilman et al. 2006; Danen-Van Oorschot et al. 2004). On the other hand, Hippi binds to apoptin only in human normal cells and seems to be involved in the normal-cell-specific neutralization of apoptin (Cheng et al. 2003; Zhang et al. 2003).

The hydrophobic region within the N-terminal half of apoptin at amino acid positions 37–46 plays a crucial role in the formation of apoptin multimers (Heilman et al. 2006; R. Ten Hoopen, personal communication). Intriguingly, this hydrophobic stretch is involved in the interaction of apoptin with its known interacting partners APC1, DEDAF, Hippi, NMI, as well as with PML (Backendorf et al. 2008; Heilman et al. 2006). This hydrophobic apoptin region seems to represent a so-called hot-spot domain (Shulman-Peleg et al. 2007), which is able to interact with a broad range of different proteins including itself. Most likely, interaction of apoptin with the observed tumor-related proteins results in prevention of protein–protein interactions essential for the survival of tumor cells. In contrast, proteins such as Hippi might be involved in the neutralization of apoptin as observed in human

**Table 1** Apoptosis characteristics of apoptin

| Cell types | p53-Independent apoptosis | Phosphorylation | Localization |
|---|---|---|---|
| Normal cells | No | No | Cytoplasmic |
| Tumor cells | Yes | Yes | Nuclear |
| Normal cells with SV40 proteins | Yes | Yes | Nuclear |
| RA-derived cells | Yes | No | Cytoplasmic |

normal cells, preventing the interference of apoptin with functional pathways in normal cells (Zhang et al. 2003).

## Apoptin-Related Proteins in TTV Population: A Flashback

Various sources have reported that CAV and TTV share similarities. Furthermore, it has been reported that TTV occurrence might be related to several diseases such as cancer and rheumatoid arthritis (Hino and Miyata 2007). An obvious question is: Does TTV harbor apoptin-like proteins? For several reasons, not much is known about the TTV proteins. TTV cannot be cultured under laboratory conditions and expression of TTV proteins is notoriously difficult. Qiu et al. (2005) have constructed a set of plasmids carrying the genome of TTV genotype 6 with hemagglutinin tags inserted in various ORFs, and examined the proteins synthesized upon transfection. Unfortunately, they only tested ORFs extending close to the 3′ terminus of the viral genome or starting at the initiation codon of ORF2.

In 1999, a publication on TTV strain TA278 by Miyata et al. (1999) drew our attention. It described a TTV virus showing some striking similarities to CAV in its genomic organization. Besides a very characteristic GC-rich region shared by both virus types, the distribution of three (putative) ORFs is very similar. One year later, Kamahora and coworkers described that one nonspliced and two spliced mRNAs are produced from this TTV genome (Kamahora et al. 2000; Hino and Miyata 2007). These mRNAs encode seven putative proteins (Fig. 1a). Remarkably, all three mRNAs encode an ORF, named ORF3 by them, which shows similarities to apoptin. Although from the primary sequences no real homology could be derived (Fig. 1b), one of the characteristics of apoptin is its relatively high content of proline residues (11.5%), which also holds true for ORF3 (10.5%). The scattered distribution of prolines within both proteins suggests a similar protein structure. In addition, both proteins contain a stretch of hydrophobic amino acids within their N-terminal region (Fig. 1a, boxed in red). Furthermore, the high level of threonines (12.4% for ORF3 and 11.5% for apoptin) also seem to be characteristic for both proteins. As described above, one of the threonines within apoptin plays an essential role in the regulation of its apoptosis activity.

## The TTV-Derived ORF TAIP Induces Apoptosis in Human Tumor Cells

Prompted by these similarities, we investigated whether ORF3 harbors an apoptosis activity like apoptin. A mammalian expression system encoding the TTV ORF3 fused to a myc-tag was constructed (Kooistra et al., 2004). Expression studies in different human tumor cell lines were carried out to analyze the cellular localization and apoptotic activity of the ORF3 protein in comparison to apoptin. The cells were fixed and the apoptin, myc-tagged ORF3 protein, and nonapoptotic control β-galactosidase were analyzed by means of indirect immunofluorescence, as described by Noteborn et al. (1990).

The percentage of apoptotic transgene-positive cells was analyzed by chromatin staining with DAPI. In apoptin-induced apoptotic cells, the chromatin is strongly condensed, the nucleus becomes fragmented and finally DAPI-staining vanishes due to nucleosomal DNA fragmentation (Danen-van Oorschot et al. 1997).

## *Hepatocarcinoma-Derived Cell Lines*

The human hepatocarcinoma-derived cell lines HepG2 (containing wild-type p53), HUH-7 (mutated, nonfunctional p53), and Hep3B (deletion of functional p53 locus) expressing apoptin, ORF3 protein, or the negative control β-galactosidase were analyzed for undergoing apoptosis.

Five days after transfection, 50%–60% of both apoptin- and ORF3-positive hepatocarcinoma-derived cells underwent apoptosis, whereas the β-galactosidase-positive cells showed only a background level of about 10%, which is due to experimental conditions (Danen-Van Oorschot et al. 1997). Apparently, besides expression of apoptin, the myc-tagged ORF3 also resulted in induction of apoptosis in the three analyzed human hepatocarcinoma-derived cell lines. Therefore, the TTV-derived ORF3 protein was renamed by us to TTV apoptosis-inducing protein, abbreviated as TAIP. Interestingly, the percentage of induced apoptosis by TAIP in the human hepatocarcinoma cell lines was not significantly different from apoptin-induced apoptosis. Moreover, apoptin and TAIP both induce apoptosis in all three cell lines, despite their different p53 status. These features imply that both apoptin and TAIP can induce apoptosis in a p53-independent way (Kooistra et al. 2004).

Two days after transfection, however, TAIP accumulated in the cytoplasm whereas apoptin was clearly present in the nucleus. Only in apoptotic TAIP-positive hepatocarcinoma-derived cells was TAIP colocalized with DAPI-stained chromatin remnants. Apoptin-induced apoptosis occurs subsequent to apoptin's transfer into the nucleus, whereas cytoplasmic TAIP seems to trigger the apoptosis process. Interestingly, in RA cells apoptin induces apoptosis as a cytoplasmic protein (see Sect. 3.2). One cannot ignore the fact, however, that nuclear localization of TAIP might be rapidly followed by apoptosis.

## *Non-hepatocarcinoma Cell Lines*

One of apoptin's striking characteristics is that it can induce apoptosis in a very broad range of human tumor cells. Therefore, we analyzed the apoptosis activity of TAIP in comparison to apoptin and the negative control β-galactosidase in three human non-hepatocarcinoma cell lines: Saos-2 and U2OS (osteosarcoma) and

H1299 (small-cell lung carcinoma). Five days after transfection with the plasmid encoding the various transgenes, approximately 60%–70% of the apoptin-positive cells had undergone apoptosis, but only 20%–30% of the TAIP-positive cells. β-Galactosidase-positive cells underwent a nonspecific background level of apoptosis of just 10%. All TAIP-positive cells contained TAIP in their cytoplasm, whereas apoptin was mainly nuclear. Therefore, one can conclude that TAIP induces apoptosis in these non-hepatocarcinoma cells, but to a significantly lower extent than apoptin (Kooistra et al. 2004).

Another remarkable feature of apoptin is that it does not induce apoptosis in normal healthy cells. Preliminary experiments with normal rat hepatocytes, injected in their nucleus with plasmid DNA encoding apoptin, TAIP, or the negative control β-galactosidase, revealed that all three transgenic products did not induce apoptosis (Y.H. Zhang, personal communication). These results indicate that TAIP causes apoptosis, like apoptin, in a tumor/transformed-related manner. However, more (human) normal cell types have to be analyzed to obtain conclusive results. The characteristics known thus far of TAIP-induced apoptosis are summarized in Table 2.

In summary, the TAIP activity resembles in certain aspects apoptin-induced apoptosis. TAIP and apoptin both induce p53-independent apoptosis in human transformed cells. In certain tumor cell types, however, TAIP is less active than apoptin, suggesting that the apoptosis range of TAIP is less broad than that of apoptin. In the six analyzed tumor cell lines, TAIP is located in the cytoplasm like apoptin is in transformed RA cells. In these RA cells, apoptin is not phosphorylated in contrast to apoptin expressed in transformed (human) tumor cells. Thus far it is unknown whether or not TAIP is phosphorylated in transformed cells. The N-terminal regions of TAIP and apoptin harbor a similar hydrophobic region, which functions in apoptin as a hot-spot domain. Possibly, the apoptotic activity of TAIP depends on this domain.

## Various Forms of TAIP in the TTV Population

Having established that there are functional and structural similarities between apoptin and the TAIP protein encoded by TTV isolate TA278, we inquired whether this protein is conserved among the many TTV isolates thus far sequenced. One of the striking differences between CAV and TTV is that the CAV genomes are highly

**Table 2** Apoptosis characteristics of TAIP

| Cell types | p53-Independent apoptosis | Localization |
|---|---|---|
| Normal cells | No | Cytoplasmic |
| Hepatocarcinoma cell lines | Yes, similar to apoptin | Cytoplasmic |
| Non-hepatocarcinoma cell lines | Yes, less than apoptin | Cytoplasmic |

conserved, whereas the TTV ones are extremely heterogeneous (Van Santen et al. 2001; He et al. 2007; Hino and Miyata 2007).

To obtain a suitable collection of sequences from each of the established TTV groups, we started from the isolates used for building two recent phylogenetic trees (Jelcic et al. 2004; Ninomiya et al. 2007). Each of the sequences was downloaded from the NCBI nucleotide collection and analyzed for the presence of an ORF3-like ORF using the Vector NTI software package (Invitrogen). The nucleotide sequence of the found ORF was then used for a database search with NCBI's Basic Local Alignment Search Tool (BLAST), so a set of sequences related to the original one was obtained. In each of these sequences, the ORF was located as before and translated to an amino-acid sequence using Vector NTI.

Both the nucleotide and the amino-acid sequences were aligned per subgroup of highly related isolates, using the alignment function of Vector NTI. Previously proposed subgroups were split or merged where necessary to arrive at meaningful consensus sequences for the encoded peptides. The complete collection of aligned sequences is available from the authors upon request.

For thc vast majority of the sequences analyzed, an ORF3-like ORF could be unambiguously assigned, but there turned out to be large differences in size (Fig. 2). The ORFs of TTV group 1 (Ninomiya et al. 2007) appeared to consist of two categories, one encoding a TAIP of 105–108 amino acids, the other encoding a mere 51–53 amino acids. The ORFs of groups 2, 3, and 5 similarly encode 50–59 amino acids, while the ORFs of group 4 appear rather variable. No ORF resembling ORF3 was found in any isolate of TTMV and in two isolates from group 3.

That the small ORFs are indeed related to the larger TAIP is evident from the nucleotide alignment in Fig. 3. Here, the sequence encoding the TAIP from TA278 is aligned with the corresponding sequence from isolates JA4 and JA1. Because these sequences can be unambiguously aligned and the various ORFs start at similar positions in the same reading frame, there is no doubt that TAIP and the small ORFs are orthologs. We therefore, tentatively, have named the peptide encoded by the small ORF as N-TAIP, for N-terminal half of TAIP. The small ORF was previously denoted as ORF XI (Leppik et al. 2007), but to our knowledge the relation to ORF3 and TAIP was not established.

We can deduce that expression of N-TAIP is of functional importance to the virus from the details of its evolutionary conservation. Figure 4 illustrates this by showing a selection of sequences surrounding the initiation and termination codons. It is clear that whenever either of these codons is mutated, an alternative arises at a nearby position, thus preserving the coding potential of the ORF. Very few sequences were found in the database where the initiation codon was mutated without compensation and we believe that these probably represent specimens that survived thanks to the presence of nonmutated helper viruses. Mutated termination codons may result in C-terminally extended proteins that are not necessarily nonfunctional, but in most cases a new termination codon was found within a short distance.

We have discussed above the presence of a hydrophobic stretch in the N-terminal half of both apoptin and TAIP, which in the first case was shown to

**Group 1**

```
TA278    (10) MiNSTLTGNGTQVYLAPTLLCAGVPTLSLILIILLLCFVPRKTHPLPVPSETCPSDGCRLSRLRQRRPEIEHHGLWLVAPKEKTVAQVETQTMEAPLEDPKTQTC
JA20      (3) MiNnTLtgNGTllYLApTLLcvavPtLSLilIILLicFVpRKiRprliiqDpypcehclLsRLptRrPVIEhHGLWVVEETpe-laqVETpTmEaPLEDPqTQTC---
JA4       (9) MNVSaSeNGSSQlFiLTlLSVAAMILSvTSVAWLLcLTaRrTRDPPeDvVLLRSGsYPLSRLLPssPVIEHHGLWvVeetpq-VeeMEekAqvETPmEDPQTQTc---
JA1       (7) MnSLFNVTGSTVAFTPTLLCVAVLILLVISItSLLCSAVRKTRTLLRHPGL
tth3     (56) MSPASRETGTRAVSDPMLLFVAVATLLAILIIWQLLWV-LHVLGPRADPAR-r
DXL1      (5) MSRGW---G-S--sv-mLLfVaVViLL-IL-IWRlVLDVLl-P--pE-R-rLQ
```

**Group 2**

```
tth6      (8) MSqDkRacGTDqfllMaLFVVAVILWvIfrdLanawvdpNhqDHrAdrRALl
PMV       (4) M-P-ek--GSep-talt-lFvvvaIllgIlLvWllI--tnlrrrv-----hl-----
                        r         i   aa                          r
```

**Group 3**

```
TCHN-A   (10) MSPPASYVGTAQFERATMLFVAVVILFFIFLVWLHVLTIRDLRRpP---RtsaRRRlPQ
TUS01    (14) M----nktGlSPasD-ThafVAVIILlFILITLLLALTICLLP-RLWdpPn-RphA-Rra
SAa-38    (5) MPPAeRDlGFRPVTNLTVLFVAVV-LFFILLAWLHVLIFRPGHRLPGVPGRRaRRf
SENV-C    (7) mGSRET-grhsfacmLqlVVvaILLAILLYWLVGMVLLlvpRppGLpDhR
SAa-01   (16) Mrqe-s-sslrAafdfTL-vavVaiLLlILlywllamv-lggrrrqvvLgr--R-
SAa-39    (2) (no ORF3 found)
SANBAN   (18) M-PassaiGtrPvSEpTl--VaVaILlctLIfWLgVm-ll----rRqVvllRaprr-
L020      (8) MSeGRSSSGFEPVCTLTGPGvVVVTLFCICLLLLHVLTFREaPLPPeaPrEREAQ
                                      a                   v
```

**Group 4**

```
JT14F     (2) MGK-LK--PR-VL-QD---CRG-RR--T--RE---GT-Q--S---LFAAA-TLLGI--SLLLALL-IL-
JT05F     (5) MGKtLKKFPRSVvRQDTKsCRGihRAttfnREKsSGyrqlLlHMLLFAAVvTLvaIf-AfLAALiTaag-Pp-Pq---TpfvpyRpyRHPkNlnsRHHGLVpVQkKaMAPkEaTeVGPTQKKi
CT23F    (14) MGKALKKFPRsVLRQDTKSCRGDRR-It--g-KinG-Q--FTrmLL-AaVVTLLgILiALLLAFLTPVpRDHLQG
                                                                          t
```

**Group 5**

```
JT33F    (20) MSRGSsACGtSPFTVAmLLFVVVGILYFTLLHLLrHMAIQQARDLLGHRE
CT44F     (7) MSRgSsTSGTSpCIVAMLLFVAVGILYAILLLLLKDMAsRgARdLLGHRE
TCHN-D1  (12) m-rgs-tsgts---VAmLL-vvVaILyatLLlllr-mA-p----Llg-R-
                              a         w
```

**TTMV**

```
         (13) (no ORF3 found)
```

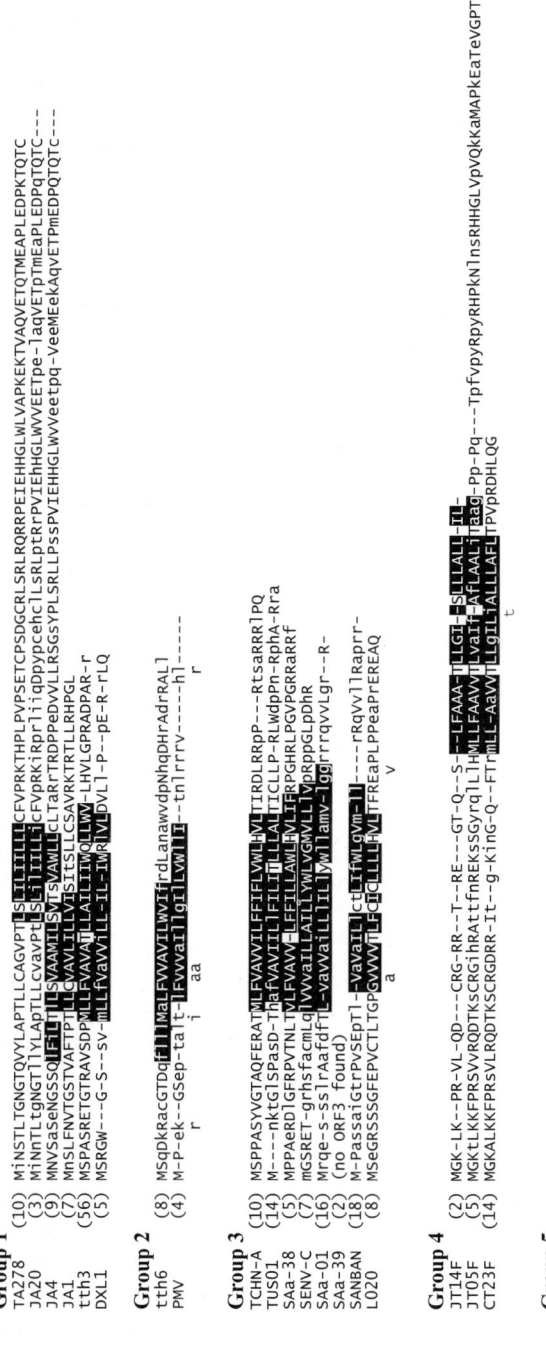

**Fig. 2** Consensus amino-acid sequences of TAIP and N-TAIP in the various groups and subgroups of TTV. For every subgroup, the name of a representative isolate is given, as well as the number of closely related sequences that were analyzed to arrive at the consensus shown. *Uppercase letters* indicate more than 75% conservation, *lowercase* 50%–75%, and less than 50% conservation is indicated by a *hyphen*. Where two amino acids are conserved at 50% each, one is shown *below* the sequence. Hydrophobic stretches are indicated in *negative print*

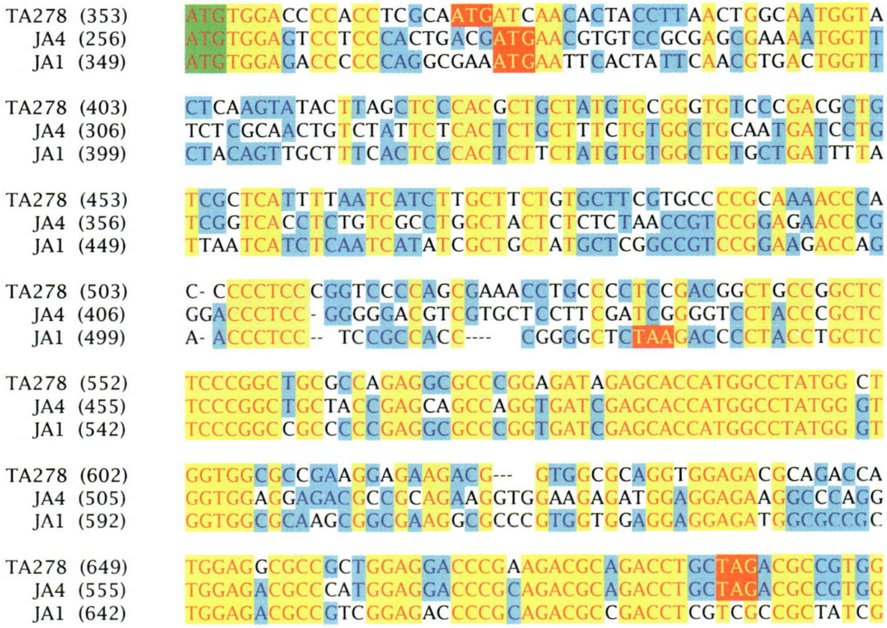


**Fig. 3** Alignment of ORF3 nucleotide sequences from group 1 TTV isolates TA278 (AB017610), JA4 (AF122917), and JA1 (AF122916). Initiation and termination codons of ORF3 are *boxed in red*, initiation codons of ORF2 in *green*. *Numbers in brackets* indicate nucleotide numbers from the respective sequences

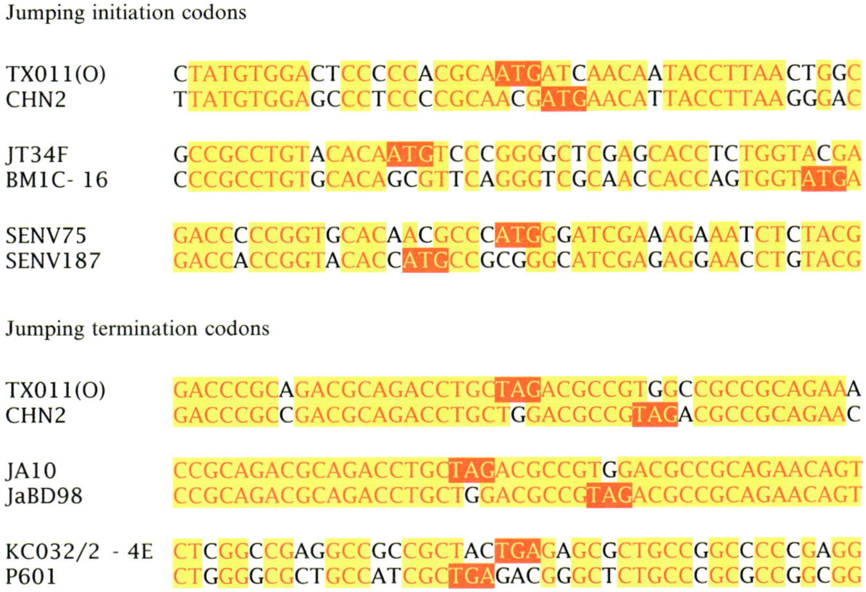


**Fig. 4** Alignment of initiation and termination regions of ORF3 from selected isolates of TTV, illustrating how mutations in initiation and termination codons are compensated by secondary mutations nearby

play an important functional role by promoting multimerization and binding a variety of host proteins. Figure 2 shows that a similar hydrophobic region is found in all of the TAIP and N-TAIP sequences that were collected, despite a remarkably weak conservation of the amino-acid sequence per se. The presence of this encoded hydrophobic stretch not only supports our proposal that TAIP and N-TAIP are indeed expressed, but also that the nature of the peptide is of importance. It is therefore unlikely that the ORF only serves a regulatory purpose, for example controlling the expression of one of the other viral genes. The high variation in the precise sequence of the hydrophobic stretch and even in its position also implies that its presence is not a mere consequence of conservation at the nucleotide level, enforced by the encoding of the overlapping ORF2. Indeed, in the two isolates from group 3 where no ORF was found (TYM9 and SAa-39), the corresponding region also no longer encodes a hydrophobic stretch.

We can thus conclude that almost all TTV variants encode a TAIP-related protein (TAIP or N-TAIP) containing a strongly conserved hydrophobic region. Both in size and in amino-acid composition, N-TAIP resembles the N-terminal half of apoptin from CAV. Most remarkably, a similar peptide of 58 amino acids is probably expressed late in infection by CAV, due to mRNA splicing (Fig. 1a).

## Diseases and Therapies

### *Relationship with Diseases*

In young chickens, CAV is an immunosuppressive pathogen (Noteborn 2004). Mutations in VP2 resulted in the cytoplasmic localization of apoptin and a reduction of cytopathogenicity in MDV-transformed chicken cells (Peters et al. 2006). As apoptin and VP2 are strongly conserved it can be assumed that these proteins will be responsible for the cytopathogenic effects observed in CAV-infected chickens worldwide.

The fact that TAIP or its truncated form N-TAIP seems to be related to apoptosis suggests that apoptosis might also play a role in the cytopathogenic effect of TTV infection. Thus far, however, only one clinicopathological study has reported that a TTV infection can induce apoptosis (Hu et al. 2002). One should examine whether TTV strains containing TAIP or its truncated version N-TAIP are inducing higher apoptosis levels in TTV-infected tissue than TTV genotypes that do not contain these elements. However, e.g., the analysis of a single heavily infiltrated spleen from a patient with Hodgkin's lymphoma showed no less than 24 different TTV genotypes (Jelcic et al. 2004). This indicates that heterogeneity within TTV populations makes it extremely difficult to relate a specific TTV genome/ORF to a specific cytopathogenic effect.

The fact that several reports have suggested that TTV is linked to certain cancers such as breast and lung cancer (De Villiers et al. 2002), hepatocarcinoma (Desai et al. 2005), colorectal cancer (De Villiers et al. 2007), leukemias, and lymphomas (Zur Hausen and De Villiers 2005), might lead to the hypothesis that TTV strains expressing TAIP-like proteins need the transforming cellular conditions to induce apoptosis. If this is the case, certain TTV genotypes might display similar behavior as known oncolytic viruses that preferentially spread in and kill tumor cells such as human parvovirus H1 (Raykov and Rommelaere 2008). Before one can draw conslusions on these hypotheses, fundamental and clinical studies have to be carried out.

## *Anticancer Therapies*

Besides being a pathogenic agent in young chickens, apoptin also harbors therapeutic potentials. One of the huge benefits of apoptin expression in human cells is that it can specifically kill transformed cells, and leaves normal, healthy ones unharmed (Backendorf et al. 2008). Clinical studies have shown that apoptin harbors the potential to become a novel anticancer agent (Visser et al., 2007; Backendorf 2008; Maddika et al. 2006; Chada and Ramesh 2007). Adenovirus delivery of apoptin in prostate cancer, breast cancer, and hepatocarcinoma xenografts in mouse models significantly reduced tumor growth (Liu et al. 2006; Pietersen 2003; Van der Eb et al. 2002; Pietersen et al. 1999). Li et al. (2006) reported antitumor effects of a recombinant fowlpox virus vector expressing apoptin in subcutaneous aggressive H22 mouse hepatoma tumors established in C57B/6 mice. Apoptin expression in combination with chemotherapeutic drugs enhanced its antitumor activity even further (Lian et al. 2007; Liu et al. 2006; Olijslagers et al. 2007). Recently, Peng et al. (2007) showed that in mice systemic delivery of apoptin via the asialoglycoprotein receptor resulted in a dramatic reduction of liver tumor tissues, whereas the surrounding normal liver cells were unaffected. The first TAIP studies in several human tumor cell lines have shown that TAIP also has the potential to become a basis for treating human tumors, and as it looks now, particularly hepatocarcinomas. Regarding the different apoptosis activities of apoptin and TAIP, one should examine whether combinatorial treatment based on both agents provides enhanced antitumor effects. Further studies have to be carried out for establishing the potential anticancer activities of TAIP.

Besides its potential as an anticancer agent, apoptin can also be valuable as a sensor of the tumor-related processes that result in its tumor-specific activation. The above-described apoptin-related kinase activity and the mentioned apoptin-interacting proteins are examples of such tumor-related factors and represent potential drug targets for the development of novel anticancer drugs. TAIP also seems to be an agent with which tumor-related processes can be identified, as it appears to have some apoptosis specificity in certain tumors such as hepatocarcinomas, although it can induce apoptosis in others.

## Conclusions

We have reported herein that the two related viruses CAV and TTV harbor (putative) apoptosis-inducing proteins, apoptin and TAIP, which display interesting similarities.

CAV-derived apoptin is synthesized in infected cells as a protein of 121 amino acids and putatively late in infection, due to splicing, also as its N-terminal half. Both apoptin protein halves harbor apoptosis activities in (human) tumor cells. The C-terminal half requires active phosphorylation at a distinct threonine.

The TTV-derived apoptin-like TAIP is present in a limited group of the heterogenic TTV population. However, a very broad panel of analyzed TTV strains contains a strongly conserved ORF representing the N-terminal half of TAIP, N-TAIP.

The N-terminal regions of both apoptin and TAIP and their variants contain a conserved hydrophobic domain, which seems to act as a hot-spot domain binding to various proteins.

Fundamental and preclinical studies have generated evidence that apoptin is a potential antitumor agent. Further studies on the TAIP-like ORFs within the TTV population might shed light on their function as a cytopathogenic agent and application as potential therapeutic agents, as was found for apoptin.

**Acknowledgements** This study was supported by the Centre for Medical Systems Biology (CMSB) in the framework of the Netherlands Genomics Initiative (NGI).

## References

Ahuja D, Saenz-Robles MT, Pipas JM (2005) SV40 large T antigen targets multiple cellular pathways to elicit cellular transformation. Oncogene 24:7729–7745

Backendorf C, Visser AE, De Boer AG, et al (2008) Apoptin: therapeutic potential of an early sensor of carcinogenic transformation. Annu Rev Pharmacol Toxicol 48:143–169

Bando M, Takahashi M, Ohno S (2008) Torque teno virus DNA titre elevated in idiopathic pulmonary fibrosis with primary lung cancer. Respirology 13:263–269

Bellon M, Nicot C (2008) Regulation of telomerase and telomeres: human tumor viruses take control. J Natl Cancer Inst 100:98–108

Chada S, Ramesh R (2007) Apoptin studies illuminate intersection between lipidomics and tumor suppressors. Mol Ther 15:7–9

Chen Y, Xu Y, Bao Q, et al (2007) Structural and biochemical insights into the regulation of protein phosphatase 2A by small t antigen of SV40. Nat Struct Mol Biol 14:527–534

Cheng CM, Huang SP, Chang YF, et al (2003) The viral death protein apoptin interacts with Hippi, the protein interactor of Huntingtin-interacting protein 1. Biochem Biophys Res Commun 305:359–364

Cho US, Morrone S, Sablina AA, et al (2007) Structural basis of PP2A inhibition by small t antigen. PLoS Biol 5:e202

Collura V, Boissy G (2007) From protein-protein complexes to interactomics. Subcell Biochem 43:135–183

Cossarizza A (2008) Apoptosis and HIV infection: about molecules and genes. Curr Pharm Des 14:237–244

Danen-Van Oorschot AA, Fischer DF, Grimbergen JM, Klein B, Zhuang S, Falkenburg JH, Backendorf C, Quax PH, Van der Eb AJ, Noteborn MH (1997) Apoptin induces apoptosis in human transformed and malignant cells but not in normal cells. Proc Natl Acad Sci USA 94:5843–5847

Danen-Van Oorschot AA, Van der Eb AJ, Noteborn MH (2000) The chicken anemia virus-derived protein apoptin requires activation of caspases for induction of apoptosis in human tumor cells. J Virol 74:7072–7078

Danen-Van Oorschot AA, Zhang YH, Leliveld SR, et al (2003) Importance of nuclear localization of apoptin for tumor-specific induction of apoptosis. J Biol Chem 278:27729–27736

Danen-Van Oorschot AA, Voskamp P, Seelen MC, et al (2004) Human death effector domain-associated factor interacts with the viral apoptosis agonist apoptin and exerts tumor-preferential cell killing. Cell Death Differ 11:564–573

De Villiers EM, Schmidt R, Delius H, et al (2002) Heterogeneity of TT virus-like sequences isolated from human tumour biopsies. J Mol Med 80:44–50

De Villiers EM, Bulajic M, Nitsch C, et al (2007) TTV infection in colorectal cancer tissues and normal tissues. Int J Cancer 121:2109–2112

Deppert W (2000) The nuclear matrix as a target for viral and cellular oncogenes. Crit Rev Eukaryot Gene Expr 10:45–61

Desai M, Pal R, Deshmukh R, et al (2005) Replication of TT virus in hepatocyte and leucocyte cell lines. J Med Virol 77:136–143

Guelen L, Paterson H, Gäken J, et al (2004) TAT-apoptin is efficiently delivered and induces apoptosis in cancer cells. Oncogene 23:1153–1165

Halazonetis TD, Gorgoulis VG, Bartek J (2008) An oncogene-induced DNA damage model for cancer development. Science 319:1352–1355

He CQ, Ding NZ, Fan W, et al (2007) Identification of chicken anemia virus putative intergenotype recombinants. Virology 366:1–7

Heilman DW, Teodoro JG, Green MR (2006) Apoptin nucleocytoplasmic shuttling is required for cell type-specific localization, apoptosis, and recruitment of the anaphase promoting complex/cyclosome to PML bodies. J Virol 80:7535–7545

Hino S, Miyata H (2007) Torque teno virus (TTV): current status. Rev Med Virol 17:45–57

Hu ZJ, Lang ZW, Zhou YS, et al (2002) Clinicopathological study on TTV infection in hepatitis of unknown etiology. World J Gastroenterol 8:288–293

Janssen K, Hofmann TG, Jans DA, et al (2007) Apoptin is modified by SUMO conjugation and targeted to promyelocytic leukemia protein nuclear bodies. Oncogene 26:1557–1566

Jelcic I, Hotz-Wagenblatt A, Hunzicker A, et al (2004) Isolation of multiple TT virus genotypes from spleen biopsy tissue from a Hodgkin's disease patient: genome reorganization and diversity in the hypervariable region. J Virol 78:7498–7507

Kamada K, Kuroishi A, Kamahora T, et al (2006) Spliced mRNAs detected during the life cycle of chicken anemia virus. J Gen Virol 87:2227–2233

Kamahora T, Hino S, Miyata H (2000) Three spliced mRNAs of TT virus transcribed from a plasmid containing the entire genome in COS1 cells. J Virol 74:9980–9986

Klanrit P, Flinterman MB, Odell EW, et al (2008) Specific isoforms of p73 control the induction of cell death induced by the viral proteins, E1A or apoptin. Cell Cycle 7:205–215

Kooistra K, Zhang YH, Henriquez NV, Weiss B, Mumberg D, Noteborn MH (2004) TT virus-derived apoptosis-inducing protein induces apoptosis preferentially in hepatocellular carcinoma-derived cells. J Gen Virol 85:1445–1450

Kooistra K, Zhang YH, Noteborn MH (2007) Viral elements sense tumorigenic processes: approaching selective cancer therapy. Mini Rev Med Chem 7:1155–1165

Landry MC, Robert A, Lavoie JN (2006) Alternative cell death pathways: lessons to be learned from a viral protein. Bull Cancer 93:921–930

Lee YH, Cheng CM, Chang YF, et al (2007) Apoptin T108 phosphorylation is not required for its tumor-specific nuclear localization but partially affects its apoptotic activity. Biochem Biophys Res Commun 354:391–395

Leliveld SR, Zhang YH, Rohn JL, et al (2003a) Apoptin induces tumor-specific apoptosis as a globular multimer. J Biol Chem 278:9042–9051

Leliveld SR, Dame RT, Mommaas MA, et al (2003b) Apoptin protein multimers form distinct higher-order nucleoprotein complexes with DNA. Nucleic Acids Res 31:4805–4813

Leppik B, Gunst K, Lehtinen M, et al (2007) In vivo and in vitro intragenomic rearrangement of TT viruses. J Virol 81:9346–9356

Li X, Jin N, Mi Z, et al (2006) Antitumor effects of a recombinant fowlpox virus expressing apoptin in vivo and in vitro. Int J Cancer 119:2948–2957

Lian H, Jin N, Li X, et al (2007) Induction of an effective anti-tumor immune response and tumor regression by combined administration of IL-18 and apoptin. Cancer Immunol Immunother 56:181–192

Liu X, Elojeimy S, El-Zawahry AM, et al (2006) Modulation of ceramide metabolism enhances viral protein apoptin's cytotoxicity in prostate cancer. Mol Ther 14:637–646

Maddika S, Mendoza FJ, Hauff K, et al (2006) Cancer-selective therapy of the future: apoptin and its mechanism of action. Cancer Biol Ther 5:10–19

Miyata H, Tsunoda H, Kazi A, et al (1999) Identification of a novel GC-rich 113-nucleotide region to complete the circular, single-stranded DNA genome of TT virus, the first human circovirus. J Virol 73:3582–3586

Nimomiya M, Takahashi M, Shimosegawa T, et al (2007) Analysis of the entire genomes of fifteen torque teno midi virus variants classifiable into a third group of genus *Anellovirus*. Arch Virol 152:1961–1975

Noteborn MH (2004) Chicken anemia virus induced apoptosis: underlying molecular mechanisms. Vet Microbiol 98:89–94

Noteborn MH (2005) Apoptin acts as a tumor-specific killer: potentials for an anti-tumor therapy. Cell Mol Biol 51:49–60

Noteborn MH, Van der Eb AJ (1999) Apoptin-induced apoptosis: potential for anti-tumor therapy. Drug Resist Updat 1:99–103

Noteborn MH, De Boer GF, Kant A, et al (1990) Expression of avian leukemia virus env-gp85 in *Spodoptera frugiperda* cells by use of a baculovirus expression vector. J Gen Virol 71:2641–2648

Noteborn MH, De Boer GF, Kant A, et al (1991) Characterization of cloned chicken anemia virus DNA that contains all elements for the infectious replication cycle. J Virol 65:3131–3139

Noteborn MH, Kranenburg O, Zantema A, et al (1992) Transcription of the chicken anemia virus (CAV) genome and synthesis of its 52-kDa protein. Gene 118:267–271

Noteborn MH, Todd D, Verschueren CA, et al (1994) A single chicken anemia virus protein induces apoptosis. J Virol 68:346–351

Noteborn MH, Danen-Van Oorschot AA, Van der Eb AJ (1998) Chicken anemia virus: induction of apoptosis by a single protein of a single-stranded DNA virus. Semin Virol 8:497–504

Nüesch JP, Rommelaere J (2007) A viral adaptor protein modulating casein kinase II activity induces cytopathic effects in permissive cells. Proc Natl Acad Sci USA 104:12482–12487

Olijslagers SJ, Zhang YH, Backendorf C, et al (2007) Additive cytotoxic effect of apoptin and chemotherapeutic agents paclitaxel and etoposide on human tumor cells. Basic Clin Pharmacol Toxicol 100:127–131

Peng DJ, Sun J, Wang YZ, et al (2007) Inhibition of hepatocarcinoma by systemic delivery of apoptin gene via the hepatic asialoglycoprotein receptor. Cancer Gene Ther 14:66–73

Peters MA, Jackson DC, Crabb BS, et al (2002) Chicken anemia virus VP2 is a novel dual specificity protein phosphatase. J Biol Chem 277:39566–39573

Peters MA, Jackson DC, Crabb BS, et al (2005) Mutation of chicken anemia virus VP2 differentially affects serine/threonine and tyrosine protein phosphatase activities. J Gen Virol 86:623–630

Peters MA, Crabb BS, Washington EA, et al (2006) Site directed mutagenesis of the VP2 gene of chicken anemia virus affects virus replication, cytopathology and host-cell MHC class 1 expression. J Gen Virol 87:823–831

Pietersen AM (2003) Preclinical studies with apoptin. Thesis, Erasmus University, Rotterdam

Pietersen AM, Van der Eb MM, Rademaker HJ, et al (1999) Specific tumor-cell killing with adenovirus vectors containing the apoptin gene. Gene Ther 6:882–892

Psyrri A, DiMaio D (2008) Human papillomavirus in cervical and head-and-neck cancer. Nat Clin Pract Oncol 5:24–31

Qiu J, Kakkola L, Cheng F, et al (2005) Human circovirus TT virus genotype 6 express six proteins following transfection of a full-length clone. J Virol 79:6505–6510

Raykov Z, Rommelaere J (2008) Potential of tumour cells for delivering oncolytic viruses. Gene Ther 15:704–710

Rohn JL, Noteborn MH (2004) The viral death effector apoptin reveals tumor-specific processes. Apoptosis 9:315–322

Rohn JL, Zhang YH, Aalbers RI, et al (2002) A tumor-specific kinase activity regulates the viral death protein apoptin. J Biol Chem 277:50820–50827

Rohn JL, Zhang YH, Leliveld SR, et al (2005) Relevance of apoptin's integrity for its functional behavior. J Virol 79:1337–1338

Russo A, Terrasi M, Agnese V, et al (2006) Apoptosis: a relevant tool for anticancer therapy. Ann Oncol 17[Suppl 7]:115–123

Shulman-Peleg A, Shatsky M, Nussinov R, et al (2007) Spatial chemical conservation of hot spot interactions in protein-protein complexes. BMC Biol 5:43

Sun GJ, Tong X, Dong Y, et al (2002) Identification of a protein interacting with Apoptin from human leucocyte cDNA library by using yeast two-hybrid screening. Sheng Wu Hua Xue Yu Sheng Wu Wu Li Xue Bao (Shanghai) 34:369–372

Tolboom TC, Zhang YH, Henriquez NV, et al (2006) Fibroblast-like synoviocytes from patients with rheumatoid arthritis are more sensitive to apoptosis induced by the viral protein, apoptin, than fibroblast-like synoviocytes from trauma patients. Clin Exp Rheumatol 24:142–147

Van der Eb MM, Pietersen AM, Speetjens FM (2002) Gene therapy with apoptin induces regression of xenografted human hepatomas. Cancer Gene Ther 9:53–61

Van Santen VL, Li L, Hoerr FJ, et al (2001) Genetic characterization of chicken anemia virus from commercial broiler chickens in Alabama. Avian Dis 45:373–378

Visser AE, Backendorf C, Noteborn MH (2007) Viral protein apoptin as a molecular tool and therapeutic bullet: implications for cancer tool. Future Virol 2:519–527

Zhang YH, Leliveld SR, Kooistra K, et al (2003) Recombinant apoptin multimers kill tumor cells but are nontoxic and epitope-shielded in a normal-cell-specific fashion. Exp Cell Res 289:36–46

Zhang YH, Kooistra K, Pietersen AM, et al (2004) Activation of the tumor-specific death effector apoptin and its kinase by an N-terminal determinant of simian virus 40 large T antigen. J Virol 78:9965–9976

Zhuang SM, Shvarts A, Van Ormondt H, et al (1995) Apoptin, a protein derived from chicken anemia virus, induces p53-independent apoptosis in human osteosarcoma cells. Cancer Res 55:486–489

Zur Hausen H, De Villiers EM (2005) Virus target cell conditioning model to explain some epidemiologic characteristics of childhood leukemias and lymphomas. Int J Cancer 115:1–5

# Chicken Anemia Virus

**K.A. Schat**

**Contents**



**Abstract** Chicken anemia virus (CAV), the only member of the genus *Gyrovirus* of the Circoviridae, is a ubiquitous pathogen of chickens and has a worldwide distribution. CAV shares some similarities with Torque teno virus (TTV) and Torque teno mini virus (TTMV) such as coding for a protein inducing apoptosis and a protein with a dual-specificity phosphatase. In contrast to TTV, the genome of CAV is highly conserved. Another important difference is that CAV can be isolated in cell culture. CAV produces a single polycistronic messenger RNA (mRNA), which is translated into three proteins. The promoter-enhancer region has four direct repeats resembling estrogen response

K.A. Schat
Department of Microbiology and Immunology, College of Veterinary Medicine, Cornell University, Ithaca, NY 14853, USA
kas24@cornell.edu

E.-M. de Villiers, H. zur Hausen (eds.), *TT Viruses: The Still Elusive Human Pathogens,* 

elements. Transcription is enhanced by estrogen and repressed by at least two other transcription factors, one of which is COUP-TF1. A remarkable feature of CAV is that the virus can remain latent in gonadal tissues in the presence or absence of virus-neutralizing antibodies. In contrast to TTV, CAV can cause clinical disease and subclinical immunosuppression especially affecting $CD8^+$ T lymphocytes. Clinical disease is associated with infection in newly hatched chicks lacking maternal antibodies or older chickens with a compromised humoral immune response.

**Abbreviations** ALV: Avian leukosis virus; CAV: Chicken (infectious) anemia virus; COUP-TF1: Chicken ovalbumin upstream promoter transcription factor 1 CTL: Cytotoxic T lymphocytes; DR: Direct repeat(s); ds: Double-stranded; EGFP: Enhanced green fluorescent protein; ERE: Estrogen response element; HRE: Hormone response elements; HVT: Herpesvirus of turkeys; IBD(V): Infectious bursal disease (virus); IB(V): Infectious bronchitis (virus); MAb: Monoclonal antibodies; MDCC: Marek's disease chicken cell line; MDV: Marek's disease virus; MHC: Major histocompatibility complex; NK: Natural killer; nt: Nucleotides; p: Passage; pi: Post infection; PCR: Polymerase chain reaction; PCV: Porcine circovirus; q(RT-)PCR: Quantitative (reverse transcription) PCR; RF: Replicative form; REV: Reticuloendotheliosis virus; SPF: Specific-pathogen-free; TCR: T cell receptor; TTV: Torque teno virus; TTMV: Torque teno mini virus; VN: Virus-neutralizing; VP: Viral protein

## Introduction

In 1979 Yuasa and coworkers (Yuasa et al. 1979) reported the presence of a new chicken pathogen causing anemia in specific-pathogen-free (SPF) chickens inoculated with herpesvirus of turkeys (HVT) to protect against Marek's disease (MD), a herpesvirus-induced T cell lymphoma. This new pathogen was first named chicken anemia agent (CAA) and is now known as chicken infectious anemia virus (CAV or CIAV). CAV is thus far the only member of the genus *Gyrovirus*, which belongs to the family of Circoviridae together with viruses of the *Circovirus* genus (ICTV 2008). In addition to these two genera, the Anelloviruses, which include the human torque teno virus (TTV) and torque teno mini virus (TTMV), are often included in the Circoviridae (Hino and Miyata 2007). As with TTV and TTMV in humans, infection with CAV is ubiquitous in its target species (chickens) and can be found in most if not all commercial poultry operations in all continents with a poultry industry (Schat and Van Santen 2008). However, unlike TTV and TTMV, CAV has been linked to specific clinical diseases and subclinical immunosuppression, although the degree of disease and immunosuppression depends on a number of factors, which will be discussed in Pathogenesis. This review will focus on the aspects that are relevant to the understanding of the pathogenesis and highlights some of the differences with Anelloviruses. A more detailed review of CAV has recently been published (Schat and van Santen 2008).

## Historical Background

The first isolation of CAV was the result of an investigation in disease problems in commercial chickens vaccinated with HVT. Yuasa et al. (1976) linked the disease problem to the presence of reticuloendotheliosis virus (REV) in HVT vaccine batches. In their subsequent study, material from one of the vaccine batches was shown to induce anemia. A highly resistant anemia-inducing agent was isolated by chick inoculation (Yuasa et al. 1979). The link to a contaminated MD vaccine suggested a link to MD vaccination as the cause of widespread distribution of CAV. However, subsequent investigations revealed that CAV was not a new pathogen associated in some way with the introduction of MDV vaccines in the early 1970s. W.C. Wellenstein (personal communication 1989, quoted by Schat and van Santen 2008) isolated the ConnB strain of CAV from an ampoule of MD tumor cells that had been stored in liquid nitrogen at least since 1969. The tumor material was obtained from chickens experimentally infected with MD virus (MDV) that had unexpectedly experienced severe anemia (Jakowski et al. 1970). This MDV strain had been passed at least 25 times in Line 7 chickens without causing anemia. Interestingly, the anemia appeared for the first time when the MDV strain was inoculated in an F2 cross of SPF×Line 7 chickens while uninfected F2 chickens remained healthy. This suggests that the SPF birds used to generate the F2 cross may have been latent carriers of CAV and that infection with MDV activated or aggravated CAV replication, resulting in anemia (see Pathogenesis of infection in SPF chickens for a discussion of latency and Interactions with Other Pathogens for interactions between CAV and other viruses).

Soiné et al. (1993, 1994) and Renshaw et al. (1996) confirmed by polymerase chain reaction (PCR) and sequencing that the ConnB strain was indeed CAV (Soiné et al. 1993). Toro et al. (2006a) confirmed that CAV infections were present in the United States as early as 1959 by analyzing banked sera from chickens used for the production of alloantisera. In Europe poultry fanciers frequently maintain old chicken breeds to show at exhibitions. These birds are not vaccinated or only vaccinated for Newcastle disease. CAV antibodies were found in 75% and 89.5% of the flocks in The Netherlands and Switzerland, respectively (Wunderwald and Hoop 2002; de Wit et al. 2004). Like TTV infections in humans, there is no indication that CAV causes any disease in these fancy breeds. The combined evidence of these studies suggests that CAV is not a new virus spread by MD vaccination, but that it is an "old" virus that has been present for a long time without causing overt problems. Based on the model proposed for the control of virus replication (see Transcriptional control for details) Miller and Schat (2004) suggested that CAV has evolved as a successful pathogen that can be maintained in successive generations of chickens without causing disease. However, in the modern high-stress environment of commercial poultry production it is a major pathogen that can cause significant economic problems (Schat and van Santen 2008).

# Characterization of CAV and Virus Replication

## *Virus Characteristics*

### Virus Structure

Viruses belonging to the Circoviridae share a number of characteristics but there are also major differences between the different genera (see also the chapter by S. Hino and A.A. Prasetyo, this volume). All viruses belonging to this family have a small covalently closed circular single-stranded DNA genome. CAV and Anelloviruses both have a negative sense genome in contrast to viruses in the genus *Circovirus*. The genome of CAV consists of 2,298 or 2,319 nucleotides (nt) (Claessens et al. 1991; Noteborn et al. 1991). The difference between these two sequences is the addition of an extra direct repeat (DR) of 21 nt in the promoter enhancer region (see Transcriptional Control). A limited number of passages ($n$=20–30) in the MDV-derived chicken cell line (MDCC)-MSB1 were reported to result in the insertion of the fifth DR without influencing the pathogenicity (Todd et al. 1995). However, other researchers, using a different isolate, were unable to demonstrate the insertion of a fifth DR after 129 passages (Chowdhury et al. 2003).

### Resistance to Physical and Chemical Treatments

Other shared characteristics of the Circoviridae are the small size of the virus particles and the extreme resistance to physical and chemical treatments. This combination complicates the elimination of these viruses from other pathogens, especially if cell culture passage is impossible or undesirable. CAV has an average size of 25 nm (Gelderblom et al. 1989; Crowther et al. 2003) while porcine circovirus type 2 (PCV-2), an important pathogen in pigs, has an average size of 20.5 nm (Crowther et al. 2003). CAV and PCV-2 are highly resistant to many common disinfectants and heat treatment (Yuasa et al. 1979; Yuasa 1992; Urlings et al. 1993; Opriessnig et al. 2007). The resistance of Circoviridae is a concern for the preparation of blood products for human use, especially in view of the widespread occurrence of TTV and TTMV. Commonly used procedures for virus inactivation in plasma products are based on the use of detergents, pasteurization of products in solution, or dry heating of lyophilized products. Welch et al. (2006) used CAV and PCV-2 as a substitute for TTV to determine the effectiveness of virus inactivation by pasteurization and dry heat. Pasteurization of CAV in heat-treated human FVIII at 60°C for 24 h reduced the titer by 1.42 log, but increasing the temperature to 75°C reduced the titer more than 3.5 log after 30 min. Heating of lyophilized material at 80°C for 72 h or 120°C for 30 min reduced the virus titers only tenfold. Inactivation of CAV in minced chicken meat required more than 10 min at 90°C (Urlings et al. 1993).

## Genetic Variation

Thus far most CAV isolates belong to the same serotype, although a second serotype has been reported (Spackman et al. 2002a, b) based on physicochemical characteristics and pathology. Interestingly, the second serotype was isolated using MDCC-CU147, which is approximately 10–100 times more susceptible to infection with serotype 1 strains than the commonly used MDCC-MSB1 cell line (Calnek et al. 2000) (see Virus Isolation and Propagation). There was no cross neutralization between the two serotypes and a commercial enzyme-linked immunosorbent assay (ELISA) test did not detect antibodies to the serotype 2 virus. CAV-specific primers failed to amplify serotype 2 DNA. Based on the lack of antibody detection in the ELISA, the negative PCR results and absence of information on the genome, it is not certain that this pathogen is even a circovirus. Many of the reported physicochemical characteristics are similar to parvoviruses and picornaviruses. As a consequence all strains are considered to belong to the same serotype in this review.

A major difference between CAV and the Anelloviruses is the genetic stability of the former. Isolates separated by time and geographical location show very little variation at the nucleotide level or the amino acid level for viral protein 1 (VP1). For example, Renshaw et al. (1996) reported that the predicted amino acid sequence for VP1 of the ConnB isolate (U69548) obtained from material frozen in 1969 (see Historical Background) has only 5 or 6 amino acid differences with the Cux-1 isolate based on the sequence provided by Noteborn et al. (1991) (M55918) and Meehan et al. (1992) (M81223), respectively. Cux-1 was isolated in Germany (von Bülow et al. 1983) while the ConnB isolate came from Connecticut in the United States. Recently a large number of isolates have been fully or partially sequenced. These isolates were obtained in Bangladesh (Islam et al. 2002), Brazil (Nogueira et al. 2007), China (He et al. 2007; Ducatez et al. 2008), Malaysia (Chowdhury et al. 2003), Nigeria (Ducatez et al. 2006), Slovenia (Krapež et al. 2006) and the United States (van Santen et al. 2001; van Santen et al. 2007). Comparison of all sequence data indicates that these isolates can be divided into 3 or 4 distinct groups, with "old" and "new" isolates scattered throughout the different groups. The overall number of nucleotide differences for the VP1 sequence is limited and most reports indicate that the maximum difference between strains at the amino acid level is less than 6% (e.g., Ducatez et al. 2006, 2008, see Fig. 1). The differences between the predicted amino acid sequences for VP2 and VP1 are even smaller and the promoter/enhancer region has very few differences. He et al. (2007) reported that recombination between genotypes can occur, which may lead to new genotypes. Thus far the minor differences between genotypes have not been linked to clear differences in pathogenicity (reviewed by Schat and van Santen 2008).

The lack of genetic diversity of CAV contrasts sharply with the high degree of genetic diversity described for TTV and TTMV. Ducatez et al. (2008) suggested that VP1 is under strong negative selection, especially in countries where chickens

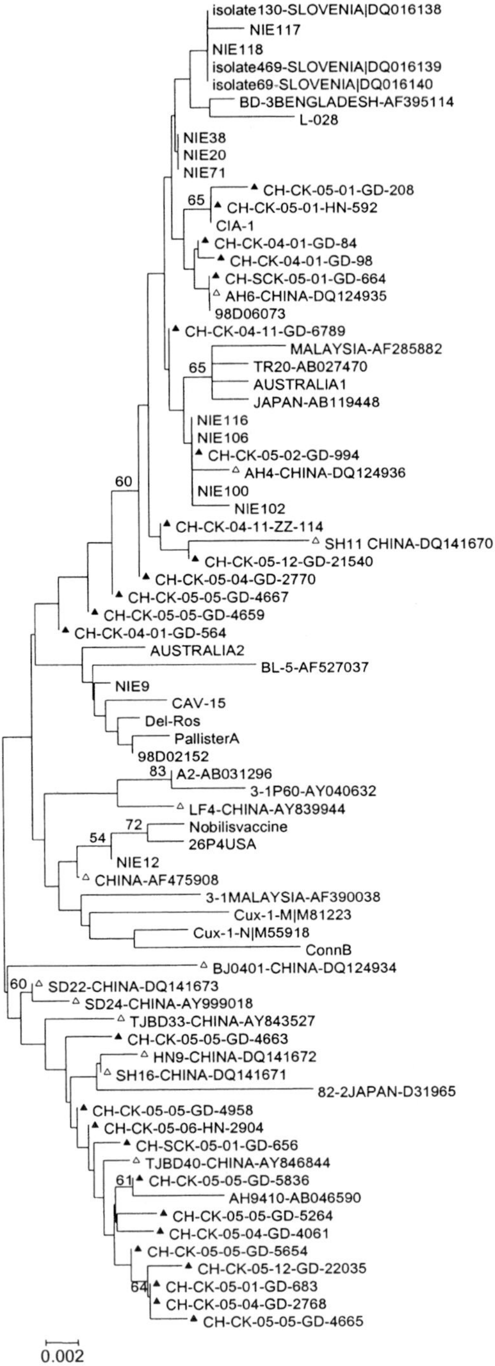


**Fig. 1** Phylogenic analysis of the predicted amino acid sequences of VP1 of 26 new CAV isolates from Guangdong and Huan provinces, China and 47 VP1 sequences available from GenBank. The Nobilis P4 vaccine strain (Intervet Schering-Plough Animal Health, Boxmeer) is also included. Numbers at nodes correspond to bootstrap values>49. Chinese strains described by Ducatez et al. (2008) are marked with a *closed triangle*, whereas other Chinese sequences from GenBank have *open triangles*. From Ducatez et al. (2008), with permission from the authors

are not vaccinated. However, humans are not vaccinated against TTV or TTMV and vaccination-driven diversity for these two virus families seems therefore unlikely. A second explanation offered by these authors is the exquisite relationship between virus and host, which may lead to vertical transmission of virus or viral DNA as proposed by Miller and Schat (2004) (see Transcriptional Control and Pathogenesis of Infection in SPF Chickens).

## Virus Isolation and Propagation

Thus far, CAV, PCV-1, and PCV-2 are the only members of the Circoviridae which can be readily isolated in cell culture. Desai et al. (2005) reported that TTV can be propagated in the Chang liver cell line and phytohemagglutinin (PHA)-stimulated peripheral blood mononuclear cells. Hino and Miyata (2007) questioned the efficacy of virus replication in these cultures.

Yuasa (1983) propagated CAV in MDCC-MSB1 and -JP2 and the avian leukosis virus-transformed B cell line LSCC-1104B1, although MDCC-RP1 and -BP1 and LSCC-1104X5 and -TLT were refractory to infection. Not all CAV strains could be isolated in MDCC-MSB1 (Soiné et al. 1994; Renshaw et al. 1996; Islam et al. 2002; Nogueira et al. 2007; van Santen et al. 2007). However, one of these isolates, CIA-1 (Lucio et al. 1990), could be propagated in the MSB1(S) subline, albeit to lower titers than Cux-1 (Renshaw et al. 1996). The difference in infectivity for MSB1 cells between CIA-1 and Cux-1 was originally thought to be related to a 316-bp fragment which included the hypervariable region of VP1 with perhaps secondary influences by differences in VP2 and VP3 (Renshaw et al. 1996). Subsequent examination of additional isolates that could not be propagated in MSB1 cells failed to reveal a consensus sequence responsible for the lack of propagation.

Calnek et al. (2000) examined 26 MDCC lines representing $CD4^+CD8^-$, $CD4^-CD8^+$, and $CD4^-CD8^-$ cells expressing either T cell receptors (TCR)$\alpha\beta1$ or TCR$\alpha\beta2$. Several cell lines representing the different phenotypes could be infected, but MDCC-CU147 ($CD4^-CD8^+$) was significantly more susceptible than MSB1(S) for the propagation of Cux-1. This cell line was approximately 100- to 1,000-fold more susceptible to infection with CIA-1 than MSB1(S). Recently van Santen et al. (2007) confirmed that CU147 can be used successfully for the propagation of CAV isolates that could not be isolated in their MSB1 subline. The reason(s) for the differences in susceptibility between cell lines has not been elucidated.

## Virus Replication

### DNA Replication and Viral Transcription

It is assumed that CAV enters cells through binding to cell surface receptors, but specific receptors have not been identified. The finding that different subsets of T cells as well as specific ALV-transformed B cell lines can be infected (see Virus

Isolation and Propagation) does not clarify this question. Claessens et al. (1991) identified a prominent hairpin structure in the sequence, which was later identified as a nuclease S1 site (Todd et al. 1996; Noteborn et al. 1992). CAV, lacking the machinery to reproduce its own DNA, depends on dividing host cells for the replication of DNA, which requires the formation of double-stranded replicative form (dsRF). The actual mechanism of DNA replication has not been resolved. Todd et al. (1996, 2001) were unable to conclusively determine if replication occurred through the rolling circle replication (RCR) model proposed for PCV-1 and PCV-2. Arguments against the RCR model for CAV are the absence of a replicase-associated (Rep) protein and the absence of sequences in VP1 needed for the binding of dinucleotide triphosphates (dNTP). In addition the conserved stem-loop structure with the nucleotide sequence TAGTATTAC, which is associated with the initiation of the RCR-based replication of DNA in PCV and geminiviruses (Bassami et al. 1998), is only semi-conserved in CAV (nt 42–50 using accession No. M81223; Meehan et al. 1992). On the other hand, three peptide motifs typically associated with RCR DNA replication are present in CAV VP1.

Once the virus or viral DNA has entered the cells, viral transcripts can be detected as early as 8 h post infection (pi). Most authors have reported the presence of a single, polycistronic message of approximately 2.1 kb (Noteborn et al. 1992; Phenix et al. 1994). Using Northern blot analysis Kamada et al. (2006) found two additional polyadenylated transcripts, the larger one is 1.6 kb and the smaller one between 1.2 and 1.3 kb. These RNA species appeared between 48 and 72 h pi in infected but not control MSB1 cells. All transcripts were amplified by RT-PCR; three different clones were obtained in addition to a clone representing the full-length transcript. The 1.3-kb clone used a splice donor and acceptor site at nt 1222 and 1814, while the smaller one of 1.2 kb used a splice donor and acceptor site at nt 994 and 1095 in addition to the ones used by the larger RNA species. Finally, the 0.8-kb clone may represent an artifact because the 5′ and 3′ flanking sequences are identical, suggesting the possibility of homologous recombination. Excluding the potential proteins from the 0.8-kb clone, the 1.3- and 1.2-kb clone could account for two additional VP1- and one additional VP2-based proteins (Fig. 2). Thus far, proteins associated with these transcripts have not been demonstrated and the relevance of these new findings for the biology of CAV is poorly understood at the present time. It will be of considerable interest to determine the fate of virus replication in vitro and in vivo when the different splice donor sites are modified and the new transcripts are no longer expressed.

## Transcriptional Control

The 5′ nontranscribed region of the CAV genome has been identified as the sole promoter/enhancer region controlling viral transcription and replication. This region contains four or five DR with a 12-bp insert after the first two DRs and several known enhancer-like elements (Fig. 3a; Noteborn et al. 1991). The first two DR and the 12-bp insert are the dominant elements based on experiments in which different deletion constructs were used to drive the expression of Cm acetyltransferase. The region encompassing the DR can enhance the thymidine kinase promoter

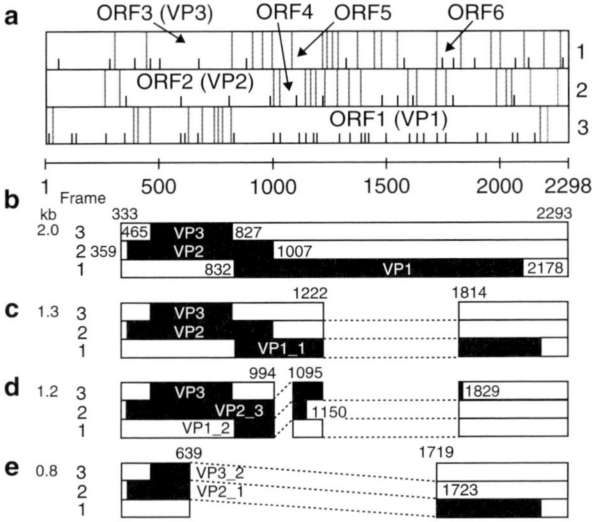


**Fig. 2 a–e** Schematic representation of the CAV transcripts and their candidate ORFs. **a** Six ORFs of the CAV genome. *Short* and *long vertical lines* indicate ATG start and stop codons, respectively. Putative ORFs encoded by the transcripts are indicated by *arrows*. ORFs evident in the 2.0-, 1.3-, 1.2-, and 0.8-kb transcripts are shown in panels **b–e**, respectively. *Dotted lines* indicate introns or deleted regions. *Black boxes* represent putative coding regions. Nucleotide *numbers* indicate the positions of ORF junctions or boundaries of ORFs. From Kamada et al. (2006), with permission from the Society for General Microbiology

independent of its orientation when inserted upstream of the promoter. The DR has a consensus sequence for the cyclic AMP response element binding protein, but it binds an unidentified T cell nuclear protein instead. The 12-bp insert binds the SP1 protein (Noteborn. et al. 1994b). Virus constructs lacking the complete DR region or containing three repeats but lacking the 12-bp insert were not viable, while additions of nucleotides (anywhere from 6 to 24) to the 12-bp insert reduced virus replication significantly (Noteborn et al. 1998b).

Cardona et al. (2000b) reported that ovaries of antibody-positive and -negative SPF hens can be positive for virus DNA (see Pathogenesis of Infection in SPF Chickens). This finding in combination with observations from the SPF poultry industry that birds often seroconvert when in production (Schat and van Santen 2008) led Miller et al. (2005) to hypothesize that CAV expression could be influenced by the hormonal regulation of the reproductive system. Inspection of the DR region suggested that the ACGTCA sequence in the repeats separated by 15 nt could function as a hormone response element (HRE). An optimal HRE typically consists of two inverted repeats separated by 15 nt, but direct repeats separated by 15 nt can also function as an HRE, especially if SP1 or NFY binding sites are located in close proximity (Wang et al. 1999; Saville et al. 2000; Safe 2001). To test the hypothesis that the CAV promoter/enhancer activity could be regulated by hormonal influences, Miller et al. (2005) generated two short (SE and S5) and two long (LE and L5) promoter constructs (Fig. 3a) to drive expression of enhanced green

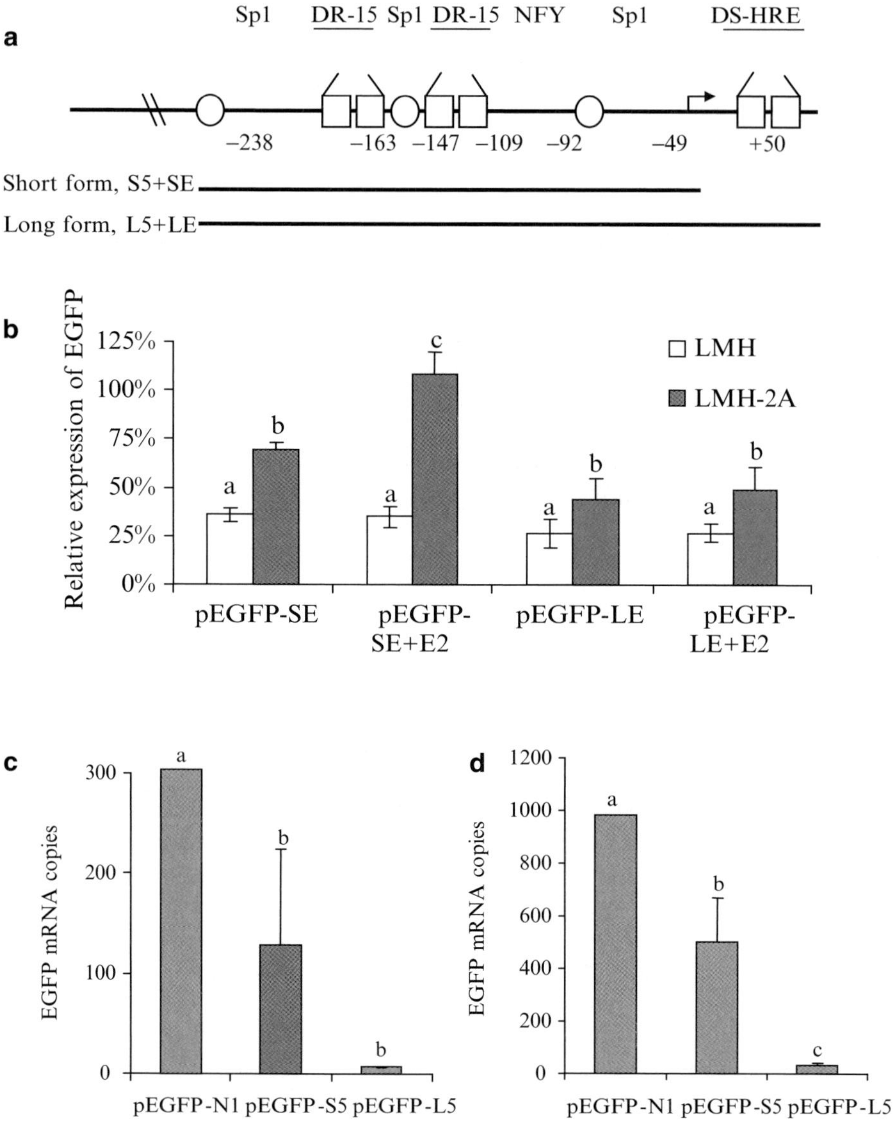


**Fig. 3** **a** Schematic diagram of the CAV promoter region with the areas used for the short and long promoter vectors indicated by the *bars* under the diagram. The CAV promoter 21-bp direct repeat sequences are TGTACAGGGGGGT**ACGTCA**CCCGTACAGGGGGGT**ACGTCA**CA, with the estrogen response element (ERE)-like sequences within the 21 bp DR in *boldface underlined type*. The ERE-like sequences arranged as direct repeats with a 15-bp separation are indicated in the diagram as *boxes* labeled *DR-15*. Known or putative transcription factor binding sites are indicated as *circles* for Sp1, *squares* for ERE-like, and *triangles* for NFY. **b** EGFP expression compared in LMH and LMH-2A cells transfected with pEGFP-SE and pEGFP-LE, with or without $10^{-7}$ M 17β-estradiol (E2). Results are the average of three or more independent experiments, each with triplicate wells. EGFP expression was measured by fluorometer and normalized as a percentage of the positive-control CMV promoter-driven vector. **c** and **d** qRT-PCR analysis of

fluorescent protein (EGFP). The differences between SE and LE versus S5 and L5 are that the former contains some potential start and stop codons from the cloning vector, which could affect translation. Transfection of the chicken liver cell line LMH and the LMH-2A, which overexpresses the estrogen receptor (Sensel et al. 1994), showed significantly enhanced expression in the presence of estrogen (Fig. 3b), suggesting that the HRE can function as an estrogen response element (ERE). It was also noted that the long promoter constructs induced a lower EGFP expression than the short forms. Real-time RT-PCR experiments indicated that the decreased response was at the transcriptional level (Fig. 3c, 3d). The decreased response was linked to an E-box-like sequence at the transcription start point (Miller et al. 2008). In addition to the upregulation of expression by estrogen, these authors also found that the chicken ovalbumin upstream promoter transcription factor 1 (COUP-TF1) can negatively regulate transcription of EGFP by binding to the short promoter form. Clearly, regulation of expression of CAV is complex and may be dose-dependent with low virus levels being controlled by a balance between COUP-TF1 and E-box-binding proteins downregulating and estrogen upregulating virus replication.

## Functions of Viral Proteins

As predicted by the initial sequence information (Claessens et al. 1991; Noteborn et al. 1991; Meehan et al. 1992) three viral proteins, VP1, VP2 and VP3, have been identified in virus-infected cells (Chandratilleke et al. 1991). These proteins are translated from a single polycistronic 2.1-kb mRNA by alternate use of initiation codons. VP1 uses the fifth AUG as the start codon (Noteborn et al. 1992), but the actual mechanism of alternate codon use has not been elucidated. Douglas et al. (1995) reported that VP2 and VP3 could be detected in infected cells as early as 12 h pi, while VP1 was not present until 30 h.

The 52-kDa VP1 is the only structural protein forming the viral capsid (Todd et al. 1990). The N-terminal region has limited similarity to protamines (Claessens et al. 1991) suggesting a DNA-binding function (Todd et al. 2001). There is a lack of information on the importance of the many differences in amino acid sequence for the tertiary structure and antigenicity of VP1. Thus far polyclonal chicken anti-CAV sera recognize all strains that have been tested, but only one neutralizing epitope has been mapped using monoclonal antibody (MAb) 2A9 (McNulty et al.

**Fig. 3** (continued) EGFP mRNA in LMH-2A (**c**) and DF-1 (**d**) cells transfected with positive control CMV promoter plasmid pEGFP-N1, or the CAV promoter constructs pEGFP-S5 or pEGFP-L5. Results are from one experiment with five separate transfected wells. The results are reported as mRNA copy numbers and are normalized for transfection efficiency by dividing by DNA copies of the expression plasmid. RNA and DNA EGFP copies were normalized to the cellular genes GAPDH and iNOS, respectively. *Error bars* in **b**, **c** and **d** indicate standard deviations, *columns with different letters* are significantly different with a $p$ value of $<0.05$ by two-tailed $t$-test analysis. From Miller et al. (2005), with permission from the American Society of Microbiology

1990; Scott et al. 2001). The formation of epitopes recognized by the polyclonal virus-neutralizing (VN) chicken antibodies requires the expression of VP1 and VP2 in the same cells (Koch et al. 1995; Noteborn et al. 1998a) suggesting that VP2 acts as a scaffold protein. Epitope mapping by Pepscan analysis of CAV proteins using a panel of MAb has only been reported for VP2. Most MAb reacted with epitopes located between amino acids 111 and 126, and one MAb reacted with an epitope between amino acids 21 and 36 (Wang et al. 2007).

Peters et al. (2002, 2005) reported a second function for the 24-kDa VP2. It has a dual-specificity phosphatase (DSP) activity catalyzing the removal of phosphatase from phosphoserine/phosphothreonine and phosphotyrosine substrates. The protein coded by ORF2 of TTMV also has DSP activity and the signature sequence for DSP activity is also present in TTV (Peters et al. 2002). The signature motif for the DSP activity in VP2 includes the amino acids ICNCGQFRKH at position 94–103. The underlined amino acids are conserved in the DSP protein family. The mutation *C95S* completely removed all DSP activity, while the mutation *C97S* reduced the phosphotyrosine activity by 30% but drastically increased the phosphoserine/phosphothreonine activity. The double mutant had no activity with either substrate. Virus titers were reduced significantly in MSB1 cells infected with the *C95S* and the *C97S* mutants. The functional importance of the DSP activity is still unknown and awaits identification of the substrate.

A key pathological finding in CAV-infected chickens is the transient depletion of thymocytes in the cortex (reviewed by Schat and van Santen 2008) caused by apoptosis (Jeurissen et al. 1992b). The apoptosis is induced by the 13.6-kDa VP3, although VP2 may also have some apoptotic activity (Noteborn 2004). Truncation of VP3 by removing amino acids 110–120 significantly reduces the apoptosis in MDCC-MSB1 cells (Noteborn et al. 1994a), but the actual mechanism of apoptosis in nontransformed chicken cells has not been elucidated. It is of interest that the cortex of virus-infected chickens can be almost completely depleted while in situ hybridization or immunohistochemistry show only a relatively small number of infected cells (Fig. 4; Allan et al. 1993; Hu et al. 1993b). Perhaps VP3 can be "recycled" into new cells, or the VP3-induced apoptosis causes a "bystander effect" affecting noninfected cells or the in situ hybridization severely underestimated the number of virus-infected cells. The finding that VP3 or apoptin selectively induces apoptosis in human cancer cells has generated considerable interest. An apoptosis-inducing protein, dubbed Torque teno virus apoptosis-inducing protein (TAIP), has also been identified in TTV (Kooistra et al. 2004). The mechanisms of apoptin and TAIP-induced apoptosis in human cells are discussed in the chapter by M.H. de Smit and M.H.M. Noteborn (this volume).

## *Attenuation*

Von Bülow and Fuchs (1986) demonstrated that Cux-1 could be attenuated by serial passage in MDCC-MSB1. Attenuation was noted after passage (p) 49, but residual pathogenicity was still present after 100 passages. Todd and collaborators (1995)

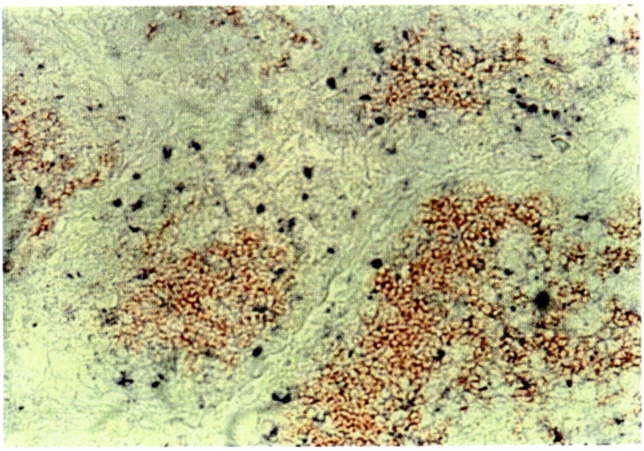

**Fig. 4** Thymus of a chicken inoculated at 1 day of age with 3,000 chicken infective doses 50% ($CID_{50}$) of the CIA-1 strain of CAV and harvested at 14 days pi. $CD4^+$ cells are stained *red* and CIA-$1^+$ cells are stained *blue*. Notice the low numbers of $CD4^+$ cells and the relatively few CIA-$1^+$ cells. From Hu (1992), MS thesis Cornell University, with permission from the author

confirmed that Cux-1 could be attenuated by passage in MDCC-MSB1, but even at p173 some residual pathogenicity was noted. Molecularly cloned viruses were obtained from p173, some of which were attenuated (e.g., clone 10) while other clones were still fully pathogenic (e.g., clone 15). Unfortunately clone 10 reverted to pathogenic virus after ten passages in 1-day-old chicks. The revertant virus contained several minor subpopulations with different mutations and it was suggested that some of the minor subpopulations could have an increased pathogenicity (Todd et al. 2003). Subsequently, Cux-1 was examined for pathogenicity and antigenicity at p310 using pooled material and clones indicating that there was still residual pathogenicity (Scott et al. 1999). Sequence analysis of VP1 indicated that 6 amino acids were changed in most of the clones compared to p13 or the original Cux-1 sequence. The relevance of these changes for the pathogenicity is not clear because clone 33 was more pathogenic than p13, while clones 31 and 34 were highly attenuated. These three clones differ only at amino acid position 89 where clone 33 has a T, like the p13 and p173 clone 10, while clones 31 and 34 have an A. Clone 33 is neutralized by MAb 2A9, which recognizes a conformational epitope. The sequence of the population that is not neutralized by MAb 2A9 is similar to clone 34 (Scott et al. 2001). Using chimeras, Todd et al. (2002) showed that amino acid 89 was responsible for the differences between clones 33 and 34 in pathogenicity and antigenicity. Site-directed mutagenesis of this amino acid from T to A did not change the pathogenicity of the cloned, low-passage Cux-1 strain. Only one other change was reported in VP3 where at position 41 clone 33 has an I, while the other two clones and Cux-1 have a T (Scott et al. 2001), but this change was not responsible for the difference in pathogenicity (Todd et al. 2002). Interestingly the three clones have a G at position 186 of VP2 instead of an E, which was also a change made by Peters et al. (2006) by site-directed mutagenesis (see below).

A low-pathogenic virus was obtained by analyzing different clones from p12 of the pathogenic AH9410 isolate (Yamaguchi et al. 2001). Only one change in VP1 (Q394H) was associated with a decrease in pathogenicity, but there were still residual lesions with the clones expressing H at amino acid position 394 such as decreased hematocrit values. Site-directed mutagenesis changing the H to Q and vice versa changed the lesion score as expected. Inspection of differences between Cux-1, its cloned apathogenic derivative C 34 (Scott et al. 2001), and the high (AH-C364) and low (AQH-C140) pathogenic clones (Yamaguchi et al. 2001) indicates that none of the changes in VP1 can be linked directly to pathogenicity (Table 1). A possible explanation is that different individual changes may affect the tertiary structure of VP1 causing subtle changes on the virion surface and hence impair the interactions with the putative virus receptor.

Peters et al. (2006) introduced specific mutations in VP2 to analyze the possibility of attenuating CAV. Mutants 1–4 were generated in the catalytic motif of DSP activity, and mutants 5–11 in the region expected to have a high degree of secondary structure (Table 2). These mutants were evaluated in cell culture (Peters et al. 2006) and by inoculation in 7-day-old embryos (Peters et al. 2007) and 1-day-old chickens (Kaffashi et al. 2008). Some of their findings are summarized in Table 2. With the exception of mutant 3, all could be propagated in MDCC-MSB1, causing relative differences in the degree of cytopathic effects. Mutant 3 did not propagate to levels allowing experimental studies. Mutants 5, 6, 7, and 9 did not cause apoptosis although VP3 could be detected in the cytoplasm, supporting the suggestion by Noteborn (2004) that VP2 may play a role in the induction of apoptosis. Unfortunately, data were not provided for the viruses with mutations in the DSP signature sequence. Interestingly, the wild-type CAV caused downregulation of major histocompatibility complex (MHC) class I on the surface of MDCC-MSB1 cells. To determine if the mutants were attenuated, 7-day-old embryos were inoculated in the yolk sac and examined for lesions in the lymphoid organs and for hemorrhages. All mutants caused significantly fewer lesions than CAU269/7, but the lesion score of mutant 11 was the only one that was not significantly different from the uninfected control (Peters et al. 2007). Infection of 1-day-old birds indicated that the relative thymus weight was decreased at 14 days pi for mutant 7 and 11, but not for mutant 6. In contrast, the thymic cortex was normal for mutant 7 and 11. A spontaneous new mutant (*S77N*) was identified in cell cultures infected with mutant 5 (*R129G*) (Kaffashi et al. 2008) raising some interesting questions about the long-term stability of the VP2 mutants. The mutation *E186G* (mutant 11) was also detected in the high passage pathogenic and apathogenic clones 33 and 34 described by Scott et al. (2001).

It is clear from these data that attenuation of CAV is complex and that many individual changes may lead to partial or complete attenuation. The approach of site-directed mutagenesis is certainly an important avenue to clarify the role of individual amino acids in the pathogenicity of CAV, but it may be necessary to introduce several changes in VP1 and VP2 and perhaps VP3 to obtain a truly attenuated stable virus that can be used as a vaccine.

**Table 1** Comparison of predicted amino acid sequence for VP1 from pathogenic Cux–1 and clone AH-C364 and the attenuated clones P310 clone 34 and AH-C140

| | Amino acid changes and position | | | | | | | | | | | | | | | | | |
|---|---|---|---|---|---|---|---|---|---|---|---|---|---|---|---|---|---|---|
| Isolate | 29 | 75 | 89 | 125 | 140 | 141 | 144 | 251 | 254 | 265 | 287 | 321 | 370 | 376 | 394 | 413 | 444 | 447 |
| Cux–1[a] | K | V | T | I | S | Q | D | Q | E | N | A | R | S | L | Q | A | Y | T |
| P310 C 34[b] | R | I | A | L | S | L | E | L | G | T | A | A | S | L | Q | A | Y | T |
| AH-C140[c] | R | V | T | L | A | Q | E | R | E | T | S | A | G | I | H | S | Y | S |
| AH-C364[c] | R | V | T | L | A | E/Q[d] | E | R | E | T | S | A | G | I | Q | S | D[e] | S |

[a]Meehan et al. 1992
[b]Scott et al. 2001
[c]Yamaguchi et al. 2001
[d]An additional pathogenic clone had Q at this position
[e]An additional pathogenic clone had Y at this position

**Table 2** Biological effects of mutations in the DSP motif (mutants 1–4) and in the sequence important for tertiary structure (mutants 5–11)

| Mutant | | Effect of mutation in: | | | | | | |
|---|---|---|---|---|---|---|---|---|
| | | MDCC-MSB1[a] | | | | Embryos[b] | Bird inoculation[c] | |
| No. | Change in amino acid | Virus rescue | Cpe in MSB1 | MHC-I decrease | Apoptosis | Lesion score | Relative thymus weight | Thymic cortex |
| | Wild-type (CAU269/7) | + | +++ | decrease | yes | 20 | decreased | decreased |
| 1 | C87R[d] | + | + | ... | ... | 9 | ... | ... |
| 2 | R101G | + | + | normal | ... | 11 | ... | ... |
| 3 | K102D | -[a] | + | ... | ... | ... | ... | ... |
| 4 | H103Y | + | + | normal | ... | 11 | ... | ... |
| 5 | R129G | + | ++ | | no | 10 | ... | ... |
| 6 | Q131P | + | ++ | intermediate | no | 12 | normal | ... |
| 7 | R/K/K150/151/152G/A/A | + | ++ | ... | no | 12 | decreased | normal |
| 8 | D/E161/162/G/G | + | + | ... | ... | 11 | ... | ... |
| 9 | L163P | + | ++ | ... | no | 15 | ... | ... |
| 10 | D169G | + | + | ... | ... | 15 | ... | ... |
| 11 | E186G | + | ++ | ... | ... | 9 | decreased | normal |
| 12 | S77N[e] | | | | | | decreased | normal[f] |
| | Negative control medium | | | | | 7 | normal | normal |

[a]Peters et al. 2006, [b]Peters et al. 2007, [c]Kaffashi et al. 2008, [d]C86R in Peters et al. 2006, although C87R is correct (G. Browning, personal communication, 23 May 2008), [e]Derived from *R129G,* [f]Normal compared to one medium control and decreased compared to a second medium control group

# Pathogenesis

Infection with CAV can have different consequences depending on whether the birds are kept as commercial production flocks, referred to as conventional chickens in this chapter, or as SPF flocks. The pathogenesis of infection and disease in the former is fairly well understood and is reviewed in Sect. 4.1. Infection with CAV is not uncommon in research or commercial SPF flocks but the maintenance of infection in these flocks is poorly understood and is discussed in Sect. 4.2. The consequences of infection for the immune response and interactions with other pathogens are detailed in Sects. 4.3 and 4.4, respectively.

## *Pathogenesis in Conventional Chickens*

The consequences of infection depend on the combination of age of infection and the presence of maternal antibodies (reviewed by Schat and van Santen 2008). Infection of young chickens, in general less than 2 weeks of age, may result in clinical disease and mortality in the absence of maternal antibodies (Yuasa et al. 1980a). Likewise clinical infections may occur in older birds when antibody responses are impaired, for example, by infectious bursal disease virus (IBDV) (Yuasa et al. 1980b) or by embryonal bursectomy (Hu et al. 1993a). In general infection cannot be established in young chickens with maternal antibodies against CAV (Yuasa et al. 1980a), which can be protective against infection until 3 or even 4 weeks of age when antibodies are no longer detectable by commercial ELISA kits (Markowski-Grimsrud and Schat 2003). Infections occur frequently in chickens once maternal antibodies wane, but infections are subclinical unless the antibody response has been damaged by infections with immunosuppressive viruses such as IBDV or MDV.

### Infection in Young Chickens Without Maternal Antibodies

Chicks hatched from eggs inoculated in the yolk-sac as 6-day-old embryos from antibody-negative hens or from eggs produced from experimentally infected, antibody-negative hens developed clinical disease (Yuasa and Yoshida 1983; Hoop 1992). Outbreaks of disease in commercial flocks have indeed been linked to late or uneven seroconversion just prior to or during egg production (Chettle et al. 1989; Engström 1999; Davidson et al. 2004) resulting in vertical transmission. Insemination with semen from viremic males may also result in vertical transmission (Hoop 1993). Testing for antibodies around 10–12 weeks of age followed by vaccination if exposure of the breeding flock has been insufficient will solve the problem the vertical transmission (Smith 2006).

In addition to vertical transmission, chickens can also become infected by horizontal infection presumably by the fecal-oral route, which is the most likely route of exposure in flocks hatched with maternal antibodies (Yuasa et al. 1983). Virus was isolated up to 49 days of age from rectal contents when chickens were inoculated at 1 day of age, even while VN antibodies were detected at 21 days pi. Another potential source of infection is transmission from feathers. Virus was detected between 7 and 14 days pi in feathers from birds infected with three different CAV isolates. Feather-extracted CAV was used to successfully infect newly hatched chickens by dripping the extract on the eye and oral mucosal surfaces (Davidson et al. 2008).

The pathogenesis of infection in 1-day-old chickens has been studied in detail by experimentally infecting chickens by injection (Taniguchi et al. 1983; Yuasa et al. 1983; Hu et al. 1993b; Smyth et al. 1993; van Santen et al. 2004) or orally (van Santen et al. 2004; Kaffashi et al. 2006). In general, high virus doses of 300–3,000 chick infective doses 50% ($CID_{50}$) or over $10^4$ tissue culture infective doses 50% ($TCID_{50}$) were used, which may represent actual exposure levels in commercial poultry operations. Virus could be isolated at low titers from many tissues as early as 1 day pi but titers increase rapidly afterwards (Yuasa et al. 1983). Within 3–4 days pi viral antigen or viral genomes can be detected initially in the lymphoid organs and bone marrow and subsequently in many other organs including the intestinal tract and skin. Viral antigens were mostly associated with lymphoid aggregates (Smyth et al. 1993). Virus shedding into the intestinal tract most likely occurs via infected lymphocytes with perhaps the cecal tonsils as an important source (van Santen et al. 2004) or perhaps through urine, as kidneys were consistently positive between 6 and 14 days pi. The finding of an occasional bird with viral antigen in the feather pulp (Smyth et al. 1993) suggested the possibility of virus shedding through feathers, which was later confirmed by Davidson et al. (2008). Once VN antibodies develop viral antigens disappear. Interestingly, virus could be reisolated from peripheral blood cells but not from serum for a long period after VN antibodies were present (K. Imai et al., unpublished data quoted by Yuasa 1994). This observation suggests that CAV can persist as a latent virus, a concept that was first proposed by McNulty (1991) and subsequently by Miller and Schat (2004) (see Pathogenesis of Infection in SPF Chickens). Recently, studies on the pathogenesis have used real-time quantitative PCR (qPCR) assays (Kaffashi et al. 2006), but the results of these studies have not changed our fundamental understanding of the pathogenesis. Unfortunately, with one exception (Markowski-Grimsrud and Schat 2003), these studies used only qPCR, which does not provide information on the actual timing and intensity of virus replication based on the quantitative analysis of transcripts by qRT-PCR.

The target cells for CAV replication are dividing cells of the hemocytoblast and T cell lineages (reviewed by Adair 2000; Miller and Schat 2004). Hemocytoblasts are the precursors for erythrocytes, thrombocytes, and heterophils (which are the mammalian equivalent of neutrophils) and infection with CAV causes a temporal decrease in these cells (Taniguchi et al. 1983). The infection in the hemocytoblasts is responsible for clinical signs such as anemia, which is often defined as hematocrit values below 27%, hemorrhages, and secondary bacterial infections due to a decrease in innate immune responses. The decrease in heterophils and thrombocytes, which have a major role in innate immune responses, contribute to the increased susceptibility to bacterial disease often leading to dermatitis or "blue-wing" disease (Engström and Luthman 1984). Infection of the T cell precursors expressing CD3 in the cortex of the thymus causes a decrease in T lymphocytes expressing CD4 and CD8 (Adair et al. 1993; Hu et al. 1993b). Infection also occurs in spleen lymphocytes especially in $CD8^+$ cells expressing TCRαβ (Adair et al. 1993). Cytotoxic T lymphocytes (CTL) express CD8 and TCRαβ, while natural killer (NK) cells express CD8 but not TCRαβ. These findings suggest that CAV infection may impair developing T cell-mediated immune responses. Markowski-Grimsrud and Schat (2003) found a significant decrease in CTL responses to REV 7 days pi with REV and CAV. Replication of CAV was demonstrated by qRT-PCR and qPCR assays at that time, presumably leading to apoptosis of the developing REV-specific CTL.

The consequences of the infection are damage to the bone marrow cells with a replacement of the normal structure by fat cells and thymus atrophy with mostly a collapse of the cortex. These changes are often reversible unless the birds die as a consequence of anemia or secondary bacterial infections. Significant loss of body weight may be noticed under experimental conditions (e.g., Hu et al. 1993b; Kaffashi ct al. 2006).

### Infection in Immunocompetent Chickens

Infection of immunocompetent chickens is often subclinical, but Smyth et al. (2006) found that experimental challenge of 3- and 6-week-old chickens resulted in damage of the thymus cortex. Interestingly, very few antigen-positive cells could be detected in the bone marrow and none of the chickens developed anemia or other clinical signs. Infection of 6-week-old birds follows the same viremia pattern in thymus, liver, and spleen as infection at 1 day of age, but lesions did not develop when birds were infected at 6 weeks of age (Kaffashi et al. 2006). However, CAV infection caused anemia and other lesions when the embryonally bursectomized birds were infected at 5 weeks of age (Hu et al. 1993a). Likewise, infection with very virulent IBDV can prolong the persistence of replicating CAV (Imai et al. 1999), which may lead to CAV-associated pathology. These experiments clearly indicate that age resistance is related to the ability to develop VN antibodies in a timely fashion and that it is not related to the disappearance of specific developmental stages of T lymphocytes as originally proposed by Jeurissen et al. (1992a).

### Factors Influencing the Pathogenesis

Genetic differences between birds, virus strains, and route of infection are possible factors that may influence the pathogenesis in addition to the presence of maternal antibodies, age-dependent maturation of the humoral immune response, or co-infection with immunosuppressive viruses. MHC-defined differences between broiler lines did not influence the susceptibility to CAV when challenged by the oral route at 4 weeks of age (Joiner et al. 2005). However, differences in antibody response were noted between three MHC-defined white leghorn lines after CAV vaccination. The S13 (MHC: $B^{13}B^{13}$) line had a significantly lower seroconversion rate than the P2a (MHC: $B^{19}B^{19}$) and N2a (MHC: $B^{21}B^{21}$) lines (Cardona et al. 2000a). Earlier Hu (1992) had already suggested that S13 chickens were more susceptible to CAV than the N2a and P2a lines. Unfortunately, the S13 line is no longer available, making it impossible to investigate the basis for the observed differences.

Thus far, significant differences in pathogenicity between field isolates have not been reported. However, comparisons between strains based on publications from different research groups are often impossible due to a lack of standardized procedures, resulting in different virus doses and routes of infection. Oral challenge often results in delayed virus replication and less severe lesions compared to intramuscular or intraabdominal injection (van Santen et al. 2004). Attenuation by cell culture passage or site-directed mutagenesis reduces the severity of lesions (von Bülow and Fuchs 1986; McKenna et al. 2003; Kaffashi et al. 2008). Antibody responses elicited by attenuated virus are similar to the responses invoked by pathogenic isolates after inoculation of 5-week-old chickens (Todd et al. 1998).

## *Pathogenesis of Infection in SPF Chickens*

The SPF flocks maintained by Cornell University were free of CAV (McNulty et al. 1989) until the mid 1990s. The flocks are maintained in a filtered-air, positive pressure house in a closed system with replacement flocks being hatched and reared in colony cages in separate rooms in the same building. These rooms have a common air source and the building is considered the unit. In 1996 the flocks became infected based on antibody testing (Cardona et al. 2000a). Since that time all birds in each flock of P2a and N2a lines are being tested for antibodies just prior to onset of egg production, between 34 and 40 weeks of age and at termination between 65 and 75 weeks of age. Seroconversion often occurs after the onset of lay and sometimes all birds within a flock become positive, but subsequent flocks may remain negative (Miller et al. 2001; Schat 2003; K.A. Schat, unpublished data 2008). During the last 12 years clinical disease has never been observed in the offspring of flocks that became antibody-positive during the laying period (K.A. Schat, unpublished data 2008), which is contrary to the experience with conventional flocks (see Infection of Young Chickens without Maternal Antibodies).

These observations suggested a major difference between the pathogenesis in SPF flocks versus conventional flocks. The seminal finding by Cardona et al. (2000b) that viral DNA could be detected in the gonadal tissues of these SPF birds in the absence or presence of antibodies suggested the possible transfer of viral DNA to the newly hatched chicks without early viral replication leading to disease. Subsequently, Miller et al. (2003) demonstrated the presence of viral DNA by nested PCR in blastodisk extracts and semen obtained from birds that were antibody negative between 432 and 494 days of age. Individual tissues from embryos

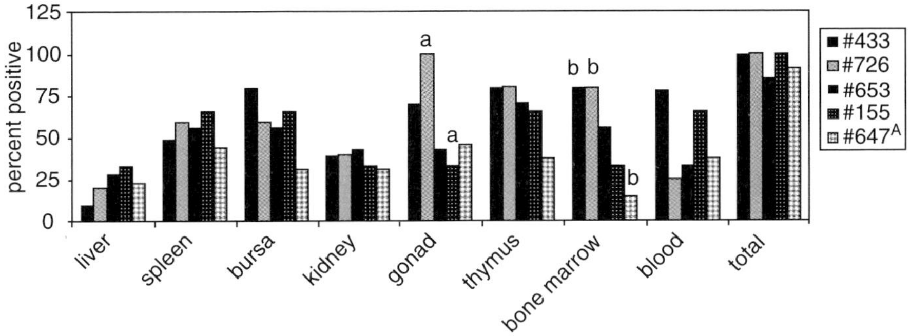


**Fig. 5** Distribution of CAV in embryos from five hens, each designated by a three-digit number. These hens were antibody-negative but had PCR-positive blastodisks. The hens were inseminated with semen from one male that was PCR-positive on 2 of 4 semen samples tested and was also positive on postmortem samples of the testicle and vas deferens. Hen No. 647 became seropositive during the period of egg collection. In total, 28 embryos were examined. Columns that have the same *lowercase superscript* are significantly different at $p<0.05$ by chi-square analysis. From Miller et al. (2003), with permission from the American Association of Avian Pathologists

obtained from antibody-negative, blastodisk-positive hens were examined for the presence of viral DNA at 18 days of embryonation. All embryos were positive but differences between tissues from individual embryos were noted indicating that the viral DNA was not transmitted through the germ-line (Fig. 5). The low levels of viral DNA in the blastodisk and the presence in most tissues at 18 days of embryonation suggested the multiplication of DNA. Embryos were examined between 0 and 12 days of incubation for the presence of viral DNA and transcripts by qPCR and qRT-PCR. Viral transcripts were detected between 4 and 7 days and between 10 and 12 days of embryonation (Fig. 6), while the number of DNA copies remained very low (data not shown) (M.M. Miller, K.A. Stucker and K.A. Schat, unpublished data). It is possible that virus is transferred from the blastoderm to the embryo through hematopoietic cells, because the first transient wave of hematopoietic precursors is derived from the yolk sac (Fellah et al. 2008). The intricate control of CAV transcription suggests a role for steroid hormones (see Transcriptional Control). Gonadal development occurs around the time of the first wave of viral transcripts and it is hypothesized that estrogen is involved in the activation of viral transcription between 4 and 7 days of embryonation. The second wave may be explained if hormone receptors develop during that period in, for example, the first wave of T lymphocytes, which start to appear around embryonation day 6 but may not be mature until shortly before the second wave of T lymphocytes appears between 10 and 12 days (Fellah et al. 2008).

The combination of strict control of viral transcription, the detection of viral DNA in gonads of adult hens in the absence of antibodies, and the presence of minute amounts of viral DNA in blastodisks and embryonal tissues led Miller and Schat (2004) to propose that the interaction between CAV and the host is an example of the ultimate host–parasite relationship. In this model, latent virus or viral DNA

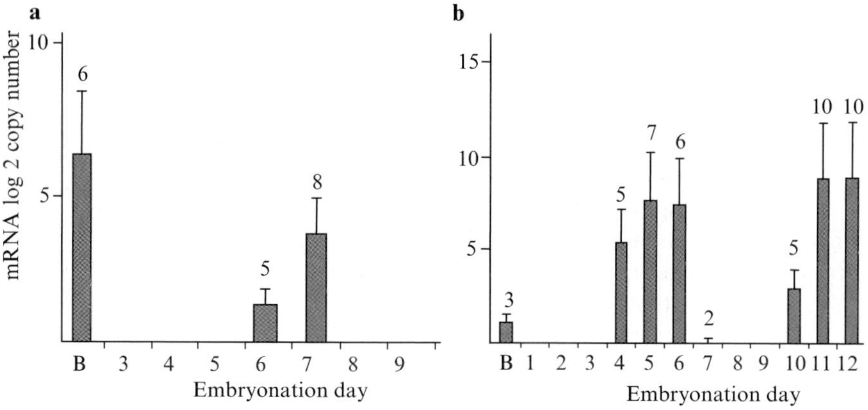


**Fig. 6 a, b** Determination of CAV transcripts in embryos using real-time quantitative RT-PCR. Total RNA was extracted from 10 fertile blastoderm (*B*) samples and 10 embryos/sampling day at embryonation day (ED) 1 and 3–9 in Expt. 1 (**a**), and from 10 blastoderm (*B*) samples and 10 embryos/sampling day at ED 1–12 in Expt. 2 (**b**). RNA copy numbers were normalized for equal loading using the cellular housekeeping gene *GAPDH*. The results are reported as log 2 of the average copy number, and *error bars* indicate the standard error of the mean. *Numbers above the bars* indicate the number positive/10 samples. Sampling days without a number had no positive samples. Unpublished data from M.M. Miller, K.A. Stucker, and K.A. Schat, with permission from Miller and Stucker

may be passed from generation to generation without showing evidence of infection and occasionally may turn on virus replication resulting in seroconversion. Miller et al. (2003) tested this model by raising three generations in isolators starting with fertile eggs from antibody-negative hens. Each generation was tested frequently by PCR and positive chicks were removed. Transmission may be intermittent: 1/9 chicks was positive from a breeding pair that had been selected for the production of the next generation based on the absence of viral DNA in 25 embryos.

Limited evidence suggests that vertical transmission of virus or viral DNA from VN antibody-positive hens to embryos may also occur in commercial breeder flocks (Brentano et al. 2005). It is currently unknown if this is important for commercial flocks. It is important to note that vaccination of pullet breeders will not prevent this type of vertical transmission (Miller et al. 2003).

## *Consequences for Immune Responses*

In Infection in Young Chickens without Maternal Antibodies it was already mentioned that infection of susceptible chickens causes a decrease in innate immune responses to bacterial infections. In addition to the decreased function of thrombocytes and heterophils, macrophage functions are also impaired. The number of macrophages in the spleen can be sharply decreased around 14 days pi (Cloud et al. 1992). Macrophages obtained from chickens infected by intramuscular injection at

1 day of age or by the oral route at 21 days of age had decreased expression of the FcR receptor, decreased phagocytic and bactericidal activity, and decreased interleukin (IL)-1 production as measured by bioassay (McConnell et al. 1993a, b). Unfortunately, these assays were performed at a time when few reagents were available to characterize the cells by macrophage-specific surface markers or by measuring IL-1β by qRT-PCR. Markowski-Grimsrud and Schat (2003) used qRT-PCR assays on whole spleen extracts and were unable to demonstrate decreased or increased transcription rates for IL-1β and IL-2 at 7 and 14 days pi, when chickens were infected between 9 and 45 days of age. They found a significant increase in interferon (IFN)-γ transcription at 7 days pi of 30-day-old birds, but the increase was also present in maternal antibody-positive birds, which did not replicate CAV. Adair et al. (1991) also reported an increase in IFN activity at 8 days pi using a bioassay, but a decrease at later sampling times.

Impairment of CTL activity is another important consequence of infection. Several groups noted a decrease in $CD8^+$ cells often with an increase in the ratio of $CD4^+$ over $CD8^+$ cells (Cloud et al. 1992; Bounous et al. 1995). As mentioned before, a marked decrease in CTL activity has been reported (Markowski-Grimsrud and Schat 2003), but NK cell activity was unaffected (Markowski-Grimsrud and Schat 2001).

## *Interactions with Other Pathogens*

Infection with immunosuppressive viruses can increase the severity of other infections or decrease the efficacy of vaccination against other pathogens. CAV is no exception in this respect, and there are many examples in which CAV infection increased the severity of other diseases in experimental infections or under field conditions. Extensive recent reviews are available for detailed information on CAV-induced enhancement of other diseases (Miller and Schat 2004; Schat and van Santen, 2008) and only a few examples will be discussed. Toro et al. (2006b) showed a clear link between CAV infection and infectious bronchitis (IB), a coronavirus-induced respiratory disease in chickens, in IB-vaccinated chickens. This may seem puzzling because IB virus (IBV) vaccines protect mostly through VN antibodies, but IBV variants appear frequently in the field, and vaccine-induced protection becomes suboptimal. In that case IBV-specific memory CTL become crucial for protection (Pei et al. 2003), which may be impaired by CAV.

MDV is another example of interaction of a second pathogen with CAV. In this case the interactions are more complicated than the interaction between CAV and IBV. Jakowski et al. (1970) noticed that birds developing anemia after challenge with MDV had a decreased incidence of MD lesions. It was subsequently shown that these birds were infected with MDV and CAV (see Historical Background). Von Bülow et al. (1983, 1986) experimentally infected 1-day-old chicks with CAV, CAV and MDV, or MDV. The dual challenge resulted in an increased degree of anemia and increased mortality, but there was no clear impact on the MD incidence. These authors also confirmed that infections with IBDV or REV increased the CAV-induced anemia and mortality and suggested that these increases were caused by immunosuppression induced by MDV, IBDV, and REV. The mechanisms of MDV-, IBDV-, and REV-induced immunosuppression are complex and have recently been reviewed (Schat and Skinner 2008). CAV can also influence MDV replication when both viruses are inoculated at the same time. For example, co-infection of CAV with the very virulent MDV strain RB-1B resulted in increased mortality and shorter mean time to death compared to chickens inoculated with RB-1B alone. Interestingly, chickens inoculated with the combination died without tumors in contrast to the birds receiving RB-1B alone. Co-infection caused an increase in transcription levels for the immediate early infected cell protein (ICP)4 of MDV in the spleen at 4, 7, and 13 days pi. If the very virulent+MDV isolate 584A was used instead of RB-1B this effect was only seen at 1 day pi. The effect of CAV on 584A-induced tumors was also less striking than with RB-1B, most likely because 584A caused a severe depletion of thymocytes and high levels of early mortality unrelated to tumor formation (Miles et al. 2001). In addition to the virulence of MDV, the challenge dose of MDV may also play a role in the interaction with CAV. Jeurissen and de Boer (1993) using a relatively low dose of MDV found that CAV significantly enhanced MD-associated lymphoproliferation in nerves and visceral organs, while co-infection with a 20-fold higher dose of MDV resulted in a decrease in lymphoproliferation compared to the high dose of MDV alone.

A more important interaction between CAV and MDV is the decrease in vaccinal immunity to MD as a consequence of CAV infection. Field observations have suggested that this may be an explanation for MD breaks as late as 24 weeks of age (Fehler and Winter 2001). Experimental data support the CAV-induced decrease in vaccinal immunity. Chickens vaccinated with HVT at 1 day of age and infected with CAV at 4 days of age and challenged with MDV were significantly less well protected than chickens that did not receive CAV. However when the MDV challenge was given at 18 days of age there was no negative effect by CAV (Yuasa and Imai 1988). Similar results were reported by Otaki et al. (1988). CAV replication decreases protective immunity by the elimination of specific CTL (see Consequences for immune responses), which can occur when CAV infects young maternal antibody-negative birds or older antibody-negative birds or when CAV is reactivated during the development of sexual maturity and subsequent egg production. The absence of an effect when chickens are challenged with MDV at 18 days of age suggests that some CTL have escaped the CAV infection and can generate a memory response in the absence of active CAV replication.

## *Control of Infection*

Vaccines against CAV have been developed using various attenuated strains (reviewed by Schat and van Santen 2008) but the need for vaccination is not universally accepted (Smith 2006). Vaccines are considered to be expensive and the fact that immunosuppression may be the only consequence of infection complicates the diagnosis of CAV as an important part of disease syndromes. There are also some technical and practical problems affecting the widespread use of vaccines. The first problem is the relative instability of the virus and relative low levels of virus titers in cell cultures. The second problem is the testing of vaccines for efficacy, which requires vaccination followed by challenge demonstrating significant protection. Challenge is typically performed 1–3 weeks after vaccination, but induction of a high level of symptoms requires challenge before 7 days of age. Markowski-Grimsrud et al. (2002) developed a qPCR assay differentiating between two strains so that one can determine if the vaccine virus protects against replication of the challenge virus. A practical problem is the timing of vaccination in young birds positive for maternal antibodies, especially if a live vaccine is used. Similar problems have been noted with vaccinations for IBDV (Eterradossi and Saif 2008). Perhaps the use of antigen-antibody complex vaccines can overcome the problems with maternal antibodies as has been shown for this type of IBDV vaccine (Haddad et al. 1997).

## Lessons for Anellovirus Infections

The TTV and TTMV group shares certain similarities with CAV such as coding for proteins with DSP and apoptotic activity, which are discussed in more detail in the chapter by S. Hino and A.A. Prasetyo (this volume). The other similarity is the ubiquitous nature of TTV and TTMV infections in humans and CAV in chickens. One of the key problems faced by the researchers studying the importance of TTV and TTMV is this high prevalence of these viruses in the human population without a clear link to one or more diseases. This situation is fairly similar to CAV, which also has a high prevalence and is not linked to a clinical disease syndrome unless the infection occurs in newly hatched, antibody-negative chicks or chickens with a severely impaired antibody response. The major impact that CAV has is the subclinical infection aggravating other diseases by causing immunosuppression. This may be very similar in TTV and TTMV infections. Vertical mother to infant transfer of TTV has been reported (e.g., Xin et al. 2004), while others have found no evidence for transmission through maternal blood. Instead they associated an increase in infection levels in infants with environmental infections (Ohto et al. 2002). The problem is that currently no methods have been widely used to determine if maternal antibodies are present to prevent virus replication either after vertical transmission or after early environmental infection. It may be of interest to use qRT-PCR in these studies to determine if and when TTV replicates, as was done in the studies on CAV-induced immunosuppression by Markowski-Grimsrud and Schat (2003).

Many research teams have linked temporal increases in TTV load in human immunodeficiency virus (HIV)-infected patients to cycles of HIV replication, which raises the question of what initiates the increase. Does an increase in HIV cause immunosuppression followed by an increase in TTV; is it the other way around; or is it a vicious circle in which each episode of HIV or TTV replication enhances the replication of the other pathogen? TTV DNA titers were significantly higher in patients with acquired immunodeficiency syndrome (AIDS), high HIV loads, or low $CD4^{+}$ T cell counts (Shibayama et al. 2001; Touinssi et al. 2001; Thom and Petrik 2007). The fact that TTV replicates in stimulated T cells (Mariscal et al. 2002) and bone marrow cells (Okamoto et al. 2000) certainly could lead to immunosuppression and thus lead to increased replication of HIV. This is similar to the replication of CAV in bone marrow cells and dividing T lymphocytes followed by enhanced replication of MDV. Girard et al. (2007) suggested that TTV may be a cofactor in the development of Kaposi's sarcoma, which supports the hypothesis that subclinical immunosuppression caused by TTV infection may increase the development of other diseases. In parallel with increased replication of CAV in immunosuppressive treatments in chickens, TTV replication increases when transplant patients receive immunosuppressive treatments (Moen et al. 2003). It will be of interest to determine if immunosuppressive treatment affecting principally the humoral or the cell-mediated responses are causing the increase in virus load and in which cell the virus replicates under those conditions.

## Conclusions

CAV is an important pathogen causing economic losses by increasing the clinical severity of diseases caused by other pathogens, mostly as a consequence of subclinical infection. This virus with a small genome has a very complex mechanism to maintain itself in low-density or low-stress populations, and the elucidation of the control of latency may unravel some fundamental aspects of gene control. The induction of apoptosis by VP3 is a crucial aspect of viral replication and subsequent pathology, but the mechanism of apoptosis in nontransformed chicken cells is poorly understood and needs additional study. Identification of the substrate(s) for the DSP activity of VP2 will be important for the further development of attenuated strains and perhaps vaccines. In general, a better understanding of the biology of CAV may lead to insight on the role of TTV in human health and disease.

**Acknowledgements** The author wishes to thank Ms. Laura Stenzler for critical reading of the manuscript. The unpublished data by Miller et al. cited in this chapter were funded in part by gifts from SPAFAS Charles River Laboratories, Wilmington, MA. Dr. Miller was supported by an Institutional NRSA award #5T32AI07618–01 and -02, by Ruth L. Kirchstein NRSA award #T32RR07059–08, and by a fellowship from the Unit of Avian Health of the Department of Microbiology and Immunology.

## References

Adair BM (2000) Immunopathogenesis of chicken anemia virus infection. Dev Comp Immunol 24:247–255

Adair BM, McNeilly F, McConnell CD, Todd D, Nelson RT, McNulty MS (1991) Effects of chicken anemia agent on lymphokine production and lymphocyte transformation in experimentally infected chickens. Avian Dis 35:783–792

Adair BM, McNeilly F, McConnell CD, McNulty MS (1993) Characterization of surface markers present on cells infected by chicken anemia virus in experimentally infected chickens. Avian Dis 37:943–950

Allan GM, Smyth JA, Todd D, McNulty MS (1993) In situ hybridization for the detection of chicken anemia virus in formalin-fixed, paraffin-embedded sections. Avian Dis 37:177–182

Bassami MR, Berryman D, Wilcox GE, Raidal SR (1998) Psittacine beak and feather disease virus nucleotide sequence analysis and its relationship to porcine circovirus, plant circoviruses, and chicken anaemia virus. Virology 249:453–459

Bounous DI, Goodwin MA, Brooks RL Jr, Lamichhane CM, Campagnoli RP, Brown J, Snyder DB (1995) Immunosuppression and intracellular calcium signaling in splenocytes from chicks infected with chicken anemia virus, CL-1 isolate. Avian Dis 39:135–140

Brentano L, Lazzarin S, Bassi SS, Klein TA, Schat KA (2005) Detection of chicken anemia virus in the gonads and in the progeny of broiler breeder hens with high neutralizing antibody titers. Vet Microbiol 105:65–72

Calnek BW, Lucio-Martinez B, Cardona C, Harris RW, Schat KA, Buscaglia C (2000) Comparative susceptibility of Marek's disease cell lines to chicken infectious anemia virus. Avian Dis 44:114–124

Cardona C, Lucio B, O'Connell P, Jagne J, Schat KA (2000a) Humoral immune responses to chicken infectious anemia virus in three strains of chickens in a closed flock. Avian Dis 44:661–667

Cardona CJ, Oswald WB, Schat KA (2000b) Distribution of chicken anaemia virus in the reproductive tissues of specific-pathogen-free chickens. J Gen Virol 81:2067–2075

Chandratilleke D, O'Connell P, Schat KA (1991) Characterization of proteins of chicken infectious anemia virus with monoclonal antibodies. Avian Dis 35:854–862

Chettle NJ, Eddy RK, Wyeth PJ, Lister SA (1989) An outbreak of disease due to chicken anaemia agent in broiler chickens in England. Vet Rec 124:211–215

Chowdhury SM, Omar AR, Aini I, Hair-Bejo M, Jamaluddin AA, Md-Zain BM, Kono Y (2003) Pathogenicity, sequence and phylogenetic analysis of Malaysian chicken anaemia virus obtained after low and high passages in MSB-1 cells. Arch Virol 148:2437–2448

Claessens JA, Schrier CC, Mockett AP, Jagt EH, Sondermeijer PJ (1991) Molecular cloning and sequence analysis of the genome of chicken anaemia agent. J Gen Virol 72:2003–2006

Cloud SS, Lillehoj HS, Rosenberger JK (1992) Immune dysfunction following infection with chicken anemia agent and infectious bursal disease virus. I. Kinetic alterations of avian lymphocyte subpopulations. Vet Immunol Immunopathol 34:337–352

Crowther RA, Berriman JA, Curran WL, Allan GM, Todd D (2003) Comparison of the structures of three circoviruses: chicken anemia virus, porcine circovirus type 2, and beak and feather disease virus. J Virol 77:13036–13041

Davidson I, Kedem M, Borochovitz H, Kass N, Ayali G, Hamzani E, Perelman B, Smith B, Perk S (2004) Chicken infectious anemia virus infection in Israeli commercial flocks: virus amplification, clinical signs, performance, and antibody status. Avian Dis 48:108–118

Davidson I, Artzi N, Shkoda I, Lublin A, Loeb E, Schat KA (2008) The contribution of feathers in the spread of chicken anemia virus. Virus Res 132:152–159

de Wit JJ, van Eck JH, Crooijmans RP, Pijpers A (2004) A serological survey for pathogens in old fancy chicken breeds in central and eastern part of The Netherlands. Tijdschr Diergeneeskd 129:324–327

Desai M, Pal R, Deshmukh R, Banker D (2005) Replication of TT virus in hepatocyte and leucocyte cell lines. J Med Virol 77:136–143

Douglas AJ, Phenix K, Mawhinney KA, Todd D, Mackie DP, Curran WL (1995) Identification of a 24-kDa protein expressed by chicken anaemia virus. J Gen Virol 76:1557–1562

Ducatez MF, Owoade AA, Abiola JO, Muller CP (2006) Molecular epidemiology of chicken anemia virus in Nigeria. Arch Virol 151:97–111

Ducatez MF, Chen H, Guan Y, Muller CP (2008) Molecular epidemiology of chicken anemia virus (CAV) in South Eastern Chinese live birds markets. Avian Dis 52:68–73

Engström BE (1999) Prevalence of antibody to chicken anaemia virus (CAV) in Swedish chicken breeding flocks correlated to outbreaks of blue wing disease (BWD) in their progeny. Acta Vet Scand 40:97–107

Engström BE, Luthman M (1984) Blue wing disease of chickens: signs, pathology and natural transmission. Avian Pathol 13:1–12

Eterradossi N, Saif YM (2008) Infectious bursal disease. In: Saif YM, Fadly AM, Glisson JR, McDougald LR, Nolan LK, Swayne DE (eds) Diseases of poultry. Blackwell Publishing, Ames, pp 185–208

Fehler F, Winter C (2001) CAV infection in older chickens: an apathogenic infection? Proc 2nd Int Symp Infect Bursal Dis Chick Infect Anaemia. Rauischholzhausen, Germany, pp 391–394

Fellah JS, Jaffredo T, Dunon D (2008) Development of the avian immune system. In: Davison F, Kaspers B, Schat KA (eds) Avian Immunology. Elsevier Academic Press, Amsterdam, pp 51–66

Gelderblom H, Kling S, Lurz R, Tischer I, von Bülow V (1989) Morphological characterization of chicken anaemia agent (CAA). Arch Virol 109:115–120

Girard C, Ottomani L, Ducos J, Dereure O, Carles MJ, Guillot B (2007) High prevalence of torque Teno (TT) virus in classical Kaposi's sarcoma. Acta Derm Venereol 87:14–17

Haddad EE, Whitfill CE, Avakian AP, Ricks CA, Andrews PD, Thoma JA, Wakenell PS (1997) Efficacy of a novel infectious bursal disease virus immune complex vaccine in broiler chickens. Avian Dis 41:882–889

He CQ, Ding NZ, Fan W, Wu YH, Li JP, Li Yl (2007) Identification of chicken anemia virus putative intergenotype recombinants. Virology 366:1–7

Hino S, Miyata H (2007) Torque teno virus (TTV): current status. Rev Med Virol 17:45–57

Hoop RK (1992) Persistence and vertical transmission of chicken anaemia agent in experimentally infected laying hens. Avian Pathol 21:493–501

Hoop RK (1993) Transmission of chicken anaemia virus with semen. Vet Rec 133:551–552

Hu L-b (1992) Role of humoral immunity and T cell subpopulations in the pathogenesis of chicken infectious anemia virus. Masters thesis. Cornell University, Ithaca, p 148

Hu L-b, Lucio B, Schat KA (1993a) Abrogation of age-related resistance to chicken infectious anemia by embryonal bursectomy. Avian Dis 37:157–169

Hu L-b, Lucio B, Schat KA (1993b) Depletion of $CD4^+$ and $CD8^+$ T lymphocyte subpopulations by CIA-1, a chicken infectious anemia virus. Avian Dis 37:492–500

ICTV (2008) The universal virus database of the International Committee on Taxonomy of Viruses. http://www.ncbi.nlm.nih.gov/ICTVdb/. Updated 15 January 2008. Cited 16 March 2008

Imai K, Mase M, Tsukamoto K, Hihara H, Yuasa N (1999) Persistent infection with chicken anaemia virus and some effects of highly virulent infectious bursal disease virus infection on its persistency. Res Vet Sci 67:233–238

Islam MR, Johne R, Raue R, Todd D, Muller H (2002) Sequence analysis of the full-length cloned DNA of a chicken anaemia virus (CAV) strain from Bangladesh: evidence for genetic grouping of CAV strains based on the deduced VP1 amino acid sequences. J Vet Med B Infect Dis Vet Public Health 49:332–337

Jakowski RM, Fredrickson TN, Chomiak TW, Luginbuhl RE (1970) Hematopoietic destruction in Marek's disease. Avian Dis 14:374–385

Jeurissen SH, de Boer GF (1993) Chicken anaemia virus influences the pathogenesis of Marek's disease in experimental infections, depending on the dose of Marek's disease virus. Vet Q 15:81–84

Jeurissen SH, Janse ME, Van Roozelaar DJ, Koch G, De Boer GF (1992a) Susceptibility of thymocytes for infection by chicken anemia virus is related to pre- and posthatching development. Dev Immunol 2:123–129

Jeurissen SH, Wagenaar F, Pol JM, van der Eb AJ, Noteborn MH (1992b) Chicken anemia virus causes apoptosis of thymocytes after in vivo infection and of cell lines after in vitro infection. J Virol 66:7383–7388

Joiner KS, Ewald SJ, Hoerr FJ, van Santen VL, Toro H (2005) Oral infection with chicken anemia virus in 4-wk broiler breeders: lack of effect of major histocompatibility B complex genotype. Avian Dis 49:482–487

Kaffashi A, Noormohammadi AH, Allott ML, Browning GF (2006) Viral load in 1-day-old and 6-week-old chickens infected with chicken anaemia virus by the intraocular route. Avian Pathol 35:471–474

Kaffashi A, Shrestha S, Browning GF (2008) Evaluation of chicken anaemia virus mutants as potential vaccine strains in 1-day-old chickens. Avian Pathol 37:109–114

Kamada K, Kuroishi A, Kamahora T, Kabat P, Yamaguchi S, Hino S (2006) Spliced mRNAs detected during the life cycle of chicken anemia virus. J Gen Virol 87:2227–2233

Koch G, van Roozelaar DJ, Verschueren CA, van der Eb AJ, Noteborn MH (1995) Immunogenic and protective properties of chicken anaemia virus proteins expressed by baculovirus. Vaccine 13:763–770

Kooistra K, Zhang YH, Henriquez NV, Weiss B, Mumberg D, Noteborn MH (2004) TT virus-derived apoptosis-inducing protein induces apoptosis preferentially in hepatocellular carcinoma-derived cells. J Gen Virol 85:1445–1450

Krapež U, Barli -Maganja D, Toplak I, Hostnik P, Rojs OZ (2006) Biological and molecular characterization of chicken anemia virus isolates from Slovenia. Avian Dis 50:69–76

Lucio B, Schat KA, Shivaprasad HL (1990) Identification of the chicken anemia agent, reproduction of the disease, and serological survey in the United States. Avian Dis 34:146–153

Mariscal LF, Lopez-Alcorocho JM, Rodriguez-Inigo E, Ortiz-Movilla N, de Lucas S, Bartolome J, Carreno V (2002) TT virus replicates in stimulated but not in nonstimulated peripheral blood mononuclear cells. Virology 301:121–129

Markowski-Grimsrud CJ, Schat KA (2001) Impairment of cell-mediated immune responses during chicken infectious anemia virus infection. Proc 2nd Int Symp Infect Bursal Dis Chick Infect Anaemia, Rauischholzhausen, Germany, pp 395–402

Markowski-Grimsrud CJ, Schat KA (2003) Infection with chicken anemia virus impairs the generation of antigen-specific cytotoxic T lymphocytes. Immunology 109:283–294

Markowski-Grimsrud CJ, Miller MM, Schat KA (2002) Development of strain-specific real-time PCR and RT-PCR assays for quantitation of chicken anemia virus. J Virol Methods 101:135–147

McConnell CD, Adair BM, McNulty MS (1993a) Effects of chicken anemia virus on cell-mediated immune function in chickens exposed to the virus by a natural route. Avian Dis 37:366–374

McConnell CD, Adair BM, McNulty MS (1993b) Effects of chicken anemia virus on macrophage function in chickens. Avian Dis 37:358–365

McKenna GF, Todd D, Borghmans BJ, Welsh MD, Adair BM (2003) Immunopathologic investigations with an attenuated chicken anemia virus in day-old chickens. Avian Dis 47:1339–1345

McNulty MS (1991) Chicken anemia agent: a review. Avian Pathol 20:187–203

McNulty MS, Connor TJ, Mc Neilly F (1989) A survey of specific pathogen-free chicken flocks for antibodies to chicken anemia agent, avian nephritis virus and group A rotavirus. Avian Pathol 18:215–220

McNulty MS, Mackie DP, Pollock DA, McNair J, Todd D, Mawhinney KA, Connor TJ, McNeilly F (1990) Production and preliminary characterization of monoclonal antibodies to chicken anemia agent. Avian Dis 34:352–358

Meehan BM, Todd D, Creelan JL, Earle JAP, Hoey EM, McNulty MS (1992) Characterization of viral DNAs from cells infected with chicken anaemia agent: sequence analysis of the cloned replicative form and transfection capabilities of cloned genome fragments. Arch Virol 124:301–319

Miles AM, Reddy SM, Morgan RW (2001) Coinfection of specific-pathogen-free chickens with Marek's disease virus (MDV) and chicken infectious anemia virus: effect of MDV pathotype. Avian Dis 45:9–18

Miller MM, Schat KA (2004) Chicken infectious anemia virus: an example of the ultimate host-parasite relationship. Avian Dis 48:734–745

Miller MM, Oswald WB, Scarlet J, Schat KA (2001) Patterns of chicken infectious anemia virus (CIAV) seroconversion in three Cornell SPF flocks. Proc 2nd Int Symp Infect Bursal Dis Chick Infect Anaemia. Rauischholzhausen, Germany, pp 410–417

Miller MM, Ealey KA, Oswald WB, Schat KA (2003) Detection of chicken anemia virus DNA in embryonal tissues and eggshell membranes. Avian Dis 47:662–671

Miller MM, Jarosinski KW, Schat KA (2005) Positive and negative regulation of chicken anemia virus transcription. J Virol 79:2859–2868

Miller MM, Jarosinski KW, Schat KA (2008) Negative modulation of the chicken infectious anemia virus promoter by COUP-TF1 and an E box-like element at the transcription start site binding δEF1. J Gen Virol (in press)

Moen EM, Sagedal S, Bjoro K, Degre M, Opstad PK, Grinde B (2003) Effect of immune modulation on TT virus (TTV) and TTV-like-mini-virus (TLMV) viremia. J Med Virol 70:177–182

Nogueira EO, A JPF, Martins Soares R, Luiz Durigon E, Lazzarin S, Brentano L (2007) Genome sequencing analysis of Brazilian chicken anemia virus isolates that lack MSB-1 cell culture tropism. Comp Immunol Microbiol Infect Dis 30:81–96

Noteborn MH, de Boer GF, van Roozelaar DJ, Karreman C, Kranenburg O, Vos JG, Jeurissen SH, Hoeben RC, Zantema A, Koch G, Van Ormondt H, Van Der Eb AJ (1991) Characterization of cloned chicken anemia virus DNA that contains all elements for the infectious replication cycle. J Virol 65:3131–3139

Noteborn MH, Kranenburg O, Zantema A, Koch G, de Boer GF, van der Eb AJ (1992) Transcription of the chicken anemia virus (CAV) genome and synthesis of its 52-kDa protein. Gene 118:267–271

Noteborn MH, Todd D, Verschueren CA, de Gauw HW, Curran WL, Veldkamp S, Douglas AJ, McNulty MS, van der Eb AJ, Koch G (1994a) A single chicken anemia virus protein induces apoptosis. J Virol 68:346–351

Noteborn MH, Verschueren CA, Zantema A, Koch G, van der Eb AJ (1994b) Identification of the promoter region of chicken anemia virus (CAV) containing a novel enhancer-like element. Gene 150:313–318

Noteborn MH, Verschueren CA, Koch G, Van der Eb AJ (1998a) Simultaneous expression of recombinant baculovirus-encoded chicken anaemia virus (CAV) proteins VP1 and VP2 is required for formation of the CAV-specific neutralizing epitope. J Gen Virol 79:3073–3077

Noteborn MH, Verschueren CA, van Ormondt H, van der Eb AJ (1998b) Chicken anemia virus strains with a mutated enhancer/promoter region share reduced virus spread and cytopathogenicity. Gene 223:165–172

Noteborn MH (2004) Chicken anemia virus induced apoptosis: underlying molecular mechanisms. Vet Microbiol 98:89–94

Ohto H, Ujiie N, Takeuchi C, Sato A, Hayashi A, Ishiko H, Nishizawa T, Okamoto H (2002) TT virus infection during childhood. Transfusion 42:892–898

Okamoto H, Takahashi M, Nishizawa T, Tawara A, Sugai Y, Sai T, Tanaka T, Tsuda F (2000) Replicative forms of TT virus DNA in bone marrow cells. Biochem Biophys Res Commun 270:657–662

Opriessnig T, Meng XJ, Halbur PG (2007) Porcine circovirus type 2-associated disease: update on current terminology, clinical manifestations, pathogenesis, diagnosis, and intervention strategies. J Vet Diagn Invest 19:591–615

Otaki Y, Nunoya T, Tajima A, Kato A, Nomura Y (1988) Depression of vaccinal immunity to Marek's disease by infection with chicken anaemia agent. Avian Pathol 17:333–347

Pei J, Briles WE, Collisson EW (2003) Memory T cells protect chicks from acute infectious bronchitis virus infection. Virology 306:376–384

Peters MA, Jackson DC, Crabb BS, Browning GF (2002) Chicken anemia virus VP2 is a novel dual specificity protein phosphatase. J Biol Chem 42:39566–39573

Peters MA, Jackson BS, Crabb BS, Browning GF (2005) Mutation of chicken anemia virus VP2 differentially affects serine/threonine and tyrosine protein phosphatase activities. J Gen Virol 86:623–630

Peters MA, Crabb BS, Washington EA, Browning GF (2006) Site-directed mutagenesis of the VP2 gene of chicken anemia virus affects virus replication, cytopathology and host-cell MHC class I expression. J Gen Virol 87:823–831

Phenix KV, Meehan BM, Todd D, McNulty MS (1994) Transcriptional analysis and genome expression of chicken anaemia virus. J Gen Virol 75:905–909

Renshaw RW, Soiné C, Weinkle T, O'Connell PH, Ohashi K, Watson S, Lucio B, Harrington S, Schat KA (1996) A hypervariable region in VP1 of chicken infectious anemia virus mediates rate of spread and cell tropism in tissue culture. J Virol 70:8872–8878
Safe S (2001) Transcriptional activation of genes by 17 beta-estradiol through estrogen receptor-Sp1 interactions. Vitam Horm 62:231–252
Saville B, Wormke M, Wang F, Nguyen T, Enmark E, Kuiper G, Gustafsson JA, Safe S (2000) Ligand-, cell-, and estrogen receptor subtype (alpha/beta)-dependent activation at GC-rich (Sp1) promoter elements. J Biol Chem 275:5379–5387
Schat KA (2003) Marek's disease, chicken infectious anemia and immunosuppression: a nasty combination. 15th Annual Australian Poultry Science Symposium, Sydney, pp 51–57
Schat KA, Skinner MA (2008) Immunosuppressive diseases and immune evasion. In: Davison F, Kaspers B, Schat KA (eds) Avian immunology. Elsevier Academic Press, Amsterdam, pp 299–322
Schat KA, Van Santen V (2008) Chicken infectious anemia. In: Saif YM, Fadly AM, Glisson JR, McDougald LR, Nolan LK, Swayne DE (eds) Diseases of poultry. Blackwell Publishing, Ames, pp 211–235
Scott AN, Connor TJ, Creelan JL, McNulty MS, Todd D (1999) Antigenicity and pathogenicity characteristics of molecularly cloned chicken anaemia virus isolates obtained after multiple cell culture passages. Arch Virol 144:1961–1975
Scott AN, McNulty MS, Todd D (2001) Characterisation of a chicken anaemia virus variant population that resists neutralisation with a group-specific monoclonal antibody. Arch Virol 146:713–728
Sensel MG, Binder R, Lazier CB, Williams DL (1994) Reactivation of apolipoprotein II gene transcription by cycloheximide reveals two steps in the deactivation of estrogen receptor-mediated transcription. Mol Cell Biol 14:1733–1742
Shibayama T, Masuda G, Ajisawa A, Takahashi M, Nishizawa T, Tsuda F, Okamoto H (2001) Inverse relationship between the titre of TT virus DNA and the CD4 cell count in patients infected with HIV. Aids 15:563–570
Smith JA (2006) Impact of subclinical immunosuppression on poultry production. Proc. Symp Impact of subclinical infection on poultry production. American Association Avian Pathologists, Athens, pp 42–48
Smyth JA, Moffett DA, McNulty MS, Todd D, Mackie DP (1993) A sequential histopathologic and immunocytochemical study of chicken anemia virus infection at one day of age. Avian Dis 37:324–338
Smyth JA, Moffett DA, Connor TJ, McNulty MS (2006) Chicken anaemia virus inoculated by the oral route causes lymphocyte depletion in the thymus in 3-week-old and 6-week-old chickens. Avian Pathol 35:254–259
Soiné C, Watson SK, Rybicki E, Lucio B, Nordgren RM, Parrish CR, Schat KA (1993) Determination of the detection limit of the polymerase chain reaction for chicken infectious anemia virus. Avian Dis 37:467–476
Soiné C, Renshaw RH, O'Connell PH, Watson SK, Lucio B, Schat KA (1994) Sequence analysis of cell culture- and non-cell culture-adapted strains of chicken infectious anemia virus. Proc Int Symp Infect Bursal Dis Chick Infect Anaemia. Rauischholzhausen, Germany, pp 364–365
Spackman E, Cloud SS, Pope CR, Rosenberger JK (2002a) Comparison of a putative second serotype of chicken infectious anemia virus with a prototypical isolate. I. Pathogenesis. Avian Dis 46:945–955
Spackman E, Cloud SS, Rosenberger JK (2002b) Comparison of a putative second serotype of chicken infectious anemia virus with a prototypical isolate. II. Antigenic and physicochemical characteristics. Avian Dis 46:956–963
Taniguchi T, Yuasa N, Maeda M, Horiuchi T (1983) Chronological observations on hemato-pathological changes in chicks inoculated with chicken anemia agent. Natl Inst Anim Health Q (Tokyo) 23:1–12
Thom K, Petrik J (2007) Progression towards AIDS leads to increased Torque teno virus and Torque teno minivirus titers in tissues of HIV infected individuals. J Med Virol 79:1–7

Todd D, Creelan JL, Mackie DP, Rixon F, McNulty MS (1990) Purification and biochemical characterization of chicken anaemia agent. J Gen Virol 71:819–823
Todd D, Connor TJ, Calvert VM, Creelan JL, Meehan B, McNulty MS (1995) Molecular cloning of an attenuated chicken anaemia virus isolate following repeated cell culture passage. Avian Pathol 24:171–187
Todd D, Creelan JL, Meehan BM, McNulty MS (1996) Investigation of the transfection capability of cloned tandemly-repeated chicken anaemia virus DNA fragments. Arch Virol 141:1523–1534
Todd D, Connor TJ, Creelan JL, Borghmans BJ, Calvert VM, McNulty MS (1998) Effect of multiple cell culture passages on the biological behaviour of chicken anaemia virus. Avian Pathol 27:74–79
Todd D, McNulty MS, Adair BM, Allan GM (2001) Animal circoviruses. Adv Virus Res 57:1–70
Todd D, Scott AN, Ball NW, Borghmans BJ, Adair BM (2002) Molecular basis of the attenuation exhibited by molecularly cloned highly passaged chicken anemia virus isolates. J Virol 76:8472–8474
Todd D, Creelan JL, Connor TJ, Ball NW, Scott ANJ, Meehan BM, McKenna GF, McNulty MS (2003) Investigation of the unstable attenuation exhibited by a chicken anaemia virus isolate. Avian Pathol 32:375–382
Toro H, Ewald S, Hoerr FJ (2006a) Serological evidence of chicken infectious anemia virus in the United States at least since 1959. Avian Dis 50:124–126
Toro H, van Santen VL, Li L, Lockaby SB, van Santen E, Hoerr FJ (2006b) Epidemiological and experimental evidence for immunodeficiency affecting avian infectious bronchitis. Avian Pathol 35:455–464
Touinssi M, Gallian P, Biagini P, Attoui H, Vialettes B, Berland Y, Tamalet C, Dhiver C, Ravaux I, De Micco P, De Lamballerie X (2001) TT virus infection: prevalence of elevated viraemia and arguments for the immune control of viral load. J Clin Virol 21:135–141
Urlings HA, de Boer GF, van Roozelaar DJ, Koch G (1993) Inactivation of chicken anaemia virus in chickens by heating and fermentation. Vet Q 15:85–88
van Santen VL, Li L, Hoerr FJ, Lauerman LH (2001) Genetic characterization of chicken anemia virus from commercial broiler chickens in Alabama. Avian Dis 45:373–388
van Santen VL, Joiner KS, Murray C, Petrenko N, Hoerr FJ, Toro H (2004) Pathogenesis of chicken anemia virus: comparison of the oral and the intramuscular routes of infection. Avian Dis 48:494–504
van Santen VL, Toro H, Hoerr FJ (2007) Biological characteristics of chicken anemia virus regenerated from clinical specimen by PCR. Avian Dis 51:66–77
von Bülow V, Fuchs B (1986) Attenuierung des Erregers der aviären infektiösen Anämie (CAA) durch Serienpassagen in Zellkulturen. J Vet Med B Infect Dis Vet Public Health 33:568–573
von Bülow V, Fuchs B, Vielitz E, Landgraf H (1983) Frühsterblichkeitssyndrom bei Küken nach Doppelinfektion mit dem Virus der Marekschen Krankheit (MDV) und einem Anämia-Erreger (CAA). Zentralbl Veterinarmed [B] 30:742–750
Wang W, Dong L, Saville B, Safe S (1999) Transcriptional activation of E2F1 gene expression by 17beta-estradiol in MCF-7 cells is regulated by NF-Y-Sp1/estrogen receptor interactions. Mol Endocrinol 13:1373–1387
Wang X, Gao H, Gao Y, Fu C, Wang Z, Lu G, Cheng Y, Wang X (2007) Mapping of epitopes of VP2 protein of chicken anemia virus using monoclonal antibodies. J Virol Methods 143:194–199
Welch J, Bienek C, Gomperts E, Simmonds P (2006) Resistance of porcine circovirus and chicken anemia virus to virus inactivation procedures used for blood products. Transfusion 46:1951–1958
Wunderwald C, Hoop RK (2002) Serological monitoring of 40 Swiss fancy breed poultry flocks. Avian Pathol 31:157–162
Xin X, Xiaoguang Z, Ninghu Z, Youtong L, Liumei X, Boping Z (2004) Mother-to-infant vertical transmission of transfusion transmitted virus in South China. J Perinat Med 32:404–406

Yamaguchi S, Imada T, Kaji N, Mase M, Tsukamoto K, Tanimura N, Yuasa N (2001) Identification of a genetic determinant of pathogenicity in chicken anaemia virus. J Gen Virol 82:1233–1238
Yuasa N (1983) Propagation and infectivity titration of the Gifu-1 strain of chicken anemia agent in a cell line (MDCC-MSB1) derived from Marek's disease lymphoma. Natl Inst Anim Health Q (Tokyo) 23:13–20
Yuasa N (1992) Effect of chemicals on the infectivity of chicken anaemia virus. Avian Pathol 21:315–319
Yuasa N (1994) Pathology and pathogenesis of chicken anemia virus infection. Proc Int Symp Infect Bursal Dis Chick Infect Anaemia. Rauischholzhausen, Germany, pp 385–389
Yuasa N, Imai K (1988) Efficacy of Marek's disease vaccine, herpesvirus of turkeys, in chickens infected with chicken anemia agent. In: Kato S, Horiuchi T, Mikami T, Hirai K (eds) Advances in Marek's disease research. Japanese Association on Marek's Disease, Osaka, pp 358–363
Yuasa N, Yoshida I (1983) Experimental egg transmission of chicken anemia agent. Natl Inst Anim Health Q (Tokyo) 23:99–100
Yuasa N, Yoshida I, Taniguchi T (1976) Isolation of a reticuloendotheliosis virus from chickens inoculated with Marek's disease vaccine. Natl Inst Anim Health Q (Tokyo) 16:141–151
Yuasa N, Taniguchi T, Yoshida I (1979) Isolation and some characteristics of an agent inducing anemia in chicks. Avian Dis 23:366–385
Yuasa N, Noguchi T, Furuta K, Yoshida I (1980a) Maternal antibody and its effect on the susceptibility of chicks to chicken anemia agent. Avian Dis 24:197–201
Yuasa N, Taniguchi T, Noguchi T, Yoshida I (1980b) Effect of infectious bursal disease virus infection on incidence of anemia by chicken anemia agent. Avian Dis 24:202–209
Yuasa N, Taniguchi T, Imada T, Hihara H (1983) Distribution of chicken anemia agent (CAA) and detection of neutralizing antibody in chicks experimentally inoculated with CAA. Natl Inst Anim Health Q (Tokyo) 23:78–81

# Geminiviruses

**H. Jeske**

**Contents**



---

H. Jeske
Institute of Biology, Department of Molecular Biology and Plant Virology, University of Stuttgart, Pfaffenwaldring 57, 70550, Stuttgart, Germany
holger.jeske@bio.uni-stuttgart.de

E.-M. de Villiers, H. zur Hausen (eds.), *TT Viruses: The Still Elusive Human Pathogens,* 

**Abstract** Plant pathogenic geminiviruses have been proliferating worldwide and have, therefore, attracted considerable scientific interest during the past three decades. Current knowledge concerning their virion and genome structure, their molecular biology of replication, recombination, transcription, and silencing, as well as their transport through plants and dynamic competition with host responses are summarized. The topics are chosen to provide a comprehensive introduction for animal virologists, emphasizing similarities and differences to the closest functional relatives, polyomaviruses and circoviruses.

## Introduction

Geminiviruses have attracted increasing scientific interest since their structural description during the 1970s. Their unusual twin geometry, their circular single-stranded (ss) DNA, unprecedented for plants, and their increasing agricultural importance have fascinated a generation of scientists already (in 2008). A growing number of publications on this subject, many of them reporting about pioneering work, have appeared, and it is beyond the scope of this chapter to acknowledge all these excellent contributions here. Comprehensive surveys of the topic are available and are recommended for studying the closer detail (Davies and Stanley 1989; Fargette et al. 2006; Gafni and Epel 2002; Gilbertson and Lucas 1996; Goodman 1981; Gutierrez 2000a, b; Gutierrez et al. 2004; Hanley-Bowdoin et al. 1999, 2004; Harrison 1985; Lazarowitz 1992; Mansoor et al. 2003; Palmer and Rybicki 1998; Rojas et al. 2005; Rybicki and Pietersen 1999; Sanderfoot and Lazarowitz 1996; Stanley 1985; Stanley and Davies 1985). The ICTV key reference on geminiviruses (Stanley et al. 2005) and recent articles on nomenclature and taxonomy (Briddon et al. 2008; Fauquet and Stanley 2005; Fauquet et al. 2008) may help the interested reader to understand the system used for geminivirus and satellite names. The family Geminiviridae is divided into the four genera, *Mastrevirus*, *Curtovirus*, *Topocuvirus*, and *Begomovirus*, the names of which were created from the abbreviations of the type members Maize streak virus (MSV), Beet curly top virus (BCTV), Tomato pseudo curly top virus (TPCTV) and Bean golden mosaic virus (BGMV), respectively. Viruses are assigned to the genera due to their sequence similarity, insect vectors, and host plants (see Table 1 for names and abbreviations).

Continuing the first discussions between plant and animal virologists about ssDNA-containing viruses (Gronenborn 2004; Gutierrez et al. 2004; Stanley 2004), here we will emphasize those topics of geminivirology that represent either fundamental differences or striking similarities between the small ssDNA viruses of mammals and plants.

## Importance of Geminiviruses

The beauty of plants was the first motif to propagate geminiviruses long before they were identified as such. A poem of the Japanese empress Koken (752 a.d.) is regarded as the most ancient description of a geminivirus-infected ornamental plant

**Table 1** Virus species and the abbreviations of their names

| Virus name | Abbreviation |
|---|---|
| Abutilon mosaic virus | AbMV |
| African cassava mosaic virus | ACMV |
| Bean dwarf mosaic virus | BDMV |
| Bean golden mosaic virus | BGMV |
| Beet curly top virus | BCTV |
| Cabbage leaf curl virus | CaLCuV |
| Indian cassava mosaic virus | ICMV |
| Maize streak virus | MSV |
| Mung bean yellow mosaic India virus | MYMIV |
| Mung bean yellow mosaic virus | MYMV |
| Pepper huasteco yellow vein virus | PHYVV |
| Satellite tobacco necrosis virus | STNV |
| Squash leaf curl virus | SLCV |
| Sri Lankan cassava mosaic virus | SLCMV |
| Tomato golden mosaic virus | TGMV |
| Tomato leaf curl Bangalore virus | ToLCBV |
| Tomato leaf curl New Delhi virus | ToLCNDV |
| Tomato pseudo curly top virus | TPCTV |
| Tomato yellow leaf curl Sardinia virus | TYLCSV |
| Tomato yellow leaf curl virus | TYLCV |
| Watermelon chlorotic stunt virus | WmCSV |
| Wheat dwarf virus | WDV |

(*Eupatorium makinoi*) (Saunders et al. 2003). During the second half of the nineteenth century, *Abutilon* plants from the West Indies became very fashionable in Europe due to their leaf mosaics (Fig. 1). Since then, the causative agent (Abutilon mosaic virus, AbMV) (Wege et al. 2000) has been disseminated world-wide by gardeners and breeding companies, a course of action which never would have been admitted by plant protection agencies nowadays. In spite of AbMV's capability to severely damage plants other than *Abutilon* under experimental conditions, including the important crop tomato (Pohl and Wege 2007), it has never been found to be involved with plant pests, presumably because it has been attenuated during vegetative propagation, and lost its insect transmissibility (Höfer et al. 1997; Höhnle et al. 2001).

Different from the two ornamental examples, many other geminiviruses have severely threatened important crop plants as a consequence of exploding insect populations and global transportation of plant material. In particular, new whitefly biotypes have been established during the extension of soybean production (Costa 1976), and have boosted the spread and evolution of geminiviruses in tropical and subtropical countries considerably. Meanwhile, geminiviruses are regarded as one of the most destructive pests for important crops such as beans, cassava, cotton, maize, pepper, sugar beet, sweet potato, and tomato (Moffat 1999). Because most of these plant species provide staple food in the subsistence agriculture of tropical and subtropical countries, combatting geminivirus epidemics today is a most

**Fig. 1** Symptomatic *Abutilon* leaf infected with AbMV

important task for social and economic stability in these areas. It is difficult to estimate the total economic losses, but by some assessments that lost value has been reported for ACMV as US $2 billion per year, and US $5 billion in the period 1995–2000 for various plants in Brazil (for detailed descriptions of the economic impact see: Hull 2002; Hull and Davies 1992; Morales 2006, 2007; Morales and Anderson 2001; Morales and Jones 2004; Rybicki and Pietersen 1999; Thresh 2006; Thresh and Cooter 2005).

In fundamental research, geminiviruses have become popular instruments to analyse the molecular biology of plant gene regulation and cell–cell communication. Due to their small genome size, studying DNA replication, transcription, mRNA processing, protein expression, and gene silencing was easier than with their host counterparts. These topics will be explored after a description of the unique geminivirus structure in the following section.

## Structure and Composition of Virions

Various devastating plant diseases have been described and analysed throughout the twentieth century, but these were only assigned to geminiviruses relatively late for several reasons. Geminiviruses are tiny in size, and hence can be mistaken for ribosomes in ultrastructural analysis. Most of them are phloem-limited and, thus, barely transmitted by mechanical inoculation. Their low titres in plants have made purification difficult. A first discrimination of the viruses from host components, however, became feasible when conspicuous pairwise particles organized in paracrystalline arrays within nuclei were detected in ultrathin sections of infected plants (for review, see Harrison 1985). The particle arrangement of AbMV was particularly remarkable, as it was composed of tubular structures that filled up the nucleus of phloem cells (Fig. 2a, b; Jeske and Werz 1980a; Jeske et al. 1977). Our current knowledge on the molecular biology of geminiviruses sheds new light on the early history, so that these tubuli may be the result of replicating viral minichromosomes in the centre of the tubuli (Fig. 2a, arrowhead) that sequester ssDNA during rolling circle replication, and concomitantly package the ssDNA into virions (Fig. 2a, arrow). Purified virions retained the conspicuous pairwise appearance (Fig. 2c) which was the reason for coining the name "geminivirus" (lat. *geminus*: twin), and contained circular ssDNA packaged in one coat protein (CP) species (Goodman 1977; Harrison et al. 1977). Although intuition often suggests the opposite, only one molecule of ssDNA is encapsidated within one twin particle (Fig. 2e). For geminiviruses with a bipartite genome, consequently, two twin particles are needed to obtain a full infection (Fig. 2f).

The fine structure of the virions has been resolved by electron cryomicroscopy in combination with image reconstruction and bioinformatic modelling of the coat protein for a *Mastrevirus* (MSV) (Zhang et al. 2001b) and a *Begomovirus* (ACMV) (Böttcher et al. 2004), using an RNA-containing isometric virus (STNV) as a reference. Both models agree in that the twin geometry (Fig. 2d) is formed by two incomplete icosahedra which are attached at a waist (W) where both halves are twisted against each other by 20 degrees. As no empty shells of geminivirions have been detected in the course of purification, it is believed that the particle structure is mainly stabilized by protein–DNA interactions and is assembled concomitantly with replication of ssDNA. Each half of the virion is composed of 11 pentameric capsomers missing one capsomer to complete a T=1 structure. The top and shoulder capsomers (Fig. 2d, T, S) can form equivalent contacts to each other, whereas the waist capsomers (Fig. 2d, W) need some deformation to fit into the geminate structure.

The geminiviral CP is a multifunctional protein that not only protects the viral DNA during vector transmission, and provides vector specificity, but is also involved in nuclear shuttling, cell-to-cell transport, and long-distance trafficking in plants. Many laboratories have dissected the genetic information of the *CP* gene using mutational analysis, either investigating spontaneous mutations or by the aid of site-directed mutagenesis. With the first three-dimensional model, we now have the unique possibility to combine the results obtained with various viruses and draw a consensus draft of the CP topology with reference to their functions (Fig. 2g–i): DNA binding capacity, insect transmission, and nuclear localization determinants are summarized immediately below and in the following sections.

CP is a very basic protein (pI of 10) and exhibits an unspecific binding affinity to nucleic acids. As many different virus species and experimental approaches have been employed in order to define the proper ssDNA binding domain, no complete consensus has been achieved about its location, but congruent results can be summarized (Fig. 2g) (Hehnle et al. 2004; Ingham et al. 1995; Kirthi and Savithri 2003; Noris et al. 1998; Qin et al. 1998). Both three-dimensional models for ACMV and MSV (Böttcher et al. 2004; Zhang et al. 2001b) agree with each other with respect

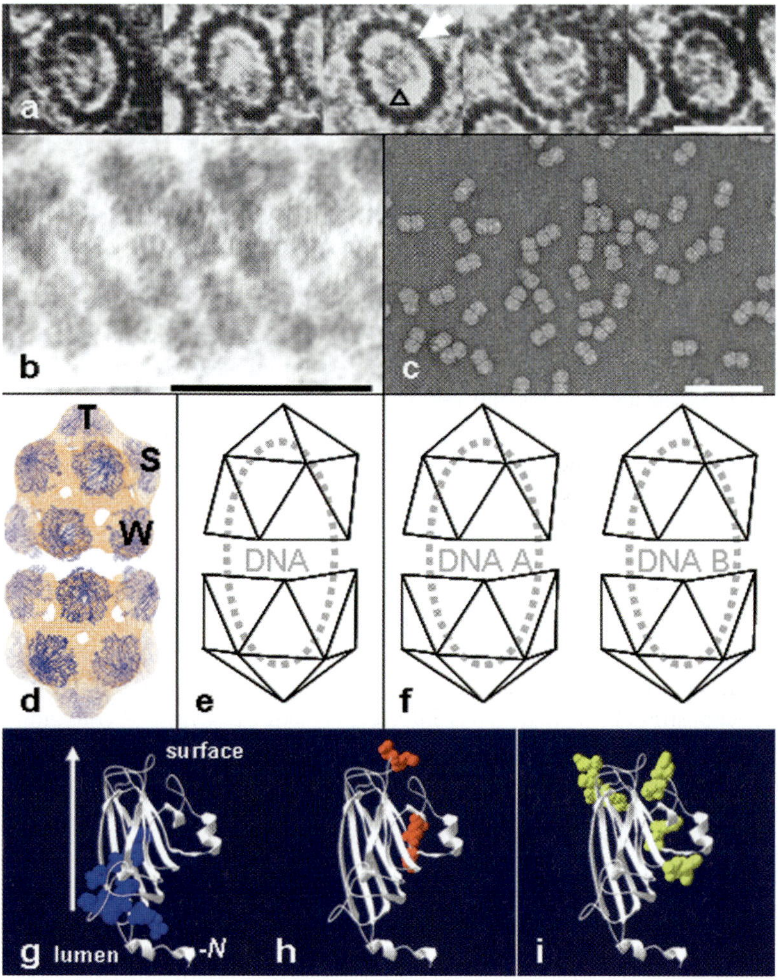


**Fig. 2 a–i** Structure of geminivirions. Cross- (**a**) and tangential (**b**) sections through tubular paracrystalline inclusion bodies in companion cell nuclei of AbMV-infected *Abutilon* plants following embedding, ultramicrotomy and positive staining (adapted from Jeske and Werz 1980b; Jeske et al. 1977, with permission from Elsevier). *Arrow* indicates geminate particles, and *arrowhead* minichromosome-like structures. Purified negatively stained African cassava mosaic virus (ACMV) particles (**c**, courtesy of K. Kittelmann). *Bar*=100 nm. Schematic drawings (**d–f**) of the geminate particle structure based on the three-dimensional reconstruction using electron cryomicroscopy (Böttcher et al. 2004). Top (*T*), shoulder (*S*), and waist (*W*) capsomers are indicated in **d** as well as the relationship of DNA to the particle for monopartite (**e**) and bipartite (**f**) geminiviruses. Current model of the coat protein (**g–i**) derived from bioinformatic comparisons (Böttcher et al. 2004). The fivefold axis (*arrow*), the surface and the lumen side of the geminate particle are indicated. Amino acid positions which were found to be relevant by mutation analysis for DNA binding (**g**), insect transmission (**h**) and nuclear import (**i**) are highlighted by *spacefilling* atoms, according to (Höhnle et al. 2001; Kheyr-Pour et al. 2000; Noris et al. 1998; Qin et al. 1998; Unseld et al. 2004; Unseld et al. 2001). The N-terminal portion of amino acids (aa) 1–57 (*N*) is not modelled and is therefore omitted here to the presence of a central eight-stranded β-barrel (Fig. 2g–i). There is some variation in the extension of the loops on the outer surface, apparently as an adaptation to the insect vector. The N-terminal portion of CP (aa 1–104), was found to be particularly important for specific DNA-binding with viruses from different genera (Liu et al. 1997; Qin et al. 1998). Amino acids at other positions additionally influenced the DNA affinity (Ingham et al. 1995; Qin et al. 1998). When projected to the structure (Fig. 2g), the relevant amino acids tend to cluster towards the lumen of the particle, which is in agreement with a function in ssDNA binding.

Some mutations may influence DNA binding indirectly by interfering with capsomer formation and inter-capsomer contacts, because DNA-binding has been demonstrated to be co-operative (TYLCV CP; Palanichelvam et al. 1998), and CP–CP interactions have been proposed to rely on the binding of N-terminal and C-terminal sequences (TYLCV CP) (Hallan and Gafni 2001).

The in vivo assembly and disassembly of geminivirions is largely unexplored. In vitro experiments have shown that geminiviral CPs bind ssDNA as well as double-stranded (ds) DNA in a sequence-nonspecific manner, with some preference for ssDNA (Hehnle et al. 2004; Ingham et al. 1995; Kirthi and Savithri 2003; Liu et al. 1997; Palanichelvam et al. 1998; Qin et al. 1998). Correspondingly, the viral DNA sequence lacks a specific origin of assembly, and probably the preferred packaging of geminiviral DNA is granted during rolling circle replication. The nuclear tubuli detected in *Abutilon* plants (Fig. 2a, b) might illustrate this process, and the protein–protein interaction of CP with a replication-associated Begomovirus Rep protein, as demonstrated for MYMIV (Malik et al. 2005) may lend additional weight to this attractive hypothesis.

Similar to the assembly process, disassembly needs further exploration. It is unknown whether geminiviruses enter the nucleus via nuclear pores or need a disintegrated nuclear envelope. If pores are used, it remains to be shown whether the whole geminivirion or only its DNA crosses the nuclear envelope. Recently, it has been shown that geminiviral particles release their DNA preferentially at the top of the virions (Kittelmann and Jeske 2008).

## Genome Structure and Evolution

In their overall genome structure, plant geminiviruses resemble animal polyomaviruses more than circoviruses (Fig. 3). Like for *Simian virus 40* (SV40), open reading frames (ORFs) are oriented bidirectionally, an intergenic region (IR or large intergenic region, LIR) contains promoters in both directions, and the origin of replication (ori), whereas termination signals (ter) are located opposite to the IR. Genes responsible for regulation of replication and transcription are combined on the left side—and those for packaging on the right side—of the genomic components. Some geminiviruses carry all necessary genes for full infection on one component and are hence classified as monopartite geminiviruses (all members of the genera *Mastre-, Curto-, and Topocuvirus* and some of *Begomovirus*). Bipartite

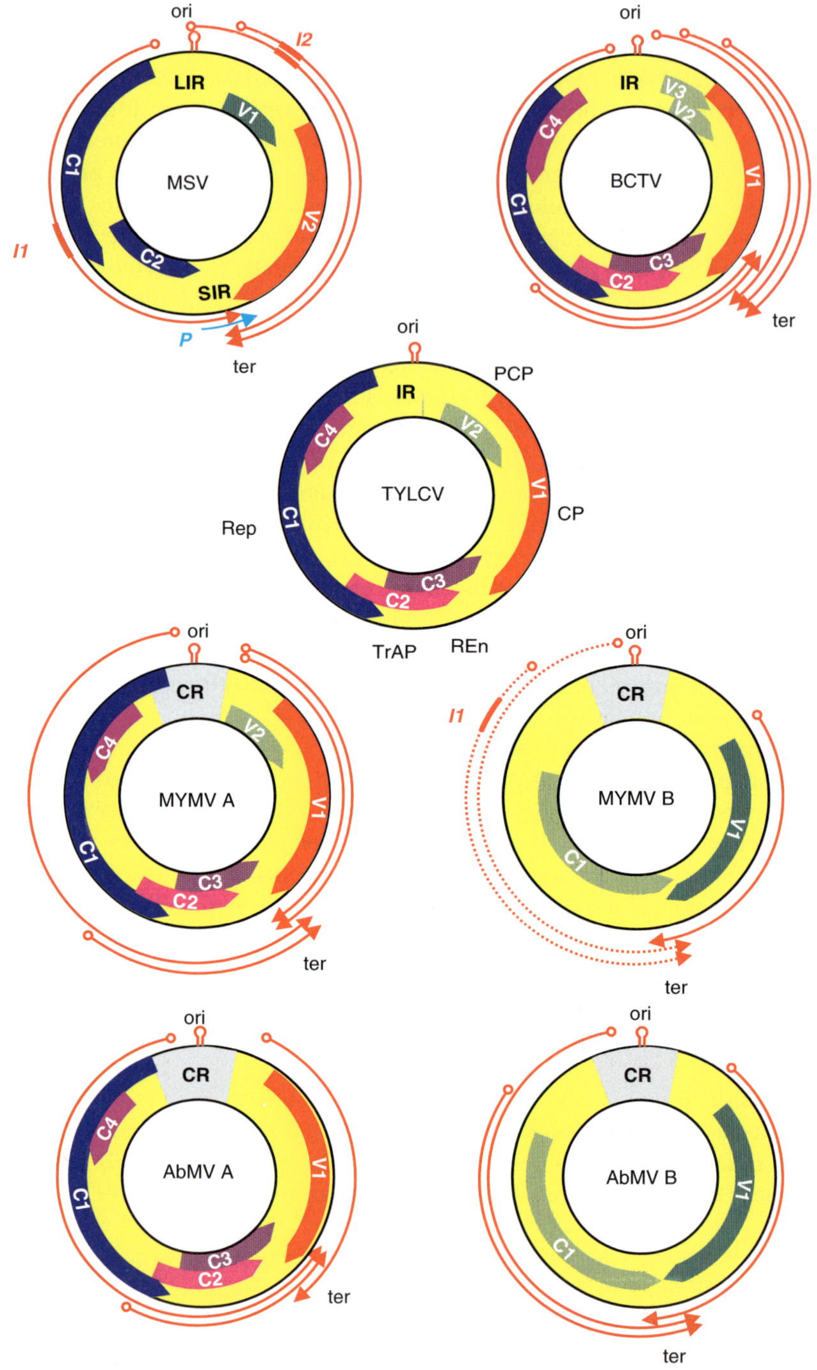


**Fig. 3** Genome maps of representative geminiviruses. For abbreviations see Table 1. *Thick arrows* indicate ORFs, *thin arrows* transcripts with introns (*Int*), or primer (*P*), and hairpins origin of replication (*ori*). Intergenic regions (*IR*) are either named long (*LIR*) or short (*SIR*) for MSV and contain common regions (*CR*) for bipartite begomoviruses of the Old World (*MYMV*) and New World

genomes with two components, called DNA A and DNA B, occur only in the genus *Begomovirus*. The viral genes were originally consecutively numbered according to their coding strand (V for viral, syn. R for right, or C for complementary, syn. L for left orientation). The numbering, however, became inconsistent to a certain extent (e.g. C4 or V2, V3), when new genes were later identified as relevant. Names for monopartite ORFs are used as such, whereas those of DNA A and B are preceded by A or B (AV1–AV2, AC1–AC4; BV1, BC1), respectively. In this review, the nomenclature (A,B)VX and (A,B)CX will be used throughout the text in order to facilitate reading, even if the synonymous naming (A,B)RX and (A,B)LX was chosen in the original literature.

With increasing knowledge about their function, characterized genes received more informative names that are widely accepted now, like *Rep* (replication-associated protein), *TrAP* (transcription activator protein), *REn* (replication enhancer), *CP* (coat protein), *PCP* (pre-coat protein), *MP* (movement protein), and NSP (nuclear shuttle protein). With the exception of PCP, which indicates the genomic position, the other names describe the initially identified functions, although it is meanwhile clear that many gene products have multiple tasks (see below).

The ORFs *C1* and *C2* in *Mastrevirus* genomes (Fig. 3, MSV) which are differentially expressed via splicing, are fused in all other geminivirus genomes to result in elongated ORFs *C1/AC1*. The corresponding new ORFs C2 (AC2) as well as C3–C4 (AC3–AC4) are unique for curto-, topocu- and begomoviruses. Additionally, although it is at the same genomic position, C2 of curto- and topocuviruses is not homologous to C2/AC2 of begomoviruses. It may serve for similar (silencing suppression) as well as different functions (transcription activation). ORF PCP (also named V1, V2 or AV2 depending on the virus species) is present in mastre-, curto-, topocu-, and Old World begomoviruses, but absent from indigenous American viruses.

Within virions, the genome is carried on completely single-stranded circular DNA. Only mastreviruses package a small complementary DNA-primer in addition (Fig. 3, MSV, P) that binds to the small intergenic region (SIR) and serves as a starter for complementary strand replication after primary infection and disassembly of the virion.

In bipartite begomovirus genomes, a stretch of about 200 bp within the intergenic region (IR) is highly conserved between two components of the same virus, but particularly diverse between different viruses. This common region (CR) harbours a small hairpin structure exposing an invariant nonanucleotide (TATAATT|AC) with highest conservation in all geminiviruses. Accordingly, this hairpin comprises the key sequence for initiating rolling circle replication (Fig. 3, ori, see below).

**Fig. 3** (continued) (*AbMV*), which are highly conserved in DNA A and DNA B of a given virus. Transcripts overlap in termination regions (*ter*) which harbour polyadenylation signals. Transcripts of MYMV B in complementary orientation are *stippled* because the termination could not be defined. Commonly used gene names are exemplified for TYLCV, but refer to the other genes at the same position correspondingly: replication-associated protein (*Rep*), transcription activator protein (*TrAP*), replication enhancer (*REn*), precoat protein (*PCP*), and coat protein (*CP*). Transcripts were mapped according to Frischmuth et al. (1991, 1993a) Shivaprasad et al. (2005) and Wright et al. (1997)

## *Defective (Interfering) DNAs*

Depending on host plant and virus, various amounts of defective (D) DNA molecules may accumulate during improper replication (reviewed in Patil and Dasgupta 2005). They may interfere with parent virus replication and attenuate symptom development, and are then called defective interfering (DI) DNAs. Usually they retain the left side of the monopartite DNA or DNA B, rarely DNA A, but in any case the viral ori. Their sizes comprise half the viral DNA component's size (1,300 bp) or may be smaller. Footprints of recombination at the deletion sites suggest that they used to be generated by illegitimate recombination.

## *Satellite DNAs*

Different from D-DNAs, circular satellite DNAs, called DNAs β (Briddon and Stanley 2006; Briddon et al. 2003) do not exhibit any extended sequence similarity to their helper viruses. Equal to D-DNAs, their size comprises about half the size of one viral genomic component, in one case less (small satellite, 682 nt; Dry et al. 1997; Li et al. 2007). The geminiviral invariant nonanucleotide is present in a hairpin sequence, and this structure is necessary for transreplication by the helper virus. The descent of DNAs β and the small satellite is unknown (Homs et al. 2008). Further circular DNAs of similar size may be acquired from *Nanovirus* genomes and are called DNA 1 (Briddon et al. 2004). DNAs β and DNA 1 code for a single ORF (βC1 and Rep, respectively). For some DNA β and helper virus combinations, the ORF βC1 is essential for symptom development, although the helper virus can independently replicate and move through plants (Gopal et al. 2007; Saeed et al. 2005; Saunders et al. 2003; Saunders et al. 2004).

## *Geographic Differentiation*

The relationships between most geminiviruses reflect their geographic descent (Harrison and Robinson 1999; Rybicki 1994). However, some species have probably been disseminated by man, even to new continents.

Nevertheless, the majority of geminiviruses show a clear geographic clustering upon sequence comparisons allowing a tentative assignment of new species to North, Central, or South America (New World), or to Europe, Africa, Middle East, India, China, and Japan (Old World). Whereas indigenous New World geminiviruses are always bipartite, monopartite viruses with or without satellite DNAs as well as bipartite geminiviruses have been found in the rest of the world. New World viruses have lost the PCP gene (Fig. 3, AbMV A) and have developed a stricter dependence on DNA B genes, correspondingly.

Recently, two begomoviruses from Vietnam have been sequenced which are more closely related to New World than to Old World geminiviruses, inferring the interesting hypothesis that New World begomoviruses have been already developed in the Old World and thereafter have been spread to the New World where they founded a new expanding population (Ha et al. 2006; Ha et al. 2008).

## *Taxonomy and Evolution*

Geminiviruses are named according to the initially described host plant, the most prevalent symptom, and the geographical origin. A demarcation threshold of 89% DNA sequence similarity for begomoviruses and 75% for mastreviruses has been set by the ICTV study group to differentiate between virus species (Fauquet and Stanley 2005; Fauquet et al. 2008). DNAs β are named with reference to their helper begomovirus (Briddon et al. 2008).

It has not been definitively proven how geminiviruses have evolved, whether from a bipartite progenitor by reduction to a monopartite descendant, or from a monopartite virus by acquiring a DNA B to result in a bipartite begomovirus. However, the latter scenario is the most parsimonious explanation on the basis of the available experimental evidence. Whereas mastre-, curto- and topocuviruses are strictly monopartite, begomoviruses have developed a series of relationships with DNAs β and DNAs B as well as with components of nanoviruses to broaden their genetic inventory.

Different from the promiscuous situation in the Old World, true New World viruses remain rather faithful to their cognate DNA partner. Reassortment (also called pseudo-recombination in plant virology) of DNA A and B components is usually restricted to closely related parents in America (Unseld et al. 2000a, and references therein; Unseld et al. 2000b). Repeated DNA elements within the CR, called "iterons" (Arguello-Astorga et al. 1994), are similar in cognate pairs of DNAs in all bipartite begomoviruses, and regulate the transreplication of DNA B by DNA A-encoded Rep proteins. This iteron-dependence is more stringent for New World (reviewed in Hanley-Bowdoin et al. 1999) than for Old World begomoviruses (Alberter et al. 2005; Li et al. 2007). Accordingly, no satellite DNAs have been found on the American continent so far.

The evolution rate of geminiviruses is remarkably high with $3–5\times10^{-4}$ nucleotides per site and year, a value close to that of plant RNA viruses (Duffy and Holmes 2008). Deamination-dependent transitions (C T, G A) were found to be overrepresented, and this result is probably due to the ssDNA nature of the genome, for which spontaneous or enzymatic deaminations cannot be repaired properly because the master template is lacking (Duffy and Holmes 2008). The rapid mutational dynamics and pseudo-recombination capabilities of geminiviruses together with their high molecular recombination frequency (see next paragraph) explain their adaptive power for new host plants and geographic areas.

## *Recombination*

With the sequencing of an increasing number of geminiviral full-length DNAs, it became obvious that molecular recombination has been a major driving force of their evolution and epidemics (Lefeuvre et al. 2007; Martin and Rybicki 2000; Martin et al. 2005; Owor et al. 2007; Padidam et al. 1999; Pita et al. 2001). The chance that two geminivirus species enter the same nucleus was found to be unexpectedly high (Morilla et al. 2004), and the discovery that geminiviruses use a recombination-dependent replication (RDR) mode in addition to rolling circle replication has explained why geminiviruses are so prone to recombination (Jeske et al. 2001). Although every part of the genome can participate in recombination, hot spots of this process have been identified (Garcia-Andres et al. 2007), which co-localize to nuclease-hypersensitive viral chromatin structures (Pilartz and Jeske 2003).

## Insect Transmission

In nature, geminiviruses are transmitted from plant to plant by either various species of leafhoppers (*Mastrevirus*, *Curtovirus*), treehoppers (*Topocuvirus*), or by a single species of whiteflies, *Bemisia tabaci* Genn. (Begomovirus, reviewed in: Byrne and Bellows 1991; Costa 1976; Czosnek et al. 2001; Morales 2007; Morales and Jones 2004). Leafhoppers may disseminate geminiviruses to temperate climate zones, north up to Sweden, and south down to South Africa, whereas *B. tabaci* is restricted to warmer climates in tropical and subtropical countries. With the climate changing and the prevalence of the B-biotype of *B. tabaci*, a further spread of begomoviruses has been observed along the northern coast of the Mediterranean and the southern states of the United States.

## *Determinants of Insect Transmission*

Irrespective of the particular insect species, geminiviruses are transmitted in a persistent circulative manner. Therefore, after uptake from the plant, they have to cross the midgut epithelium, are transported through the haemocoel, and enter salivary glands to be injected into a new host plant. As the vector insects are phloem-feeders, geminiviruses are directly deposited into phloem cells (sieve elements, companion cells, phloem parenchyma cells).

For efficient transmission, capsids have to protect the DNA in insect gut and haemocoel, and probably need to attach specifically to receptors in the midgut and salivary gland epithelia. Insect virus receptors have not been identified so far, but whiteflies accumulate groEL-like chaperons in their haemocoel which specifically stabilize geminiviruses and are produced by endosymbiotic bacteria (Akad et al. 2004; Morin et al. 1999).

The specificity between virus and insect is solely determined by the coat protein (Briddon et al. 1990; Höhnle et al. 2001). A domain of the coat protein that is important for whitefly transmission has been narrowed down by mutational analyses for several begomoviruses [AbMV, WmCSV, Tomato yellow leaf curl Sardinia virus (TYLCSV): Höfer et al. 1997; Höhnle et al. 2001; Kheyr-Pour et al. 2000; Noris et al. 1998].

The amino acid positions which have been implicated in insect transmission are highlighted in the model (Fig. 2 h), and they cluster at the outer surface as expected for a domain that has to be recognized by an insect receptor, just at the distal vertex of the pentameric capsomer.

## Tissue Tropism

In plants, the main entrances for viruses are the veins of leaves, which are composed of two main tissues, the water-conducting xylem and the carbohydrates and other nutrients transporting phloem. Among those, only the latter tissue contains a continuous protoplasm and is therefore directly connected to the symplast of the whole plant via plasmodesmata. As their insect vectors are strict phloem-feeders there is no obvious need for geminiviruses to get out from this tissue for further spread from plant to plant. Nevertheless, some geminiviruses were found in other tissues of the leaf mesophyll (spongy and palisade parenchyma), as well as the epidermis, although the majority of them remain phloem-limited (reviewed in Wege 2007). After well-aimed injection into phloem cells by the aid of the insect's stylet, geminivirions can move inside the sieve elements but they are not able to replicate therein, because these cells do not contain nuclei. For further propagation they need to enter nucleus-containing phloem cells, like companion and phloem parenchyma cells. These cells are connected to each other and sieve elements by specialized cytoplasmic bridges (plasmodesmata, reviewed in Robards and Lucas 1990). In order to eventually exit from the phloem, geminiviruses have to pass through further plasmodesmata connecting the adjacent tissues. Each plasmodesma has its own architecture and is therefore regarded as a selective gate allowing or inhibiting the passage of small and macro-molecules. Its function typically changes with leaf development during the transition from sink to source status for carbohydrates. Correspondingly, plant viruses in general infect sink leaves much more easily than source leaves with their narrow plasmodesmata.

A failure to leave the phloem may be the consequence of disabled geminiviral transport proteins, phloem-specific promoters, or an insufficient activation of the host's replication machinery in mesophyll cells (Levy and Czosnek 2003; Morra and Petty 2000). These explanations, however, do not completely satisfy the available experimental evidence. The transport proteins of strictly phloem-limited geminiviruses (AbMV, ICMV) efficiently moved from cell to cell in the epidermis of sink leaves if the encoding DNA constructs were delivered by biolistic bombardment (Rothenstein et al. 2007; Zhang et al. 2001a). Moreover, the rate of mechanical infection was increased in transgenic plants that contained tandem copies of AbMV DNA

B and thus were able to express the transport proteins MP and NSP in all cells (Wege and Pohl 2007). A lack of mechanical inoculation in plants is often regarded as an indication for phloem-limitation of the virus, because it is believed that it is more difficult to productively infect phloem cells by this means. Despite its capability to reach the phloem in the inoculated leaf, AbMV remained phloem-limited in the systemically infected leaves of transgenic DNA B plants (Wege and Pohl 2007). These observations are best explained by taking into consideration the resistance response of the plant, in particular transcriptional and post-transcriptional gene silencing (TGS, PTGS) in systemically invaded organs. If geminiviruses are more efficiently defeated in the mesophyll than in the phloem, the latter tissue may serve as a retreat for the virus. Occasional emigrations from the phloem (Rothenstein et al. 2007) and promotion of geminiviruses to invade mesophyll cells by co-infected unrelated RNA viruses with a strong silencing suppressor (Wege and Siegmund 2007) are observations in favour of this interpretation (see Sects. 8.4 and 9).

All the tissues discussed so far are composed of highly differentiated cells. On the contrary, geminiviruses have never been found to accumulate in meristematic tissues (Horns and Jeske 1991; Lucy et al. 1996). Accordingly, meristem tissue cultures are suitable means for obtaining geminivirus-free plants (Kartha and Gamborg 1975).

Although most investigations have concentrated on the fate of geminiviruses in foliage leaves, they are spread through roots and shoots as well, may enter flowers and fruits, but not seeds (reviewed in Wege 2007). Correspondingly, progeny plants are purged of geminiviruses.

## Replication, Recombination, and Repair

After having entered the nucleus following injection of virions by the insect and disassembly of the capsid, the circular ssDNA is converted into circular dsDNA which is packaged into host nucleosomes. Viral minichromosomes remain extrachromosomal and serve as templates for transcription, further replication and finally the production of progeny ssDNA that is encapsidated by coat protein again. The replication processes have attracted the most interest of geminivirologists during the past two decades and have been extensively reviewed (Gutierrez 2000a; Gutierrez et al. 2004; Hanley-Bowdoin et al. 2004; Hanley-Bowdoin et al. 1999; Jeske 2007).

### *Complementary Strand Replication*

In addition to the viral genome, mastrevirus capsids contain a small primer complementary to the short intergenic region (Fig. 3, MSV, P) that serves to initiate complementary strand synthesis by host enzymes, presumably repair polymerases

(Fig. 4, CSR). All other geminiviruses need de novo generation of RNA primers which are extended by DNA polymerase, and are removed by RNase H. The resulting copy strand is ligated by ligases, and twisted by topoisomerases to wrap the dsDNA around nucleosome cores. During these processes, the viral DNA completely relies on already present host enzymes.

## *Activation of Host DNA Synthesis*

As the small geminiviral genomes do not encode for their own DNA polymerase and host repair polymerases may not be efficient enough to propagate the viral DNA, they have evolved genes that re-activate S phase in differentiated cells. The key player in this concert is the gene product of *C1* (RepA) for mastreviruses and *C1/AC1* for all other geminiviruses (Rep, Fig. 3). Rep is a multifunctional protein with many similarities to SV40 T antigen, but an additional nicking-closing activity, similar to that of ssDNA phage and conjugative bacterial plasmids. Mastrevirus RepA contains the motif LXCXE which enabled the binding of human retinoblastoma (RB) protein, a result that has initiated the discovery of plant Rb-related proteins (pRBR) (Gutierrez 2000a). The LXCXE motif is not conserved in other geminiviral genera, so that different binding domains for pRBR have evolved for begomoviruses (Hanley-Bowdoin et al. 2004). Although both RepA encoded by *C1* ORF and Rep encoded by the spliced *C1:C2* ORF of mastreviruses contain the LXCXE motif, pRBR binding activity is masked in Rep (Liu et al. 1999).

Mutations of the pRBR binding domains reduced replication efficiency to lower levels, but did not abolish geminiviral replication completely. Interestingly, these mutations led to phloem-limitation of otherwise non-restricted viruses like MSV

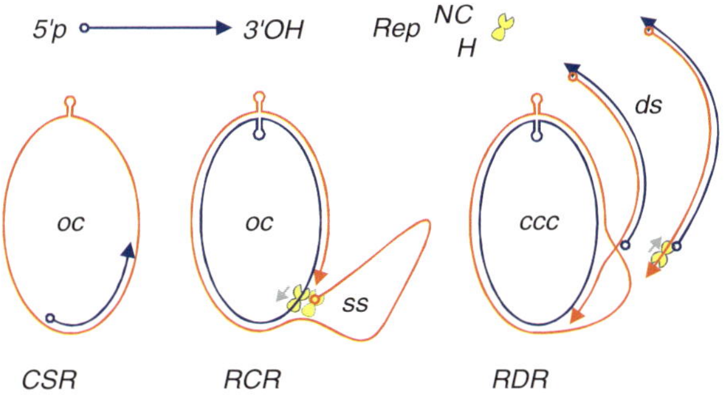


**Fig. 4** Schematic planar representation of replication modes of geminiviruses with open circular (*oc*) or covalently closed circular (*ccc*) templates, and single-stranded (*ss*) or double-stranded (*ds*) products, for complementary strand replication (*CSR*), rolling circle replication (*RCR*) and recombination-dependent replication (*RDR*). The replication-associated protein (*Rep*) with its nicking-closing (*NC*) and helicase (*H*) domain is shown during different putative functions, and *small arrows* indicate the direction of movement

and TGMV (Kong et al. 2000; McGivern et al. 2005; Shepherd et al. 2005, 2006). Which host mechanisms allow a replication of the mutated viruses in differentiated phloem cells has to be determined.

## *Rolling Circle Replication*

Because of its circular ssDNA packaged in isometric capsids, bacteriophage ΦX174 served as an early reference model for geminivirus replication. Indeed, the origin for viral strand synthesis of this bacteriophage (TG|ATATT AATAAC; NC_001422) shows some similarity to the later identified viral strand origin of geminiviruses (TAATATT|AC) and nanoviruses (TAKTATT|AC). The symbol "|" indicates the cleavage site, at which RepA/Rep opens the viral strand and thereby initiates rolling circle replication (RCR) (Fig. 4). RepA/Rep is covalently linked to the newly generated 5′ end (Rep-5′-AC...) and the conserved energy of this bond is used to re-ligate a new 3′OH end to the 5′ phosphate after one or more rounds of replication. The resulting circular ssDNA may be immediately packaged into virions (see above) or complemented by CSR (Fig. 4). The protein domain with nicking-closing activity is located in RepA or the corresponding N-terminal half of Rep (aa 1–180), a portion that also harbours the specific DNA-binding domain for origin recognition and transcriptional auto-repression (see below).

The C-terminal half of Rep (aa 180—362) was postulated early on to contain a helicase domain (Koonin and Ilyina 1992), which has been identified only recently (Choudhury et al. 2006; Clerot and Bernardi 2006). Both examined helicases proceeded in 3′ to 5′ direction, but differed with respect to their template requirements. It is not clear whether this divergence was due to the virus species analysed (TYLSCV, MYMIV) or the applied technical approaches. The MYMIV Rep-helicase required more than 5 nt free space of ssDNA for efficient activity. A displaced 5′ or 3′ end was not necessary for its action, but a 5′ flap was more efficiently unwound. These features of MYMIV-Rep were expected for a helicase involved in RCR, which slides along the complementary strand in order to remove the viral strand after nicking (Fig. 4, RCR). The requirement of a 3′ overhang for TYLCSV Rep-helicase combined with its 3′-to-5′ polarity does not fit to an RCR model and would be more suitable for providing primers in the context of an RDR model (Fig. 4, RDR, see below).

The structure of the N-terminal portion of Rep has been analysed in closer detail (Campos-Olivas et al. 2002; Vega-Rocha et al. 2007a, b), but the C-terminal portion of Rep awaits further characterization.

The function of Rep is dependent on various oligomerization states, homo-oligomerization as well as hetero-oligomerization with other viral proteins, namely REn, or host proteins (Table 2; reviewed in Jeske 2007). A central interaction platform (aa 120–180) is responsible for both purposes. MYMIV ΔRep-helicase (aa 120–362) has been purified in a 24mer complex with highest activity

(Choudhury et al. 2006), whereas TYLCSV ΔRep (aa 122–359) formed hexamers to dodecamers, and higher order complexes in vitro (Clerot and Bernardi 2006).

## *Recombination-Dependent Replication*

Using two-dimensional agarose gel electrophoresis followed by blotting and hybridization (reviewed in Jeske 2007), it became clear that major amounts of geminiviral DNA intermediates are not readily explained by CSR and RCR. A third fitting mode of replication was identified as RDR (Fig. 4) for at least begomoviruses, their satellites and curtoviruses (Alberter et al. 2005; Jeske et al. 2001; Jovel et al. 2007; Morilla et al. 2006; Preiss and Jeske 2003). RDR does not rely on an origin of replication, but rather can use any free 3′ end of ssDNA or of ssDNA overhangs in a dsDNA to initiate replication in a homologous stretch of supercoiled master DNA (Fig. 4). Whether geminiviral proteins are directly involved in RDR (AC2 and AC3 are not necessary) remains to be investigated. However, the helicase activity described above for TYLCSV Rep would neatly fit into this scenario, acting like RecBCD helicase in bacterial recombination (Fig. 4). Thus, extended 3′ ends of viral ssDNA could easily invade homologous dsDNA with the help of host recombination enzymes, using the RecA-like eukaryotic protein Rad51 of plants (Sung et al. 2003; Symington 2002). In this way, fragments of geminiviral DNA that may result from incomplete synthesis or from nucleolytic attack can be repaired with high fidelity. Moreover, RDR may lead to recombination if two partially related viruses have entered the same nucleus, a scenario which would help to explain the recombinogenic potential of geminiviruses.

## *Integration into Host DNA*

Like most plant viruses, geminiviruses are excluded from meristems, and therefore integration into host DNA has no evolutionary advantage. Even if integration occurred, this event would not be propagated to larger tissues or progeny plants and thus would remain undetected. Nonetheless, rare evolutionary footprints of partial geminiviral DNAs (Ashby et al. 1997; Bejarano et al. 1996; Kenton et al. 1995; Lim et al. 2000; Murad et al. 2004) have been discovered in chromosomes of *Nicotiana tabacum* and related species (section *Tomentosae*) which originate from the Americas. By contrast, endogenous geminivirus-related DNA was absent from *N. benthamiana*, the most frequently used experimental test plant from Australia. With these few examples, there is much less evidence for accidentally integrated copies in plant chromosomes than for plant pararetroviruses (reviewed in Harper et al. 2002).

**Table 2** Host proteins which interact with viral proteins

| VP | Virus | Host protein | S | Context | Reference |
|---|---|---|---|---|---|
| RepA | WDV | Plant retinoblastoma-like protein (pRBR) | H.s. Z.m. | Replication | Xie et al. 1995, 1996 |
| RepA | WDV | Geminivirus Rep A-binding (GRAB): NAC domain protein | T.a. | Development, senescence | Xie et al. 1999 |
| Rep | WDV | Replication factor C (TmRFC–1) | T.a. | Replication, clamp loading | Luque et al. 2002 |
| Rep | TGMV | Plant retinoblastoma-related protein (pRBR) | Z.m. | Replication | Ach et al. 1997 |
| Rep | TGMV | Ser/Thr kinase; kinesin; histone H3 | A.t. | Plant cell division, development; activation of SNF1-related kinases | Kong and Hanley-Bowdoin 2002; Shen and Hanley-Bowdoin 2006 |
| Rep | TYLCSV | SUMO-conjugating enzyme (NbSCE1) | N.b. | Replication repair | Castillo et al. 2004 |
| C2 AC2 | TGMV, BCTV | Adenosine kinase (ADK) | A.t. | Susceptibility reduction[a]; adenosine salvage and methyl cycle maintenance | Wang et al. 2003, 2005 |
| C2 AC2 | TGMV, BCTV | SNF1 kinase | A.t. | Susceptibility reduction[a] | Hao et al. 2003 |
| Rep REn | TYLCSV | Proliferating cell nuclear antigen (PCNA) | A.t. S.l. | Replication | Castillo et al. 2003 |
| REn | TGMV, TYLCV | pRBR | | Replication, cell cycle | Settlage et al. 2001, 2005 |
| REn | TYLCV | PCNA | | Replication processivity | Settlage et al. 2005 |
| REn | ToLCV | NAC domain protein (SlNAC1) | S.l. | Replication | Selth et al. 2005 |
| AC4, C4 | TGMV, BCTV | Leucine-rich repeat receptor-like kinase (LRR-RLK): shaggy-related protein kinase (AtSKeta, AtSKzeta); | A.t. | Symptom development; brassinosteroid signalling pathway | Piroux et al. 2007 |
| CP | TYLCV | Karyopherin α | S.l. | Nuclear import | Kunik et al. 1999 |
| CP | MYMV | Importin α | O.s. C.a. | Nuclear import | Guerra-Peraza et al. 2005 |
| PCP | ToLCV | UDPglucose:protein transglucosylase (SlUPTG1) | S.l. | Cell wall synthesis, transport | Selth et al. 2006 |
| V2 | TYLCV | SlSGS3 | S.l. | Silencing | Glick et al. 2008 |
| NSP | TGMV, TCrLYV, CaLCuV | Leucine-rich repeat receptor-like kinase (LRR-RLK): NSP-interacting kinase; (LeNIK, GmNIK, AtNIK1–3) | S.l. G.m. A.t. | Plant development, susceptibility reduction[a] | Fontes et al. 2004; Mariano et al. 2004 |

| | | | | | |
|---|---|---|---|---|---|
| NSP | CaLCuV | Proline-rich extensin-like receptor protein kinase (PERK): NSP-associated kinase (NsAK) | A.t. | Susceptibility enhancement[a] | Florentino et al. 2006 |
| NSP | CaLCuV | Acetyltransferase (AtNSI) | A.t. | Nuclear export, susceptibility reduction[a] | Carvalho and Lazarowitz 2004; Carvalho et al. 2006; McGarry et al. 2003 |

VP, viral protein. S, source organisms: A.t., *Arabidopsis thaliana*; C.a., pepper, *Capsicum annuum*; G.m., soybean,*Glycine max*; H.s., *Homo sapiens*; N.b., *Nicotiana benthamiana*; O.s., rice, *Oryza sativa*; S.l., tomato, *Solanum lycopersicum=Lycopersicon esculentum*, L.e.; T.a., common wheat, *Triticum aestivum*; Z.m., maize, *Zea mays*.

[a]Refers to the function of the host protein, if not impaired by the viral proteins which inhibit this function in some cases

## Transcription

Transcription of geminiviral DNA (reviewed in Hanley-Bowdoin et al. 1999) is very reminiscent of SV40 in exploiting classic eukaryotic control elements such as TATA and CAAT boxes as well as polyadenylation signals for a bidirectional transcription (Fig. 3). Promoters are located within the LIRfor mastreviruses, or within the IR for begomoviruses, and the terminator regions overlap at the opposite side of the genomes (ter). The corresponding RNA Pol II-dependent transcripts have been mapped for several geminiviruses of different genera (Fig. 3).

Different from the SV40 model, a further promoter is located in front of the ORFs C2 and C3 for curtoviruses or AC2 and AC3 for begomoviruses, each of which are expressed from bicistronic transcripts (TAC2–3). Splicing is only rarely used as a differential expression strategy for geminiviruses. The only constitutive splicing serves to fuse C1 and C2 ORFs for Rep expression in the case of mastreviruses (Fig. 3, MSV, Int1). Further introns within the V1 ORF of MSV (Fig. 3, Int2) or in front of BC1 ORF of MYMV (Fig. 3, Int1) are not conserved in other geminivirus genomes of the respective genera.

### *Transactivation and Repression*

Geminiviruses regulate their particular promoters by the help of their own proteins as well as host transcription factors (reviewed in Hanley-Bowdoin et al. 1999). Different key proteins are relevant for the distinct genera. Mastreviruses use Rep A to transactivate viral sense transcripts (Horvath et al. 1998; Munoz-Martin et al. 2003), and bipartite begomoviruses AC2 (TrAP) to promote transcription of AV1 as well as of BV1 (Sunter and Bisaro 1991; Sunter and Bisaro 1992; Sunter and Bisaro 1997; Sunter et al. 1994; Sunter et al. 1990). BCTV C2 ORF, although a positional homologue of AC2, is not involved in transcriptional control (Hormuzdi and Bisaro 1995). For WDV, a stretch of bent DNA has been identified and implicated in the regulation of virus-sense transcription (Suarez-Lopez et al. 1995), and RepA as well as Rep binding sites have been mapped upstream and downstream of the ori (Castellano et al. 1999; Missich et al. 2000). On the contrary, TrAP obviously lacks a specific DNA binding sequence. Therefore, it is believed that its function is mediated by host factors.

Rep binds to virus-specific iterated sequences upstream of the ori, referred to as "iterons" (Arguello-Astorga et al. 1994). This interaction is responsible for autorepression of complementary-sense TAC1–4 transcription, in addition to its role in replication control (Eagle et al. 1994; Sunter et al. 1993). Interestingly, the repression of TGMV AC1 enhanced the promoter strength for the downstream transcript TAC2–3 (Shung and Sunter 2007). A larger upstream activation element of this promoter has been identified in curto- and begomoviruses (Shung et al. 2006) including a highly conserved binding site of nine nucleotides (AACGTCATC) with a potential signature for host transcription factor binding (ATF/CREB, TGA2/ASF-1, v-Myb) (Tu and Sunter 2007). TGA transcription factor binding is particularly remarkable, because these factors are involved in the regulation of plant pararetroviruses and salicylate-mediated stress response as well (Dong 2004; Durrant and Dong 2004; van Loon et al. 2006). Thus, geminiviruses could profit from the first defence line of the plant to give a boost to their silencing suppressors AC2 and replication enhancers AC3.

The bicistronic mRNA was shown to programme relatively low AC2 and high AC3 abundance, probably as a consequence of leaky ribosome scanning for the AC2 ATG in a suboptimal Kozak context (Shung et al. 2006).

In comparison to Rep itself, the AC4 protein of TGMV exerted a minor down-regulation of TAC1–4 and used other DNA binding sites (Eagle and Hanley-Bowdoin 1997; Gröning et al. 1994; Orozco and Hanley-Bowdoin 1998; Shung and Sunter 2007). However, it is in question whether C4/AC4 functions in this manner for other geminiviruses.

Like in SV40, the genome organization of geminiviruses reflects a temporal regulation by expressing the early genes via complementary sense and late genes via viral sense transcripts (PHYVV) (Shimada-Beltran and Rivera-Bustamante 2007). Recombinant green fluorescent protein (GFP) constructs and transcripts were activated as early as 2 h after transfection for the complementary sense and at 6 h after transfection for the viral sense genes.

## *Chromatin Structure and Transcriptional Silencing of Viral Genes (TGS)*

Following CSR, the geminiviral covalently closed circular (ccc) dsDNA is packaged into host histones, thus forming a viral minichromosome (Abouzid et al. 1988; Pilartz and Jeske 1992; Pilartz and Jeske 2003). One genomic component of 2,600 bp offers space for 13 nucleosomes, but minichromosomes with 11 and 12 nucleosomes have been detected as well. One nucleosome-free gap was located in the common region of the bipartite AbMV, another in the promoter region for the TAC2–3 or for the TBC1 transcript for DNA A or DNA B, respectively (Pilartz and Jeske 2003). Positioning of nucleosomes may, therefore, exert transcription control on a structural level in order to facilitate transcription factor binding.

In agreement with the idea that geminiviral dsDNA is regulated at the chromatin level, de novo methylation of both symmetric and asymmetric cytosines has been detected for transgenic and extrachromosomal copies of ToLCV DNA (Bian et al. 2006; Seemanpillai et al. 2003). Because geminiviral DNA replication was found to be impaired by methylation (Brough et al. 1988; Ermak et al. 1993), TGS involving methylation of DNA may be regarded as a resistance mechanism of the plant. In this context, Bian et al. (2006) proposed that replication via RCR may be an advantage in order to escape from the counter-defence. The same might hold true for RDR, because a methylated parental strand will be separated from the daughter strand during both replication modes, and thus cannot serve as a template for maintenance methylase. Similarly, depletion of methyl group supply could promote viral replication and transcription, as has been discussed for a downstream effect of adenosine kinase inactivation by geminiviral AC2/C2 (Bisaro 2006).

In further concurrence with the importance of chromatin remodelling is the finding that the geminiviral nuclear shuttle protein (NSP) interacted with an acetyltransferase (AtNSI, Table 2), which was able to acetylate histones as well as geminiviral CP. Overexpression of this acetyltransferase increased the susceptibility of *A. thaliana* plants for CaLCuV (Carvalho and Lazarowitz 2004; McGarry et al. 2003). As this protein, however, failed to co-activate transcription in vitro, it has to be determined whether it is directly responsible for geminiviral chromatin opening *in planta*. In addition, its acetyltransferase activity on CP as well as on histones was inhibited by the presence of NSP (Carvalho et al. 2006), thus suggesting that it may be spatially and temporally regulated during the course of infection.

## *Post-transcriptional Gene Silencing of Viral Genes*

PTGS is now a well-elucidated network of the eukaryotic RNA metabolism that specifically down-regulates genes in development, removes overexpressed and aberrant RNA, or controls transposons and viruses. Triggered either by micro (mi) RNA, mRNA folded into hairpin structures, or linear dsRNA, specialized RNase III-like enzymes (DICER) cleave these substrates and generate short interfering (si) RNAs. The resulting 21- to 24-nt oligonucleotides of both polarities are unwound and one strand enables an RNA-induced silencing complex (RISC) to specifically cut the complementary mRNA by an activity called "Slicer". In addition to this primary inactivation, the generated small oligonucleotides may serve as primers on complementary RNA to start synthesis of new dsRNA by RNA-dependent RNA polymerase (RdRP), thus amplifying the dsRNA trigger for further PTGS rounds. Once established, PTGS can induce TGS by RNA-directed DNA methylation (RdDM).

In plants (*A. thaliana*), four genes for DICER-like enzymes (*DCL1* to *DCL4*) have been characterized, all of which fulfil different tasks in miRNA processing (DCL1), anti-viral defence (DCL2, DCL4), or control of retroelements and transposons as well as TGS (DCL3). Six genes encode RdRPs, and *RDR6* (detected as *suppressor of gene silencing2/silencing defective1 SGS2/SDE1*) is possibly involved in relay amplification of the systemic silencing signal.

With respect to geminiviruses, the interplay of TGS and PTGS processes and their roles in virus resistance have been reviewed recently (Bisaro 2006; Vanitharani et al. 2005). Although geminiviruses do not generate dsRNA during their replication, do multiply in nuclei, and produce only moderate amounts of transcripts, siRNAs corresponding to all parts of the viral genome have been detected during infection. The overlapping bidirectional transcription (Fig. 3, at ter) possibly predestines the generation of aberrant dsRNAs and thereby may be a hotspot of PTGS initiation followed by transitive spread of the silencing signal to other parts of the viral genome. Surprisingly, even siRNAs corresponding to the intergenic region are generated (Pooggin and Hohn 2004; Pooggin et al. 2003).

Geminivirus-specific siRNAs comprise different size classes (21, 22, and 24 nt), each of which is indicative for the action of a particular DCL enzyme (Akbergenov et al. 2006; Blevins et al. 2006). These results show that the plant counteracts geminiviral transcription at different levels, including PTGS with the aid of *DCL1* within the nucleus as well as *DCL2*, *DCL3* and *DCL4* in the cytoplasm. This contrasts with RNA viruses, which are mainly affected by *DCL4*. Moreover, siRNAs can initiate TGS and increase methylation of the viral minichromosome DNA (Bian et al. 2006).

In summary, geminiviruses encounter a rather complex response in plants that efficiently substitutes for the lacking immune system. Future investigations will show whether these defence mechanisms work equally well in different tissues or whether geminiviruses may at least partially escape from this battle in the phloem. Furthermore, it will be challenging to explore the enforcement of these silencing pathways with the help of artificial transgenes in order to provide sustainable plant resistance against geminiviruses (reviewed in Pooggin and Hohn 2004; Vanderschuren et al. 2007; Vanitharani et al. 2005).

## *Suppression of Silencing*

To this end, one has to recognize that geminiviruses have encountered the plants' protective walls for millions of years already, and that they have evolved efficient strategies to escape or combat the hosts' responses. The current view on geminiviral suppression of silencing is not uniform, as different viruses seem to use different genes for this purpose and their gene products may interfere with various steps of basal and specific resistance. Moreover, experimental techniques used to unravel the putative functions of silencing suppressors during generation, processing, or systemic spread of the silencing signals may have an influence of the conclusions drawn.

So far, three geminiviral genes have attended most interest in this context: AC2/C2, which is characterized as TrAP for begomoviruses, and AC4/C4, identified early as a symptom determinant for Old World geminiviruses (reviewed in Bisaro 2006; Vanitharani et al. 2005), as well as V2/PCP, which is involved in cell-to-cell spread (Glick et al. 2008; Zrachya et al. 2007). In addition, the *βC1* gene product of satellite DNA β interferes with silencing as well, and is thereby an essential determinant of symptom development (Cui et al. 2005; Gopal et al. 2007; Saeed et al. 2005; Saunders et al. 2004).

Some geminiviral suppressors of silencing can bind directly to siRNAs and miRNA (Chellappan et al. 2005), while others cannot (Wang et al. 2005), and therefore it is the common view that they may exert their effects also by interacting with host proteins and disturb basal as well as more specific defence pathways indirectly (see Table 2, reviewed in Bisaro 2006; Trinks et al. 2005; Yang et al. 2007, and reference therein).

Again, different geminiviruses seem to prefer different strategies in this context. The combination of geminiviruses in a plant may consequently induce synergistic effects in symptom development and pathogenesis (Morilla et al. 2004; Vanitharani et al. 2004). Moreover, geminivirus virulence may be increased by the co-infection with unrelated viruses (Pohl and Wege 2007; Wege 2007; Wege and Siegmund 2007).

## *Virus-Induced Gene Silencing*

In spite of their silencing suppressors, geminiviruses can, nevertheless, be used as vectors for efficient virus-induced gene silencing (VIGS), and numerous constructs have been tested experimentally to selectively knock down host genes by TGS or PTGS (Atkinson et al. 1998; Carrillo-Tripp et al. 2006; Fofana et al. 2004; Kjemtrup et al. 1998; Li et al. 2007; Muangsan and Robertson 2004; Muangsan et al. 2004; Peele et al. 2001; Seemanpillai et al. 2003; Tao and Zhou 2004). An attenuated vector based on AbMV DNA A has recently been developed which generated the silencing signal in phloem cells but knocked down genes in the mesophyll as well (Krenz et al. 2008). This VIGS vector can be combined with plants transgenic for dimers of AbMV DNA B, which do not exert any adverse effect on plant development in contrast to some DNA Bs of other geminiviruses (Wege and Pohl 2007).

## Transport Within the Plant

The movement of viral nucleoprotein complexes from cell to cell differs fundamentally between higher plants and animals. Plant cells are usually connected by cytoplasmic bridges (plasmodesmata), and the plant cormus can be considered as a huge symplast. Correspondingly, there is no need for a plant virus to bud through a membrane and recognize a target cell from the intercellular environment. If plant viruses have, nevertheless, evolved proteins that attach to the outer surface of plasma membranes, this is regarded to be an adaptation to their transmitting insect. Plant viruses, instead, face the challenging task of snaking through tiny tunnels between plasma membrane and endoplasmic reticulum of plasmodesmata (Fig. 5, PM, ER). Their normal diameter is much smaller than that of virus particles, of other nucleoprotein complexes, or of unstructured larger nucleic acids. Accordingly, plant viruses have adopted several strategies in order to widen plasmodesmata, assemble secondary plasmodesmata, and stretch their nucleic acids (reviewed in Waigmann et al. 2004).

Geminiviruses, in addition, have to pass another small hole, the nuclear pore, in both directions. This is accomplished by active processes using nuclear localization and nuclear export signals (NLS, NES) in their proteins. There is now increasing consensus that pairs of geminiviral proteins are responsible for the two tasks (Fig. 5). One protein has to facilitate nuclear shuttling and binding of viral DNA, the other has to open the gate of plasmodesmata and promote the transfer of viral DNA to the adjacent cell. During evolution of geminiviruses, two such pairs of proteins have been adapted for these purposes. For the first variant, V1/AV1/CP functions as a nuclear shuttle protein and V2/AV2/PCP as a gate-keeper and membrane anchor. This arrangement is typical for monopartite mastreviruses and curtoviruses, but has been maintained in Old World bipartite geminiviruses. The second variant is limited to bipartite begomoviruses, for which DNA B genes have successively taken over transport tasks, BV1/NSP for nuclear shuttling and BC1/MP for plasmodesmata transfer. Various aspects of geminiviral transport and some controversial interpretations have been extensively reviewed (Gafni and Epel 2002; Lazarowitz 1999; Lazarowitz and Beachy 1999; Rojas et al. 2005; Wege 2007).

If regarded completely bipartite, geminiviruses need both DNA components for full systemic and symptomatic infection. Nevertheless, on the condition that DNA A of bipartite geminiviruses was delivered alone by agro-infection, it was able to move systemically in *N. benthamiana* plants, but without inducing symptoms (Briddon and Markham 2001; Evans and Jeske 1993; Haible et al. 2006; Klinkenberg and Stanley 1990). In contrast to these results, biolistically inoculated DNA A needed the complementation of, at least transiently expressed, NSP and movement protein (MP) in order to reach upper leaves (Jeffrey et al. 1996; Saeed et al. 2007).

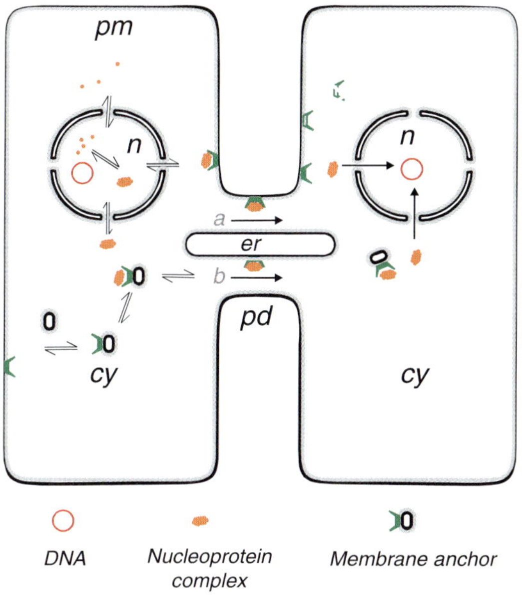


**Fig. 5** Cell-to-cell transport model for geminiviruses. Two cells with nuclei (*n*), cytoplasm (*cy*), plasma membrane (*pm*), endoplasmic reticulum (*er*), and connecting plasmodesmos (*pd*) are schematically drawn in order to differentiate the inner (protoplasmic) and outer (exterior) leaflet of membranes in *grey* and *black*, respectively. The location and transport of viral DNA, nucleoprotein complexes, and membrane anchor protein are shown along the plasma membrane (*a*) or the ER (*b*) that traverse plasmodesmata

The relative independence of DNA A has been exploited to judge the promiscuity of DNA B genes in mobilizing unrelated geminiviruses (Briddon and Markham 2001; Frischmuth et al. 1993b; Petty et al. 1995). The results demonstrated that *trans*-complementation of geminiviral movement proteins is possible over continental as well as genus barriers, although Old World viruses mobilized other viruses more efficiently than their counterparts from the New World.

The movement-complementing gene products have recently been extended to the βC1 protein encoded by satellite DNA (Saeed et al. 2007). The DNA A of the bipartite ToLCNDV could be mobilized by transient expression of βC1 protein encoded by the unrelated Cotton leaf curl disease (CLCuD) satellite, and in contrast to the complementation by DNA B genes, full symptomatic and systemic infection was obtained.

## *Coat Protein*

After translation in the cytoplasm, CP has to be imported into the nucleus. Although the monomer is small enough to penetrate the nuclear pore passively, multimers may assemble already in the cytoplasm. Accordingly, several nuclear localization signals (NLS) have been mapped using GFP fusion proteins at the N-terminus, within the central portion, and the C-terminus (Guerra-Peraza et al. 2005; Kunik et al. 1998; Rojas et al. 2001; Unseld et al. 2004; Unseld et al. 2001). In addition, the central portion contains nuclear export sequences that direct fusion proteins to the cytoplasm and the cell periphery in a time-dependent manner (Unseld et al. 2001). Confirming an active nuclear import, CP was found to interact with host karyopherin (importin) α (Table 2).

The orientation of the N-terminus towards the lumen of the capsid should mask the N-terminal NLS in virions. If geminiviruses, therefore, entered new nuclei via an active mechanism through nuclear pores, an additional NLS exposed at the surface of the particle would be needed. Meeting this prerequisite, amino acids which were pinpointed by mutational analysis as relevant for nuclear localization (SLCV CP) (Qin et al. 1998) cluster on the outer surface in the three-dimensional model of capsomers (Fig. 2i).

Although for bipartite geminiviruses the tasks of cell-to-cell movement and long distance transport in plants are taken over by NSP and MP, the CP nevertheless seems to keep some function in this context. If NSP was disabled by mutations, CP could complement some of the respective defects (Qin et al. 1998). Accordingly, it has been suggested that an intimate co-operation of both proteins for efficient trafficking still exists (Carvalho and Lazarowitz 2004; Carvalho et al. 2006; McGarry et al. 2003).

## *Precoat Protein*

Calling the gene located in front of *CP precoat protein* (*PCP*) has been proposed due to its genomic position, but without implying structural tasks (Padidam et al.

1996). This unifying nomenclature has been widely used, and similar functions have been assigned to this gene for members of different geminivirus genera. For mastreviruses, MSV PCP has been detected at the cell periphery and in plasmodesmata (Dickinson et al. 1996), and it was found to move both intra- and intercellularly (Kotlizky et al. 2000). The PCPs of monopartite [TYLCV (Rojas et al. 2001) and ToLCV (Selth et al. 2006)] and bipartite begomoviruses (ICMV) (Rothenstein et al. 2007) were targeted to the cell periphery when expressed as GFP fusion proteins, and appeared in punctate fluorescent spots that indicate plasmodesmal localization particularly in plasmolysed cells (Selth et al. 2006). In addition, TYLCV as well as ICMV PCP moved between epidermal cells, but to a lesser extent than MSV PCP or BDMV MP. TYLCV PCP was also able to increase the size exclusion limit of plasmodesmata in a low proportion of cells (Rojas et al. 2001). Interestingly, ICMV PCP in contrast to TYLCV PCP was able to enter nuclei and formed conspicuous pairs of globular inclusions, before it reached the cells' periphery and the adjacent cells in a temporally regulated fashion (Fig. 6a–d; Rothenstein et al. 2007).

Discrepancies in movement efficiencies have been discussed with reference to the tissue specificities of the viruses under investigation, e.g. phloem-limitation for TYLCV and AbMV, or phloem-emigration for BDMV (Levy and Czosnek 2003; Rojas et al. 2001). Thorough microinjection studies (Noueiry et al. 1994; Rojas et al. 2001) have revealed that cells may be permissive for cell-to-cell movement and gate-opening to various degrees even within a given tissue. During leaf development from sink to source status, cells undergo a gradual transition in physiology and structure, which is rather a patchwork than a linear process in the leaf lamina. The higher percentage of transport-permissive cells in sink compared to source leaves supports this notion for TYLCV PCP, as detected with fluorescently labelled or GFP-fusion proteins (Rojas et al. 2001). Similarly, occasional emigrations to the mesophyll of the otherwise phloem-limited ICMV as detected by in situ hybridization show the capability of ICMV transport proteins to act outside of the phloem, in principle (Rothenstein et al. 2007).

## *C4/AC4*

Mutational and movement complementation analyses have suggested that ORF C4 may be directly involved in cell-to-cell movement (Jupin et al. 1994; Rojas et al. 2001; Saeed et al. 2007). Considering the pleiotropic role of C4/AC4 in transcriptional regulation, silencing and pathogenicity (Chellappan et al. 2004; Piroux et al. 2007; Vanitharani et al. 2004), the movement effect may also be indirect by inhibiting the spread of a systemic silencing signal. The localization of C4 at the plasma membrane, presumably managed by anchoring via myristoyl residues (Fondong et al. 2007; Piroux et al. 2007; Rojas et al. 2001), would agree with a function in cell–cell transport as well as cell–cell communication. At present, the variable effects of C4/AC4 of different geminivirus genera and origins make a general conclusion difficult (Fondong et al. 2007; Piroux et al. 2007, and references therein).

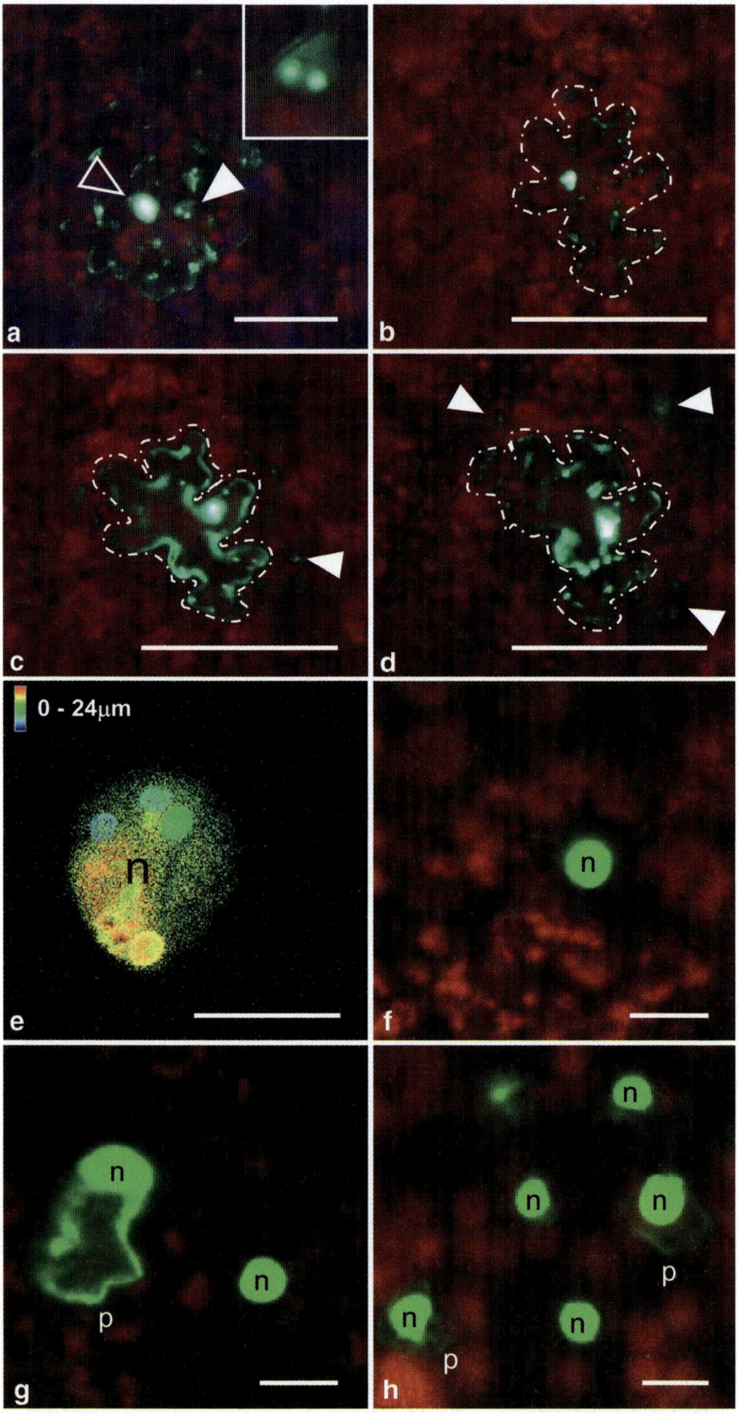


**Fig. 6** Intra- and intercellular transport of geminiviral proteins monitored by transiently expressed GFP-fusion proteins after biolistic inoculation of epidermis cells. ICMV GFP:PCP in *N. benthaminana*

## *Nuclear Shuttle Protein*

Similar to its evolutionary predecessor CP, NSP binds ssDNA and dsDNA without sequence-specificity, and several signatures for nuclear import and export have been mapped on the NSP amino acid sequence (reviewed in Lazarowitz 1999). Depending on the virus under investigation and the experimental technique used, there is some debate in the literature as to whether NSP physically interacts with BC1/MP and is transported to the next cell, or whether it hands viral DNA to MP for intercellular transport (reviewed in Rojas et al. 2005).

In plants, GFP-fused AbMV NSP was translocated to the adjacent cell if coexpressed with MP, or in the presence of complementing wt virus (Fig. 6; Zhang et al. 2001a). Whereas the NSP fluorescence was distributed in the nucleus as well as at the cell periphery of the donor cell, it accumulated in the nucleus of the next, recipient cell upon mobilization by transiently expressed MP.

The same intracellular translocation has been shown in a fission yeast (*Schizosaccharomyces pombe*) model, which allowed the localization of transiently expressed geminiviral proteins with highest resolution with the aid of freeze-fracture immunolabelling (FreeFI) (Aberle et al. 2002). This technique facilitates the assignment of a protein to the two faces of a membrane and yields a higher density of immunolabelling for membrane-bound proteins, because membranes are rather exposed from their surface than from a cross-section, as after ultramicrotomy. By this method, it was possible to follow the fate of NSP in yeast cells after co-expression with MP, unravelling its path from the interior of the nucleus to a tight attachment at the cytoplasmic face of the plasma membrane (Frischmuth et al. 2007). These results are more compatible with a co-operative than with a successive action of NSP and MP in case of AbMV.

## *Movement Protein*

Microinjection experiments with plants using bacterially expressed and fluorescently labelled begomovirus MP revealed its most remarkable feature, namely to increase the size selection limit of plasmodesmata (BDMV; Noueiry et al. 1994;

**Fig. 6** (continued) leaves at 18 h post-bombardment (hpb, **a**) show prominent cytoplasmic inclusion (*open arrowhead*), a pair of globular nuclear inclusion (*filled arrowhead*), and peripheral localization of the protein; *inset*: the nucleus at higher magnification. Transport of ICMV GFP:PCP to adjacent cells (*filled arrowheads*) was monitored at the same tissue site at 10 (**b**), 24 (**c**), and 48 (**d**) hpb. The initially bombarded cell is marked by *white dotted lines* in **b–d**, *bar* represents 5 μm. AbMV GFP:NSP in the nucleus of an onion cell with globular inclusions (**e**), or in *N. tabacum* cells (**f–h**), either expressed alone (**f**), with unlabelled AbMV MP (**g**), or with AbMV DNA A and B. Nuclear (**n**) and peripheral (**p**) NSP localization is shown in false colours (**e**) of *stacked images* or *green* colour (**f–h**), and *red fluorescence* is due to chlorophyll in chloroplasts. *Bars* represent 10 μm in **e–h**. Images were adapted from Rothenstein et al. (2007) and Zhang et al. (2001a), with permission of Elsevier

Rojas et al. 2001). In accordance with this task, MP was localized at the cell periphery of various plant cells and protoplasts (Rojas et al. 2001; Sanderfoot and Lazarowitz 1995; von Arnim et al. 1993; Zhang et al. 2001a; Zhang et al. 2002), and in specialized endoplasmic reticulum-derived tubular structures of protophloem cells (Ward et al. 1997). In yeast cells, AbMV MP was assigned to the cytoplasmic leaflet of ER vesicles and plasma membranes (Aberle et al. 2002), where it is thought to be inserted as monotopic membrane proteins due to a central amphipathic helix (Zhang et al. 2002). The amount of MP in infected plants is usually low and its expression is temporally regulated. Moreover, it appeared in several isoforms due to various post-translational modifications including phosphorylation (Kleinow et al. 2008). Upon transient overexpression of GFP-fused AbMV MP in plant cells, substantial amounts of fluorescent signals were detected in the vicinity of nuclei and as punctate spots at the cell periphery of the donor cell, but only faint signals were detected in the recipient cells after mobilization of NSP and MP (Zhang et al. 2001a).

There is some discrepancy in the literature how far MP directly can bind DNA, which may have its origin in differently chosen viruses or experimental approaches. SLCV MP bound only low amounts of ssDNA (Pascal et al. 1994), whereas BDMV MP was found to select ssDNA as well as dsDNA in a form- and size-specific manner (Rojas et al. 1998). AbMV MP exhibited only weak true binding to dimeric cccDNA (Hehnle et al. 2004). As outlined by Hehnle et al., electrophoretic mobility assays with MP should be interpreted with great care, because electrophoretic interference may lead one to believe in complex formation. To be convincing, the presence of a protein in a putative complex has to be shown by immunological means in addition to gel-retarded DNA (Hehnle et al. 2004).

In addition to or as a consequence of their role as movement proteins, geminiviral MPs determine the viral pathogenicity (Duan et al. 1997; Hou et al. 2000; Pascal et al. 1993; Saunders et al. 2001). In transgenic plants, their gene insertion favoured the selection of truncated alleles, or led to silencing of the transgene, a genetic consequence which has been also observed in fission yeast cells (Frischmuth et al. 2004). Nevertheless, it became possible recently to establish plants transgenic for AbMV DNA B which developed normally and activated their silenced DNA B component upon infection with the cognate DNA A (Wege and Pohl 2007).

## *Interactions of Geminiviral Transport Proteins with Each Other*

In vitro assays have shown a physical interaction between mastrevirus PCP and CP (Liu et al. 2001). For TYLCV, both proteins were found to co-operate in cell-to-cell movement and DNA transfer in microinjection studies (Rojas et al. 2001). Yeast two-hybrid tests verified a binding between begomovirus NSP and MP in one case (TGMV) (Mariano et al. 2004), but not in another (AbMV) (Frischmuth et al. 2004), although a homo-oligomerization of the C-terminus of AbMV MP could be detected. A plasma membrane-located yeast interaction assay (Cytotrap) exhibited autoinduction of the signal cascade with full-length AbMV MP confirming its tight association with the cytoplasmic leaflet of the plasma membrane and explaining why intact MP cannot meet NSP in the Gal4-based nuclear test (Frischmuth et al. 2004). In fission yeast cells (*S. pombe*), AbMV NSP and MP co-operated the same way as in plants and, thus, NSP was redirected to the cytoplasmic leaflet of plasma membranes after co-expression with MP (Frischmuth et al. 2007). These results demonstrate that the co-operation of both proteins is direct and does not need any plant-specific protein, although a general eukaryotic adaptor may be necessary to mediate physical contacts of NSP and MP. A conspicuous ternary complex of DNA, NSP and MP could be obtained by in vitro assembly of these components and observed by electron microscopic visualization (Hehnle et al. 2004). Whether and how these structures are formed *in planta* is unknown, and in particular the conformation of viral nucleic acid within the movement complex has yet to be determined.

### *Interaction of Geminiviral and Host Proteins in the Context of Spread in Plants*

Interactions between geminiviral and host proteins may have two evolutionary advantages for the virus, in order to either exploit host functions for its own multiplication and spread, or to escape the plant's counter-defence. Several host proteins have been identified by yeast two-hybrid screens during the past decade that may serve the one or other purpose (Table 2). Surprisingly, more detected interactions are interpreted to play roles in combatting host responses and increasing susceptibility than in utilizing host proteins for viral movement. Correspondingly, several plasma membrane-resident (LRR-RLKs, PERK-RLKs, Table 1) and cytoplasmic and/or nuclear kinases (ADK, SNF1 K) have been identified that act in the basal host's signalling network. Only one of these (NsAK) has been reported to phosphorylate a geminiviral protein (CaLCuV NSP) (Florentino et al. 2006). A further host enzyme (acetyltransferase, AtNSI), which is also involved in basal signal regulation, is the second example of a protein that is able to modify a geminiviral partner (CaLCuV CP) (Carvalho et al. 2006). The authors propose that NSP may recruit an acetyltransferase that modifies CP, thereby directly regulating the interplay of NSP and CP during transport.

## Conclusions and Perspectives

The biology of geminiviruses has been explored extensively and in great detail over the past three decades, revealing a multitude of genetic capabilities and the flexibility of these tiny viruses. Having understood many aspects of their life cycle, it is of utmost importance to use this knowledge to engineer sustainable resistant crop plants in order to combat worldwide epidemics. Diagnostic tools have been developed to prevent further spread of these viruses, but new unconventional means to protect plants are necessary. Pathogen-derived approaches using our knowledge of antisense RNA and dominant-negative protein effects, silencing, and DI DNA have to be validated. Moreover, the ample repertoire of resistance mechanisms and genes in wild plant species needs further exploration and transfer to cultivated plants. For being successful in this long-lasting battle, it will become inevitable to understand the temporal and spatial dynamics as well as the genetic flexibility and complexity of geminiviruses, demanding a systems biology-based analysis of the particular virus–host interactions.

**Acknowledgements** The many critical discussions on this manuscript by Katharina Kittelmann, Tatjana Kleinow and Christina Wege are greatly appreciated. I have to apologize to the authors of many excellent publications that could not be cited here because of space limitations.

## References

Aberle HJ, Rütz ML, Karayavuz M, et al (2002) Localizing BC1 movement proteins of Abutilon mosaic geminivirus in yeasts by subcellular fractionation and freeze-fracture immunolabelling. Arch Virol 147:103–107

Abouzid AM, Frischmuth T, Jeske H (1988) A putative replicative form of the Abutilon mosaic virus (gemini group) in a chromatin-like structure. Mol Gen Genet 212:252–258

Ach RA, Durfee T, Miller AB, et al (1997) RRB1 and RRB2 encode maize retinoblastoma-related proteins that interact with a plant D-type cyclin and geminivirus replication protein. Mol Cell Biol 17:5077–5086

Akad F, Dotan N, Czosnek H (2004) Trapping of Tomato yellow leaf curl virus (TYLCV) and other plant viruses with a GroEL homologue from the whitefly *Bemisia tabaci*. Arch Virol 149:1481–1497

Akbergenov R, Si-Ammour A, Blevins T, et al (2006) Molecular characterization of geminivirus-derived small RNAs in different plant species. Nucleic Acids Res 34:462–471

Alberter B, Rezaian AM, Jeske H (2005) Replicative intermediates of ToLCV and its satellite DNAs. Virology 331:441–448

Arguello-Astorga GR, Guevara-Gonzalez RG, Herrera-Estrella LR, et al (1994) Geminivirus replication origins have a group-specific organization of iterative elements: a model for replication. Virology 203:90–100

Ashby MK, Warry A, Bejarano ER, et al (1997) Analysis of multiple copies of geminiviral DNA in the genome of four closely related *Nicotiana* species suggests a unique integration event. Plant Mol Biol 35:313–321

Atkinson RG, Bieleski LR F, Gleave AP, et al (1998) Post-transcriptional silencing of chalcone synthase in petunia using a geminivirus-based episomal vector. Plant J 15:593–604

Bejarano ER, Khashoggi A, Witty M, et al (1996) Integration of multiple repeats of geminiviral DNA into the nuclear genome of tobacco during evolution. Proc Natl Acad Sci USA 93:759–764

Bian XY, Rasheed MS, Seemanpillai M, et al (2006) Analysis of silencing escape of Tomato leaf curl virus: an evaluation of the role of DNA methylation. Mol Plant Microbe Interact 19:614–624

Bisaro DM (2006) Silencing suppression by geminivirus proteins. Virology 344:158–168

Blevins T, Rajeswaran R, Shivaprasad PV, et al (2006) Four plant Dicers mediate viral small RNA biogenesis and DNA virus induced silencing. Nucleic Acids Res 34:6233–6246

Böttcher B, Unseld S, Ceulemans H, et al (2004) Geminate structures of African cassava mosaic virus. J Virol 78:6709–6714
Briddon RW, Markham PG (2001) Complementation of bipartite begomovirus movement functions by topocuviruses and curtoviruses. Arch Virol 146:1811–1819
Briddon RW, Stanley J (2006) Subviral agents associated with plant single-stranded DNA viruses. Virology 344:198–210
Briddon RW, Pinner MS, Stanley J, et al (1990) Geminivirus coat protein gene replacement alters insect specificity. Virology 177:85–94
Briddon RW, Bull SE, Amin I, et al (2003) Diversity of DNA beta, a satellite molecule associated with some monopartite begomoviruses. Virology 312:106–121
Briddon RW, Bull SE, Amin I, et al (2004) Diversity of DNA 1: a satellite-like molecule associated with monopartite begomovirus-DNA beta complexes. Virology 324:462–474
Briddon RW, Brown JK, Moriones E, et al (2008) Recommendations for the classification and nomenclature of the DNA-beta satellites of begomoviruses. Arch Virol 153:763–781
Brough CL, Hayes RJ, Coutts RH A, et al (1988) Effect of mutagenesis in vitro on the ability of cloned tomato golden mosaic virus DNA to infect *Nicotiana* benthamiana plants. J Gen Virol 69:481–492
Byrne DN, Bellows TS Jr (1991) Whitefly biology. Annu Rev Entomol 36:431–457
Campos-Olivas R, Louis JM, Clerot D, Gronenborn B, Gronenborn AM (2002) The structure of a replication initiator unites diverse aspects of nucleic acid metabolism. Proc Natl Acad Sci USA 99:10310–10315
Carrillo-Tripp J, Shimada-Beltran H, Rivera-Bustamante R (2006) Use of geminiviral vectors for functional genomics. Curr Opin Plant Biol 9:209–215
Carvalho MF, Lazarowitz SG (2004) Interaction of the movement protein NSP and the Arabidopsis acetyltransferase AtNSI is necessary for Cabbage leaf curl geminivirus infection and pathogenicity. J Virol 78:11161–11171
Carvalho MF, Turgeon R, Lazarowitz SG (2006) The geminivirus nuclear shuttle protein NSP inhibits the activity of AtNSI, a vascular-expressed Arabidopsis acetyltransferase regulated with the sink-to-source transition. Plant Physiol 140:1317–1330
Castellano MM, Sanz-Burgos AP, Gutierrez C (1999) Initiation of DNA replication in a eukaryotic rolling-circle replication: identification of multiple DNA-protein complexes at the geminivirus origin. J Mol Biol 290:639–652
Castillo AG, Collinet D, Deret S, et al (2003) Dual interaction of plant PCNA with geminivirus replication accessory protein (Ren) and viral replication protein (Rep). Virology 312:381–394
Castillo AG, Kong LJ, Hanley-Bowdoin L, et al (2004) Interaction between a geminivirus replication protein and the plant sumoylation system. J Virol 78:2758–2769
Chellappan P, Vanitharani R, Fauquet CM (2004) Short interfering RNA accumulation correlates with host recovery in DNA virus-infected hosts, and gene silencing targets specific viral sequences. J Virol 78:7465–7477
Chellappan P, Vanitharani R, Fauquet CM (2005) MicroRNA-binding viral protein interferes with Arabidopsis development. Proc Natl Acad Sci USA 102:10381–10386
Choudhury NR, Malik PS, Singh DK, et al (2006) The oligomeric Rep protein of Mung bean yellow mosaic India virus (MYMIV) is a likely replicative helicase. Nucleic Acids Res 34:6362–6377
Clerot D, Bernardi F (2006) DNA helicase activity is associated with the replication initiator protein rep of tomato yellow leaf curl geminivirus. J Virol 80:11322–11330
Costa AS (1976) Whitefly transmitted plant diseases. Annu Rev Phytopathol 16:429–449
Cui X, Li G, Wang D, et al (2005) A begomovirus DNAbeta-encoded protein binds DNA, functions as a suppressor of RNA silencing, and targets the cell nucleus. J Virol 79:10764–10775
Czosnek H, Ghanim M, Morin S, et al (2001) Whiteflies: vectors, and victims (?), of geminiviruses. Adv Virus Res 57:291–322
Davies JW, Stanley J (1989) Geminivirus genes and vectors. Trends Genet 5:77–81

Dickinson VJ, Halder J, Woolston CJ (1996) The product of maize streak virus ORF V1 is associated with secondary plasmodesmata and is first detected with the onset of viral lesions. Virology 220:51–59

Dong X (2004) NPR1, all things considered. Curr Opin Plant Biol 7:547–552

Dry IB, Krake LR, Rigden JE, et al (1997) A novel subviral agent associated with a geminivirus: the first report of a DNA satellite. Proc Natl Acad Sci USA 94:7088–7093

Duan YP, Powell CA, Purcifull DE, et al (1997) Phenotypic variation in transgenic tobacco expressing mutated geminivirus movement/pathogenicity (BC1) proteins. Mol Plant Microbe Interact 10:1065–1074

Duffy S, Holmes EC (2008) Phylogenetic evidence for rapid rates of molecular evolution in the single-stranded DNA begomovirus tomato yellow leaf curl virus. J Virol 82:957–965

Durrant WE, Dong X (2004) Systemic acquired resistance. Annu Rev Phytopathol 42:185–209

Eagle PA, Hanley-Bowdoin L (1997) Cis elements that contribute to geminivirus transcriptional regulation and the efficiency of DNA replication. J Virol 71:6947–6955

Eagle PA, Orozco BM, Hanley-Bowdoin L (1994) A DNA sequence required for geminivirus replication also mediates transcriptional regulation. Plant Cell 6:1157–1170

Ermak G, Paszkowski U, Wohlmuth M, et al (1993) Cytosine methylation inhibits replication of African cassava mosaic virus by two distinct mechanisms. Nucleic Acids Res 25:3445–3450

Evans D, Jeske H (1993) DNA B facilitates, but is not essential for, the spread of Abutilon mosaic virus in agroinoculated *Nicotiana benthamiana*. Virology 194:752–757

Fargette D, Konate G, Fauquet C, et al (2006) Molecular ecology and emergence of tropical plant viruses. Annu Rev Phytopathol 44:235–260

Fauquet CM, Stanley J (2005) Revising the way we conceive and name viruses below the species level: a review of geminivirus taxonomy calls for new standardized isolate descriptors. Arch Virol 150:2151–2179

Fauquet CM, Briddon RW, Brown JK, et al (2008) Geminivirus strain demarcation and nomenclature. Arch Virol 153:783–821

Florentino LH, Santos AA, Fontenelle MR, et al (2006) A PERK-like receptor kinase interacts with the geminivirus nuclear shuttle protein and potentiates viral infection. J Virol 80:6648–6656

Fofana IB, Sangare A, Collier R, et al (2004) A geminivirus-induced gene silencing system for gene function validation in cassava. Plant Mol Biol 56:613–624

Fondong VN, Reddy RV, Lu C, et al (2007) The consensus N-myristoylation motif of a geminivirus AC4 protein is required for membrane binding and pathogenicity. Mol Plant Microbe Interact 20:380–391

Fontes EP, Santos AA, Luz DF, et al (2004) The geminivirus nuclear shuttle protein is a virulence factor that suppresses transmembrane receptor kinase activity. Genes Dev 18:2545–2556

Frischmuth S, Frischmuth T, Jeske H (1991) Transcript mapping of Abutilon mosaic virus, a geminivirus. Virology 185:596–604

Frischmuth S, Frischmuth T, Latham J, et al (1993a) Transcriptional analysis of the virus-sense genes of the geminivirus beet curly top virus. Virology 197:312–319

Frischmuth S, Kleinow T, Aberle HJ, et al (2004) Yeast two-hybrid systems confirm the membrane-association and oligomerization of BC1 but do not detect an interaction of the movement proteins BC1 and BV1 of Abutilon mosaic geminivirus. Arch Virol 149:2349–2364

Frischmuth S, Wege C, Hülser D, et al (2007) The movement protein BC1 promotes redirection of the nuclear shuttle protein BV1 of Abutilon mosaic geminivirus to the plasma membrane in fission yeast. Protoplasma 230:117–123

Frischmuth T, Roberts S, von Arnim A, et al (1993b) Specificity of bipartite geminivirus movement proteins. Virology 196:666–673

Gafni Y, Epel BL (2002) The role of host and viral proteins in intra- and inter-cellular trafficking of geminiviruses. Mol Plant Pathol 60:231–241

Garcia-Andres S, Tomas DM, Sanchez-Campos S, et al (2007) Frequent occurrence of recombinants in mixed infections of tomato yellow leaf curl disease-associated begomoviruses. Virology 365:210–219

Gilbertson R, Lucas WJ (1996) How do viruses traffic on the 'vascular highway'? Trends Plant Sci 1:260–268

Glick E, Zrachya A, Levy Y, et al (2008) Interaction with host SGS3 is required for suppression of RNA silencing by tomato yellow leaf curl virus V2 protein. Proc Natl Acad Sci USA 105:157–161

Goodman RM (1977) Single-stranded DNA genome in a whitefly-transmitted plant virus. Virology 83:171–179

Goodman RM (1981) Geminiviruses. J Gen Virol 54:9–21

Gopal P, Pravin Kumar P, Sinilal B, et al (2007) Differential roles of C4 and betaC1 in mediating suppression of post-transcriptional gene silencing: evidence for transactivation by the C2 of Bhendi yellow vein mosaic virus, a monopartite begomovirus. Virus Res 123:9–18

Gröning BR, Hayes RJ, Buck KW (1994) Simultaneous regulation of tomato golden mosaic virus coat protein and AL1 gene expression: expression of the AL4 gene may contribute to suppression of the AL1 gene. J Gen Virol 75:721–726

Gronenborn B (2004) Nanoviruses: genome organisation and protein function. Vet Microbiol 98:103–109

Guerra-Peraza O, Kirk D, Seltzer V, et al (2005) Coat proteins of Rice tungro bacilliform virus and Mung bean yellow mosaic virus contain multiple nuclear-localization signals and interact with importin alpha. J Gen Virol 86:1815–1826

Gutierrez C (2000a) DNA replication and cell cycle in plants: learning from geminiviruses. EMBO J 19:792–799

Gutierrez C (2000b) Geminiviruses and the plant cell cycle. Plant Mol Biol 43:763–772

Gutierrez C, Ramirez-Parra E, Mar Castellano M, et al (2004) Geminivirus DNA replication and cell cycle interactions. Vet Microbiol 98:111–119

Ha C, Coombs S, Revill P, et al (2006) Corchorus yellow vein virus, a New World geminivirus from the Old World. J Gen Virol 87:997–1003

Ha C, Coombs S, Revill P, et al (2008) Molecular characterization of begomoviruses and DNA satellites from Vietnam: additional evidence that the New World geminiviruses were present in the Old World prior to continental separation. J Gen Virol 89:312–326

Haible D, Kober S, Jeske H (2006) Rolling circle amplification revolutionizes diagnosis and genomics of geminiviruses. J Virol Methods 135:9–16

Hallan V, Gafni Y (2001) Tomato yellow leaf curl virus (TYLCV) capsid protein (CP) subunit interactions: implications for viral assembly. Arch Virol 146:1765–1773

Hanley-Bowdoin L, Settlage SB, Orozco BM, et al (1999) Geminiviruses: models for plant DNA replication, transcription, and cell cycle regulation. CRC Crit Rev Plant Sci 18:71–106

Hanley-Bowdoin L, Settlage SB, Robertson D (2004) Reprogramming plant gene expression: a prerequisite to geminivirus DNA replication. Mol Plant Pathol 5:149–156

Hao L, Wang H, Sunter G, et al (2003) Geminivirus AL2 and L2 proteins interact with and inactivate SNF1 kinase. Plant Cell 15:1034–1048

Harper G, Hull R, Lockhart B, et al (2002) Viral sequences integrated into plant genomes. Annu Rev Phytopathol 40:119–136

Harrison BD (1985) Advances in geminivirus research. Annu Rev Phytopathol 23:55–82

Harrison BD, Robinson DJ (1999) Natural genomic and antigenic variation in whitefly-transmitted geminiviruses (begomoviruses). Annu Rev Phytopathol 37:369–398

Harrison BD, Barker H, Bock KR, et al (1977) Plant-viruses with circular single-stranded DNA. Nature 270:760–762

Hehnle S, Wege C, Jeske H (2004) The interaction of DNA with the movement proteins of geminiviruses revisited. J Virol 78:7698–7706

Höfer P, Bedford ID, Markham PG, et al (1997) Coat protein gene replacement results in whitefly transmission of an insect nontransmissible geminivirus isolate. Virology 236:288–295

Höhnle M, Höfer P, Bedford ID, et al (2001) Exchange of three amino acids in the coat protein results in efficient whitefly transmission of a nontransmissible Abutilon mosaic virus isolate. Virology 290:164–171

Homs M, Kober S, Kepp G, Jeske H (2008) Mitochondrial plasmids of sugar beet amplified via rolling circle method detected during curtovirus screening. Virus Res 136:124–129

Hormuzdi SG, Bisaro DM (1995) Genetic analysis of beet curly top virus: examination of the roles of L2 and L3 genes in viral pathogenesis. Virology 206:1044–1054

Horns T, Jeske H (1991) Localization of Abutilon mosaic virus DNA within leaf tissue by in-situ hybridization. Virology 181:580–588

Horvath GV, Pettko-Szandtner A, Nikovics K, et al (1998) Prediction of functional regions of the maize streak virus replication-associated proteins by protein-protein interaction analysis. Plant Mol Biol 38:699–712

Hou YM, Sanders R, Ursin VM, et al (2000) Transgenic plants expressing geminivirus movement proteins: abnormal phenotypes and delayed infection by Tomato mottle virus in transgenic tomatoes expressing the Bean dwarf mosaic virus BV1 or BC1 proteins. Mol Plant Microbe Interact 13:297–308

Hull R (2002) Matthews' plant virology, 4th edn. Academic Press, San Diego

Hull R, Davies JW (1992) Approaches to nonconventional control of plant virus diseases. CRC Crit Rev Plant Sci 11:17–33

Ingham DJ, Pascal E, Lazarowitz SG (1995) Both bipartite geminivirus movement proteins define viral host range, but only BL1 determines viral pathogenicity. Virology 207:191–204

Jeffrey JL, Pooma W, Petty IT (1996) Genetic requirements for local and systemic movement of tomato golden mosaic virus in infected plants. Virology 223:208–218

Jeske H (2007) Replication of geminiviruses and the use of rolling circle amplification for their diagnosis. In: Czosnek H (ed) Tomato yellow leaf curl virus disease. Springer, Dordrecht, pp 141–156

Jeske H, Werz G (1980a) Ultrastructural and biochemical investigations on the whitefly transmitted Abutilon mosaic virus (AbMV). J Phytopathol 97:43–55

Jeske H, Werz G (1980b) Cytochemical characterization of plastidal inclusions in Abutilon mosaic infected *Malva parviflora* mesophyll cells. Virology 106:155–158

Jeske H, Menzel D, Werz G (1977) Electron microscopic studies on intranuclear virus-like inclusions in mosaic-diseased *Abutilon sellowianum* Reg. J Phytopathol 89:289–295

Jeske H, Lütgemeier M, Preiss W (2001) Distinct DNA forms indicate rolling circle and recombination-dependent replication of Abutilon mosaic geminivirus. EMBO J 20:6158–6167

Jovel J, Preiß W, Jeske H (2007) Characterization of DNA-intermediates of an arising geminivirus. Virus Res 130:63–70

Jupin I, De Kouchkovsky F, Jouanneau F, et al (1994) Movement of tomato yellow leaf curl geminivirus (TYLCV): involvement of the protein encoded by ORF C4. Virology 204:82–90

Kartha KK, Gamborg OL (1975) Elimination of cassava mosaic disease by meristem culture. Phytopathology 65:826–828

Kenton A, Khashoggi A, Parokonny A, et al (1995) Chromosomal location of endogenous geminivirus-related DNA sequences in *Nicotiana tabacum L.* Chromosome Res 3:346–350

Kheyr-Pour A, Bananej K, Dafalla GA, et al (2000) Watermelon chlorotic stunt virus from the Sudan and Iran: sequence comparison and identification of a whitefly-transmission determinant. Phytopathology 90:629–635

Kirthi N, Savithri HS (2003) A conserved zinc finger motif in the coat protein of Tomato leaf curl Bangalore virus is responsible for binding to ssDNA. Arch Virol 148:2369–2380

Kittelmann K, Jeske H (2008) Disassembly of African cassava mosaic virus. J Gen Virol 89:2029–2036

Kjemtrup S, Sampson KS, Peele CG, et al (1998) Gene silencing from plant DNA carried by a geminivirus. Plant J 14:91–100

Kleinow T, Holeiter G, Nischang M, et al (2008) Post-translational modifications of Abutilon mosaic virus movement protein (BC1) in fission yeast. Virus Res 131:86–94

Klinkenberg FA, Stanley J (1990) Encapsidation and spread of African cassava mosaic virus DNA A in the absence of DNA B when agroinoculated to *Nicotiana benthamiana*. J Gen Virol 71:1409–1412

Kong LJ, Hanley-Bowdoin L (2002) A geminivirus replication protein interacts with a protein kinase and a motor protein that display different expression patterns during plant development and infection. Plant Cell 14:1817–1832

Kong LJ, Orozco BM, Roe JL, et al (2000) A geminivirus replication protein interacts with the retinoblastoma protein through a novel domain to determine symptoms and tissue specificity of infection in plants. EMBO J 19:3485–3495

Koonin EV, Ilyina TV (1992) Geminivirus replication proteins are related to prokaryotic plasmid rolling circle DNA replication initiator proteins. J Gen Virol 73:2763–2766

Kotlizky G, Boulton MI, Pitaksutheepong C, et al (2000) Intracellular and intercellular movement of maize streak geminivirus V1 and V2 proteins transiently expressed as green fluorescent protein fusions. Virology 274:32–38

Krenz B, Wege C, Jeske H (2008) Abutilon mosaic virus as a stable and attenuated vector for virus-induced gene silencing and limited phloem-specific protein expression. (submitted).

Kunik T, Palanichelvam K, Czosnek H, et al (1998) Nuclear import of the capsid protein of tomato yellow leaf curl virus (TYLCV) in plant and insect cells. Plant J 13:393–399

Kunik T, Mizrachy L, Citovsky V, et al (1999) Characterization of a tomato karyopherin alpha that interacts with the tomato yellow leaf curl virus (TYLCV) capsid protein. J Exp Bot 50:731–732

Lazarowitz SG (1992) Geminiviruses: genome structure and gene function. CRC Crit Rev Plant Sci 11:327–349

Lazarowitz SG (1999) Probing plant cell structure and function with viral movement proteins. Curr Opin Plant Biol 2:332–338

Lazarowitz SG, Beachy RN (1999) Viral movement proteins as probes for intracellular and intercellular trafficking in plants. Plant Cell 11:535–548

Lefeuvre P, Martin DP, Hoareau M, et al (2007) Begomovirus 'melting pot' in the south-west Indian Ocean islands: molecular diversity and evolution through recombination. J Gen Virol 88:3458–3468

Levy A, Czosnek H (2003) The DNA-B of the non-phloem-limited bean dwarf mosaic virus (BDMV) is able to move the phloem-limited Abutilon mosaic virus (AbMV) out of the phloem, but DNA-B of AbMV is unable to confine BDMV to the phloem. Plant Mol Biol 53:789–803

Li D, Behjatnia SA, Dry IB, et al (2007) Genomic regions of tomato leaf curl virus DNA satellite required for replication and for satellite-mediated delivery of heterologous DNAs. J Gen Virol 88:2073–2077

Lim KY, Matyasek R, Lichtenstein CP, et al (2000) Molecular cytogenetic analyses and phylogenetic studies in the *Nicotiana* section Tomentosae. Chromosoma 109:245–258

Liu H, Boulton MI, Oparka KJ, et al (2001) Interaction of the movement and coat proteins of Maize streak virus: implications for the transport of viral DNA. J Gen Virol 82:35–44

Liu HT, Boulton MI, Davies JW (1997) Maize streak virus coat protein binds single- and double-stranded DNA in vitro. J Gen Virol 78:1265–1270

Liu L, Saunders K, Thomas CL, et al (1999) Bean yellow dwarf virus RepA, but not Rep, binds to maize retinoblastoma protein, and the virus tolerates mutations in the consensus binding motif. Virology 256:270–279

Lucy AP, Boulton MI, Davies JW, et al (1996) Tissue specificity of *Zea mays* infection by maize streak virus. Mol Plant Microbe Interact 9:22–31

Luque A, Sanz-Burgos AP, Ramirez-Parra E, et al (2002) Interaction of geminivirus Rep protein with replication factor C and its potential role during geminivirus DNA replication. Virology 302:83–94

Malik PS, Kumar V, Bagewadi B, et al (2005) Interaction between coat protein and replication initiation protein of Mung bean yellow mosaic India virus might lead to control of viral DNA replication. Virology 337:273–283

Mansoor S, Briddon RW, Zafar Y, et al (2003) Geminivirus disease complexes: an emerging threat. Trends Plant Sci 8:128–134

Mariano AC, Andrade MO, Santos AA, et al (2004) Identification of a novel receptor-like protein kinase that interacts with a geminivirus nuclear shuttle protein. Virology 318:24–31

Martin D, Rybicki E (2000) RDP: detection of recombination amongst aligned sequences. Bioinformatics 16:562–563

Martin DP, van der Walt E, Posada D, et al (2005) The evolutionary value of recombination is constrained by genome modularity. PLoS Genet 1:475–479

McGarry RC, Barron YD, Carvalho MF, et al (2003) A novel Arabidopsis acetyltransferase interacts with the geminivirus movement protein NSP. Plant Cell 15:1605–1618

McGivern DR, Findlay KC, Montague NP, et al (2005) An intact RBR-binding motif is not required for infectivity of Maize streak virus in cereals, but is required for invasion of mesophyll cells. J Gen Virol 86:797–801

Missich R, Ramirez-Parra E, Gutierrez C (2000) Relationship of oligomerization to DNA binding of Wheat dwarf virus RepA and Rep proteins. Virology 273:178–188

Moffat A (1999) Geminiviruses emerge as serious crop threat. Science 286:1835

Morales FJ (2006) History and current distribution of begomoviruses in Latin America. Adv Virus Res 67:127–162

Morales FJ (2007) Tropical Whitefly IPM Project. Adv Virus Res 69:249–311

Morales FJ, Anderson PK (2001) The emergence and dissemination of whitefly-transmitted geminiviruses in Latin America. Arch Virol 146:415–441

Morales FJ, Jones PG (2004) The ecology and epidemiology of whitefly-transmitted viruses in Latin America. Virus Res 100:57–65

Morilla G, Krenz B, Jeske H, et al (2004) Tête à tête of Tomato yellow leaf curl virus (TYLCV) and Tomato yellow leaf curl Sardinia virus (TYLCSV) in single nuclei. J Virol 78:10715–10723

Morilla G, Castillo AG, Preiß W, et al (2006) A versatile transreplication-based system to identify cellular proteins involved in geminivirus replication. J Virol 80:3624–3633

Morin S, Ghanim M, Zeidan M, et al (1999) A GroEL homologue from endosymbiotic bacteria of the whitefly *Bemisia tabaci* is implicated in the circulative transmission of tomato yellow leaf curl virus. Virology 256:75–84

Morra MR, Petty IT (2000) Tissue specificity of geminivirus infection is genetically determined. Plant Cell 12:2259–2270

Muangsan N, Robertson D (2004) Geminivirus vectors for transient gene silencing in plants. Methods Mol Biol 265:101–115

Muangsan N, Beclin C, Vaucheret H, et al (2004) Geminivirus VIGS of endogenous genes requires SGS2/SDE1 and SGS3 and defines a new branch in the genetic pathway for silencing in plants. Plant J 38:1004–1014

Munoz-Martin A, Collin S, Herreros E, et al (2003) Regulation of MSV and WDV virion-sense promoters by WDV nonstructural proteins: a role for their retinoblastoma protein-binding motifs. Virology 306:313–323

Murad L, Bielawski JP, Matyasek R, et al (2004) The origin and evolution of geminivirus-related DNA sequences in *Nicotiana*. Heredity 92:352–358

Noris E, Vaira AM, Caciagli P, et al (1998) Amino acids in the capsid protein of tomato yellow leaf curl virus that are crucial for systemic infection, particle formation, and insect transmission. J Virol 72:10050–10057

Noueiry AO, Lucas WJ, Gilbertson RL (1994) Two proteins of a plant DNA virus coordinate nuclear and plasmodesmatal transport. Cell 76:925–932

Orozco BM, Hanley-Bowdoin L (1998) Conserved sequence and structural motifs contribute to the DNA binding and cleavage activities of a geminivirus replication protein. J Biol Chem 273:24448–24456

Owor BE, Martin DP, Shepherd DN, et al (2007) Genetic analysis of maize streak virus isolates from Uganda reveals widespread distribution of a recombinant variant. J Gen Virol 88:3154–3165

Padidam M, Beachy RN, Fauquet CM (1996) The role of AV2 ("precoat") and coat protein in viral replication and movement in tomato leaf curl geminivirus. Virology 224:390–404

Padidam M, Sawyer S, Fauquet CM (1999) Possible emergence of new geminiviruses by frequent recombination. Virology 285:218–225

Palanichelvam K, Kunik T, Citovsky V, et al (1998) The capsid protein of tomato yellow leaf curl virus binds cooperatively to single-stranded DNA. J Gen Virol 79:2829–2833
Palmer KE, Rybicki EP (1998) The molecular biology of mastreviruses. Adv Virus Res 50:183–234
Pascal E, Goodlove PE, Wu LC, et al (1993) Transgenic tobacco plants expressing the geminivirus BL1 protein exhibit symptoms of viral disease. Plant Cell 5:795–807
Pascal E, Sanderfoot AA, Ward BM, et al (1994) The geminivirus BR1 movement protein binds single-stranded DNA and localizes to the cell nucleus. Plant Cell 6:995–1006
Patil BL, Dasgupta I (2005) Defective interfering DNAs of plant viruses. CRC Crit Rev Plant Sci 24:1–18
Peele C, Jordan CV, Muangsan N, et al (2001) Silencing of a meristematic gene, proliferating cell nuclear antigen (PCNA), using geminivirus-derived vectors. Plant J 271:357–366
Petty IT, Miller CG, Meade-Hash TJ, et al (1995) Complementable and noncomplementable host adaptation defects in bipartite geminiviruses. Virology 212:263–267
Pilartz M, Jeske H (1992) Abutilon mosaic geminivirus double-stranded DNA is packed into minichromosomes. Virology 189:800–802
Pilartz M, Jeske H (2003) Mapping of Abutilon mosaic geminivirus minichromosomes. J Virol 77:10808–10818
Piroux N, Saunders K, Page A, et al (2007) Geminivirus pathogenicity protein C4 interacts with Arabidopsis thaliana shaggy-related protein kinase AtSKeta, a component of the brassinosteroid signalling pathway. Virology 362:428–440
Pita JS, Fondong VN, Sangare A, et al (2001) Recombination, pseudorecombination and synergism of geminiviruses are determinant keys to the epidemic of severe cassava mosaic disease in Uganda. J Gen Virol 82:655–665
Pohl D, Wege C (2007) Synergistic pathogenicity of a phloem-limited begomovirus and tobamoviruses despite negative interference. J Gen Virol 88:1034–1040
Pooggin M, Hohn T (2004) Fighting geminiviruses by RNAi and vice versa. Plant Mol Biol 55:149–152
Pooggin M, Shivaprasad PV, Veluthambi K, et al (2003) RNAi targeting of DNA virus in plants. Nat Biotechnol 21:131–132
Preiss W, Jeske H (2003) Multitasking in replication is common among geminiviruses. J Virol 77:2972–2980
Qin SW, Ward BM, Lazarowitz SG (1998) The bipartite geminivirus coat protein aids BR1 function in viral movement by affecting the accumulation of viral single-stranded DNA. J Virol 72:9247–9256
Robards AW, Lucas WJ (1990) Plasmodesmata. Annu Rev Plant Physiol Plant Mol Biol 41:369–419
Rojas MR, Noueiry AO, Lucas WJ, et al (1998) Bean dwarf mosaic geminivirus movement proteins recognize DNA in a form- and size-specific manner. Cell 95:105–113
Rojas MR, Jiang H, Salati R, et al (2001) Functional analysis of proteins involved in movement of the monopartite begomovirus, Tomato yellow leaf curl virus. Virology 291:110–125
Rojas MR, Hagen C, Lucas WJ, et al (2005) Exploiting chinks in the plant's armor: evolution and emergence of geminiviruses. Annu Rev Phytopathol 43:361–394
Rothenstein D, Krenz B, Selchow O, et al (2007) Tissue and cell tropism of Indian cassava mosaic virus (ICMV) and its AV2 (precoat) gene product. Virology 359:137–145
Rybicki EP (1994) A phylogenetic and evolutionary justification for three genera of Geminiviridae. Arch Virol 139:49–77
Rybicki EP, Pietersen G (1999) Plant virus disease problems in the developing world. Adv Virus Res 53:127–175
Saeed M, Behjatnia SA A, Mansoor S, et al (2005) A geminiviral DNA beta satellite modulates pathogenesis by a single complementary-sense transcript. Mol Plant Microbe Interact 18:7–14

Saeed M, Zafar Y, Randles JW, et al (2007) A monopartite begomovirus-associated DNA beta satellite substitutes for the DNA B of a bipartite begomovirus to permit systemic infection. J Gen Virol 88:2881–2889

Sanderfoot AA, Lazarowitz SG (1995) Cooperation in viral movement: the geminivirus BL1 movement protein interacts with BR1 and redirects it from the nucleus to the cell periphery. Plant Cell 7:1185–1194

Sanderfoot AA, Lazarowitz SG (1996) Getting it together in plant virus movement: cooperative interactions between bipartite geminivirus movement proteins. Trends Cell Biol 6:353–358

Saunders K, Wege C, Karuppannan V, et al (2001) The distinct disease phenotypes of the common and yellow vein strains of Tomato golden mosaic virus are determined by nucleotide differences in the 3′-terminal region of the gene encoding the movement protein. J Gen Virol 82:45–51

Saunders K, Bedford ID, Yahara T, et al (2003) Aetiology: the earliest recorded plant virus disease. Nature 422:831

Saunders K, Norman A, Gucciardo S, et al (2004) The DNA beta satellite component associated with Ageratum yellow vein disease encodes an essential pathogenicity protein (beta C1). Virology 324:37–47

Seemanpillai M, Dry I, Randles J, et al (2003) Transcriptional silencing of geminiviral promoter-driven transgenes following homologous virus infection. Mol Plant Microbe Interact 16:429–438

Selth LA, Dogra SC, Rasheed MS, et al (2005) A NAC domain protein interacts with tomato leaf curl virus replication accessory protein and enhances viral replication. Plant Cell 17:311–325

Selth LA, Dogra SC, Rasheed MS, et al (2006) Identification and characterization of a host reversibly glycosylated peptide that interacts with the Tomato leaf curl virus V1 protein. Plant Mol Biol 61:297–310

Settlage SB, Miller AB, Gruissem W, et al (2001) Dual interaction of a geminivirus replication accessory factor with a viral replication protein and a plant cell cycle regulator. Virology 279:570–576

Settlage SB, See RG, Hanley-Bowdoin L (2005) Geminivirus C3 protein: replication enhancement and protein interactions. J Virol 79:9885–9895

Shen W, Hanley-Bowdoin L (2006) Geminivirus infection up-regulates the expression of two Arabidopsis protein kinases related to yeast SNF1- and mammalian AMPK-activating kinases. Plant Physiol 142:1642–1655

Shepherd DN, Martin DP, McGivern DR, et al (2005) A three-nucleotide mutation altering the Maize streak virus Rep pRBR-interaction motif reduces symptom severity in maize and partially reverts at high frequency without restoring pRBR-Rep binding. J Gen Virol 86:803–813

Shepherd DN, Martin DP, Varsani A, et al (2006) Restoration of native folding of single-stranded DNA sequences through reverse mutations: an indication of a new epigenetic mechanism. Arch Biochem Biophys 453:108–122

Shimada-Beltran H, Rivera-Bustamante RF (2007) Early and late gene expression in pepper huasteco yellow vein virus. J Gen Virol 88:3145–3153

Shivaprasad PV, Akbergenov R, Trinks D, et al (2005) Promoters, transcripts, and regulatory proteins of mung bean yellow mosaic geminivirus. J Virol 79:8149–8163

Shung CY, Sunter G (2007) AL1-dependent repression of transcription enhances expression of Tomato golden mosaic virus AL2 and AL3. Virology 364:112–122

Shung CY, Sunter J, Sirasanagandla SS, et al (2006) Distinct viral sequence elements are necessary for expression of Tomato golden mosaic virus complementary sense transcripts that direct AL2 and AL3 gene expression. Mol Plant Microbe Interact 19:1394–13405

Stanley J (1985) The molecular biology of geminiviruses. Adv Virus Res 31:139–177

Stanley J (2004) Subviral DNAs associated with geminivirus disease complexes. Vet Microbiol 98:121–129

Stanley J, Davies JW (1985) Structure and function of the DNA genome of geminiviruses. In: Davies J (ed) Molecular plant virology. CRC Press, Boca Raton, pp 191–218

Stanley J, Bisaro DM, Briddon RW, et al (2005) Geminiviridae. In: Fauquet CM, Mayo MA, Maniloff J, et al (eds) Virus taxonomy. VIIIth report of the International Committee on Taxonomy of Viruses. Elsevier/Academic Press, London, pp 301–326

Suarez-Lopez P, Martinez-Sals E, Hernandez P, et al (1995) Bent DNA in the large intergenic region of wheat dwarf geminivirus. Virology 208:303–311

Sung P, Krejci L, Van Komen S, et al (2003) Rad51 recombinase and recombination mediators. J Biol Chem 278:42729–42732

Sunter G, Bisaro DM (1991) Transactivation in a geminivirus: AL2 gene product is needed for coat protein expression. Virology 180:416–419

Sunter G, Bisaro DM (1992) Transactivation of geminivirus AR1 and BR1 gene expression by the viral AL2 gene product occurs at the level of transcription. Plant Cell 4:1321–1331

Sunter G, Bisaro DM (1997) Regulation of a geminivirus coat protein promoter by AL2 protein (TrAP): evidence for activation and derepression mechanisms. Virology 232:269–280

Sunter G, Hartitz MD, Hormuzdi SG, et al (1990) Genetic analysis of tomato golden mosaic virus: ORF AL2 is required for coat protein accumulation while ORF AL3 is necessary for efficient DNA replication. Virology 179:69–77

Sunter G, Hartitz MD, Bisaro DM (1993) Tomato golden mosaic virus leftward gene expression: autoregulation of geminivirus replication protein. Virology 195:275–280

Sunter G, Stenger DC, Bisaro DM (1994) Heterologous complementation by geminivirus AL2 and AL3 genes. Virology 203:203–210

Symington LS (2002) Role of RAD52 epistasis group genes in homologous recombination and double-strand break repair. Microbiol Mol Biol Rev 66:630–670

Tao X, Zhou X (2004) A modified viral satellite DNA that suppresses gene expression in plants. Plant J 38:850–860

Thresh JM (2006) Control of tropical plant virus diseases. Adv Virus Res 67:245–295

Thresh JM, Cooter RJ (2005) Strategies for controlling cassava mosaic virus disease in Africa. Plant Pathol 54:587–614

Trinks D, Rajeswaran R, Shivaprasad PV, et al (2005) Suppression of RNA silencing by a geminivirus nuclear protein, AC2, correlates with transactivation of host genes. J Virol 79:2517–2527

Tu J, Sunter G (2007) A conserved binding site within the Tomato golden mosaic virus AL-1629 promoter is necessary for expression of viral genes important for pathogenesis. Virology 367:117–125

Unseld S, Ringel M, Konrad A, et al (2000a) Virus-specific adaptations for the production of a pseudorecombinant virus formed by two distinct bipartite geminiviruses from Central America. Virology 274:179–188

Unseld S, Ringel M, Höfer P, et al (2000b) Host range and symptom variation of pseudorecombinant virus produced by two distinct bipartite geminiviruses. Arch Virol 145:1449–1454

Unseld S, Höhnle M, Ringel M, et al (2001) Subcellular targeting of the coat protein of African cassava mosaic geminivirus. Virology 286:373–383

Unseld S, Frischmuth T, Jeske H (2004) Short deletions in nuclear targeting sequences of African cassava mosaic virus coat protein prevent geminivirus twinned particle formation. Virology 318:89–100

van Loon LC, Rep M, Pieterse CM (2006) Significance of inducible defense-related proteins in infected plants. Annu Rev Phytopathol 44:135–162

Vanderschuren H, Akbergenov R, Pooggin MM, et al (2007) Transgenic cassava resistance to African cassava mosaic virus is enhanced by viral DNA-A bidirectional promoter-derived siRNAs. Plant Mol Biol 64:549–557

Vanitharani R, Chellappan P, Pita JS, et al (2004) Differential roles of AC2 and AC4 of cassava geminiviruses in mediating synergism and suppression of posttranscriptional gene silencing. J Virol 78:9487–9498

Vanitharani R, Chellappan P, Fauquet CM (2005) Geminiviruses and RNA silencing. Trends Plant Sci 10:144–151

Vega-Rocha S, Byeon IL, Gronenborn B, Gronenborn AM, Campos-Olivas R (2007a) Solution structure, divalent metal and DNA binding of the endonuclease domain from the replication initiation protein from porcine circovirus. J Mol Biol 367:473–487

Vega-Rocha S, Gronenborn B, Gronenborn AM, Campos-Olivas R (2007b) Solution structure of the endonuclease domain from the master replication initiator protein of the nanovirus Faba bean necrotic yellow virus and comparison with the corresponding geminivirus and circovirus structures. Biochemistry 46:6201–6212

von Arnim A, Frischmuth T, Stanley J (1993) Detection and possible functions of African cassava mosaic virus DNA B gene products. Virology 192:264–272

Waigmann E, Ueki S, Trutnyeva K, et al (2004) The ins and outs of nondestructive cell-to-cell and systemic movement of plant viruses. CRC Crit Rev Plant Sci 23:195–250

Wang H, Hao L, Shung CY, et al (2003) Adenosine kinase is inactivated by geminivirus AL2 and L2 proteins. Plant Cell 15:3020–3032

Wang H, Buckley KJ, Yang X, et al (2005) Adenosine kinase inhibition and suppression of RNA silencing by geminivirus AL2 and L2 proteins. J Virol 79:7410–7418

Ward BM, Medville R, Lazarowitz SG, et al (1997) The geminivirus BL1 movement protein is associated with endoplasmic reticulum-derived tubules in developing phloem cells. J Virol 71:3726–3733

Wege C (2007) Movement and localization of Tomato yellow leaf curl viruses in the infected plant. In: Czosnek H (ed) Tomato yellow leaf curl virus disease. Springer, Dordrecht, pp 185–206

Wege C, Pohl D (2007) Abutilon mosaic virus DNA B component supports mechanical virus transmission, but does not counteract begomoviral phloem limitation in transgenic plants. Virology 365:173–186

Wege C, Siegmund D (2007) Synergism of a DNA and an RNA virus: enhanced tissue infiltration of the begomovirus Abutilon mosaic virus (AbMV) mediated by Cucumber mosaic virus (CMV). Virology 357:10–28

Wege C, Gotthardt RD, Frischmuth T, et al (2000) Fulfilling Koch's postulates for Abutilon mosaic virus. Arch Virol 145:2217–2225

Wright EA, Heckel T, Groenendijk J, et al (1997) Splicing features in maize streak virus virion- and complementary-sense gene expression. Plant J 12:1285–1297

Xie Q, Suarez-Lopez P, Gutierrez C (1995) Identification and analysis of a retinoblastoma binding motif in the replication protein of a plant DNA virus: requirement for efficient viral DNA replication. EMBO J 14:4073–4082

Xie Q, Sanz-Burgos AP, Hannon GJ, et al (1996) Plant cells contain a novel member of the retinoblastoma family of growth regulatory proteins. EMBO J 15:4900–4908

Xie Q, Sanz-Burgos AP, Guo H, et al (1999) GRAB proteins, novel members of the NAC domain family, isolated by their interaction with a geminivirus protein. Plant Mol Biol 39:647–656

Yang X, Baliji S, Buchmann RC, et al (2007) Functional modulation of the geminivirus AL2 transcription factor and silencing suppressor by self-interaction. J Virol 81:11972–11981

Zhang SC, Wege C, Jeske H (2001a) Movement proteins (BC1 and BV1) of Abutilon mosaic geminivirus are cotransported in and between cells of sink but not of source leaves as detected by green fluorescent protein tagging. Virology 290:249–260

Zhang SC, Ghosh R, Jeske H (2002) Subcellular targeting domains of Abutilon mosaic geminivirus movement protein BC1. Arch Virol 147:2349–2363

Zhang W, Olson NH, Baker TS, et al (2001b) Structure of the maize streak virus geminate particle. Virology 279:471–477

Zrachya A, Glick E, Levy Y, et al (2007) Suppressor of RNA silencing encoded by Tomato yellow leaf curl virus-Israel. Virology 358:159–165

# Index

# Current Topics in Microbiology and Immunology

## Volumes published since 2002

Vol. 271: **Koehler, Theresa M. (Ed.):** Anthrax. 2002. 14 figs. X, 169 pp. ISBN 3-540-43497-6

Vol. 272: **Doerfler, Walter; Böhm, Petra (Eds.):** Adenoviruses: Model and Vectors in Virus-Host Interactions. Virion and Structure, Viral Replication, Host Cell Interactions. 2003. 63 figs., approx. 280 pp. ISBN 3-540-00154-9

Vol. 273: **Doerfler, Walter; Böhm, Petra (Eds.):** Adenoviruses: Model and Vectors in VirusHost Interactions. Immune System, Oncogenesis, Gene Therapy. 2004. 35 figs., approx. 280 pp. ISBN 3-540-06851-1

Vol. 274: **Workman, Jerry L. (Ed.):** Protein Complexes that Modify Chromatin. 2003. 38 figs., XII, 296 pp. ISBN 3-540-44208-1

Vol. 275: **Fan, Hung (Ed.):** Jaagsiekte Sheep Retrovirus and Lung Cancer. 2003. 63 figs., XII, 252 pp. ISBN 3-540-44096-3

Vol. 276: **Steinkasserer, Alexander (Ed.):** Dendritic Cells and Virus Infection. 2003. 24 figs., X, 296 pp. ISBN 3-540-44290-1

Vol. 277: **Rethwilm, Axel (Ed.):** Foamy Viruses. 2003. 40 figs., X, 214 pp. ISBN 3-540-44388-6

Vol. 278: **Salomon, Daniel R.; Wilson, Carolyn (Eds.):** Xenotransplantation. 2003. 22 figs., IX, 254 pp. ISBN 3-540-00210-3

Vol. 279: **Thomas, George; Sabatini, David; Hall, Michael N. (Eds.):** TOR. 2004. 49 figs., X, 364 pp. ISBN 3-540-00534X

Vol. 280: **Heber-Katz, Ellen (Ed.):** Regeneration: Stem Cells and Beyond. 2004. 42 figs., XII, 194 pp. ISBN 3-540-02238-4

Vol. 281: **Young, John A. T. (Ed.):** Cellular Factors Involved in Early Steps of Retroviral Replication. 2003. 21 figs., IX, 240 pp. ISBN 3-540-00844-6

Vol. 282: **Stenmark, Harald (Ed.):** Phosphoinositides in Subcellular Targeting and Enzyme Activation. 2003. 20 figs., X, 210 pp. ISBN 3-540-00950-7

Vol. 283: **Kawaoka, Yoshihiro (Ed.):** Biology of Negative Strand RNA Viruses: The Power of Reverse Genetics. 2004. 24 figs., IX, 350 pp. ISBN 3-540-40661-1

Vol. 284: **Harris, David (Ed.):** Mad Cow Disease and Related Spongiform Encephalopathies. 2004. 34 figs., IX, 219 pp. ISBN 3-540-20107-6

Vol. 285: **Marsh, Mark (Ed.):** Membrane Trafficking in Viral Replication. 2004. 19 figs., IX, 259 pp. ISBN 3-540-21430-5

Vol. 286: **Madshus, Inger H. (Ed.):** Signalling from Internalized Growth Factor Receptors. 2004. 19 figs., IX, 187 pp. ISBN 3-540-21038-5

Vol. 287: **Enjuanes, Luis (Ed.):** Coronavirus Replication and Reverse Genetics. 2005. 49 figs., XI, 257 pp. ISBN 3-540- 21494-1

Vol. 288: **Mahy, Brain W. J. (Ed.):** Foot-and-Mouth-Disease Virus. 2005. 16 figs., IX, 178 pp. ISBN 3-540-22419X

Vol. 289: **Griffin, Diane E. (Ed.):** Role of Apoptosis in Infection. 2005. 40 figs., IX, 294 pp. ISBN 3-540-23006-8

Vol. 290: **Singh, Harinder; Grosschedl, Rudolf (Eds.):** Molecular Analysis of B Lymphocyte Development and Activation. 2005. 28 figs., XI, 255 pp. ISBN 3-540-23090-4

Vol. 291: **Boquet, Patrice; Lemichez Emmanuel (Eds.):** Bacterial Virulence Factors and Rho GTPases. 2005. 28 figs., IX, 196 pp. ISBN 3-540-23865-4

Vol. 292: **Fu, Zhen F. (Ed.):** The World of Rhabdoviruses. 2005. 27 figs., X, 210 pp. ISBN 3-540-24011-X

Vol. 293: **Kyewski, Bruno; Suri-Payer, Elisabeth (Eds.):** CD4+CD25+ Regulatory T Cells: Origin, Function and Therapeutic Potential. 2005. 22 figs., XII, 332 pp. ISBN 3-540-24444-1

Vol. 294: **Caligaris-Cappio, Federico, Dalla Favera, Ricardo (Eds.):** Chronic Lymphocytic Leukemia. 2005. 25 figs., VIII, 187 pp. ISBN 3-540-25279-7

Vol. 295: **Sullivan, David J.; Krishna Sanjeew (Eds.):** Malaria: Drugs, Disease and Post-genomic Biology. 2005. 40 figs., XI, 446 pp. ISBN 3-540-25363-7

Vol. 296: **Oldstone, Michael B. A. (Ed.):** Molecular Mimicry: Infection Induced Autoimmune Disease. 2005. 28 figs., VIII, 167 pp. ISBN 3-540-25597-4

Vol. 297: **Langhorne, Jean (Ed.):** Immunology and Immunopathogenesis of Malaria. 2005. 8 figs., XII, 236 pp. ISBN 3-540-25718-7

Vol. 298: **Vivier, Eric; Colonna, Marco (Eds.):** Immunobiology of Natural Killer Cell Receptors. 2005. 27 figs., VIII, 286 pp. ISBN 3-540-26083-8

Vol. 299: **Domingo, Esteban (Ed.):** Quasispecies: Concept and Implications. 2006. 44 figs., XII, 401 pp. ISBN 3-540-26395-0

Vol. 300: **Wiertz, Emmanuel J.H.J.; Kikkert, Marjolein (Eds.):** Dislocation and Degradation of Proteins from the Endoplasmic Reticulum. 2006. 19 figs., VIII, 168 pp. ISBN 3-540-28006-5

Vol. 301: **Doerfler, Walter; Böhm, Petra (Eds.):** DNA Methylation: Basic Mechanisms. 2006. 24 figs., VIII, 324 pp. ISBN 3-540-29114-8

Vol. 302: **Robert N. Eisenman (Ed.):** The Myc/Max/Mad Transcription Factor Network. 2006. 28 figs., XII, 278 pp. ISBN 3-540-23968-5

Vol. 303: **Thomas E. Lane (Ed.):** Chemokines and Viral Infection. 2006. 14 figs. XII, 154 pp. ISBN 3-540-29207-1

Vol. 304: **Stanley A. Plotkin (Ed.):** Mass Vaccination: Global Aspects – Progress and Obstacles. 2006. 40 figs. X, 270 pp. ISBN 3-540-29382-5

Vol. 305: **Radbruch, Andreas; Lipsky, Peter E. (Eds.):** Current Concepts in Autoimmunity. 2006. 29 figs. IIX, 276 pp. ISBN 3-540-29713-8

Vol. 306: **William M. Shafer (Ed.):** Antimicrobial Peptides and Human Disease. 2006. 12 figs. XII, 262 pp. ISBN 3-540-29915-7

Vol. 307: **John L. Casey (Ed.):** Hepatitis Delta Virus. 2006. 22 figs. XII, 228 pp. ISBN 3-540-29801-0

Vol. 308: **Honjo, Tasuku; Melchers, Fritz (Eds.):** Gut-Associated Lymphoid Tissues. 2006. 24 figs. XII, 204 pp. ISBN 3-540-30656-0

Vol. 309: **Polly Roy (Ed.):** Reoviruses: Entry, Assembly and Morphogenesis. 2006. 43 figs. XX, 261 pp. ISBN 3-540-30772-9

Vol. 310: **Doerfler, Walter; Böhm, Petra (Eds.):** DNA Methylation: Development, Genetic Disease and Cancer. 2006. 25 figs. X, 284 pp. ISBN 3-540-31180-7

Vol. 311: **Pulendran, Bali; Ahmed, Rafi (Eds.):** From Innate Immunity to Immunological Memory. 2006. 13 figs. X, 177 pp. ISBN 3-540-32635-9

Vol. 312: **Boshoff, Chris; Weiss, Robin A. (Eds.):** Kaposi Sarcoma Herpesvirus: New Perspectives. 2006. 29 figs. XVI, 330 pp. ISBN 3-540-34343-1

Vol. 313: **Pandolfi, Pier P.; Vogt, Peter K. (Eds.):** Acute Promyelocytic Leukemia. 2007. 16 figs. VIII, 273 pp. ISBN 3-540-34592-2

Vol. 314: **Moody, Branch D. (Ed.):** T Cell Activation by CD1 and Lipid Antigens, 2007, 25 figs. VIII, 348 pp. ISBN 978-3-540-69510-3

Vol. 315: **Childs, James, E.; Mackenzie, John S.; Richt, Jürgen A. (Eds.):** Wildlife and Emerging Zoonotic Diseases: The Biology, Circumstances and Consequences of Cross-Species Transmission. 2007. 49 figs. VII, 524 pp. ISBN 978-3-540-70961-9

Vol. 316: **Pitha, Paula M. (Ed.):** Interferon: The 50th Anniversary. 2007. VII, 391 pp. ISBN 978-3-540-71328-9

Vol. 317: **Dessain, Scott K. (Ed.):** Human Antibody Therapeutics for Viral Disease. 2007. XI, 202 pp. ISBN 978-3-540-72144-4

Vol. 318: **Rodriguez, Moses (Ed.):** Advances in Multiple Sclerosis and Experimental Demyelinating Diseases. 2008. XIV, 376 pp. ISBN 978-3-540-73679-9

Vol. 319: **Manser, Tim (Ed.):** Specialization and Complementation of Humoral Immune Responses to Infection. 2008. XII, 174 pp. ISBN 978-3-540-73899-2

Vol. 320: **Paddison, Patrick J.; Vogt, Peter K. (Eds.):** RNA Interference. 2008. VIII, 273 pp. ISBN 978-3-540-75156-4

Vol. 321: **Beutler, Bruce (Ed.):** Immunology, Phenotype First: How Mutations Have Established New Principles and Pathways in Immunology. 2008. XIV, 221 pp. ISBN 978-3-540-75202-8

Vol. 322: **Romeo, Tony (Ed.):** Bacterial Biofilms. 2008. XII, 299. ISBN 978-3-540-75417-6

Vol. 323: **Tracy, Steven; Oberste, M. Steven; Drescher, Kristen M. (Eds.):** Group B Coxsackieviruses. 2008. ISBN 978-3-540-75545-6

Vol. 324: **Nomura, Tatsuji; Watanabe, Takeshi; Habu, Sonoko (Eds.):** Humanized Mice. 2008. ISBN 978-3-540-75646-0

Vol. 325: **Shenk, Thomas E.; Stinski, Mark F.; (Eds.):** Human Cytomegalovirus. 2008. ISBN 978-3-540-77348-1

Vol. 326: **Reddy, Anireddy S.N; Golovkin, Maxim (Eds.):** Nuclear pre-mRNA processing in plants. 2008. ISBN 978-3-540-76775-6

Vol. 327: **Manchester, Marianne; Steinmetz, Nicole F. (Eds.):** Viruses and Nanotechnology. 2008. ISBN 978-3-540-69376-5

Vol. 328: **van Etten, (Ed.):** Lesser Known Large dsDNA Viruses. 2008. ISBN 978-3-540-68617-0

Vol. 329: **Diane E. Griffin; Michael B.A. Oldstone (Eds.):** Measles. 2009. ISBN 978-3-540-70522-2

Vol. 330: **Diane E. Griffin; Michael B.A. Oldstone (Eds.):** Measles. 2009. ISBN 978-3-540-70616-8

Vol. 331: **Ethel-Michele de Villiers; Harald zur Hausen (Eds.):** TT Viruses. 2009. ISBN 978-3-540-70971-8

Printing: Krips bv, Meppel, The Netherlands
Binding: Stürtz, Würzburg, Germany